NEW QUESTS IN STELLAR ASTROPHYSICS:
THE LINK BETWEEN STARS AND COSMOLOGY

ASTROPHYSICS AND SPACE SCIENCE LIBRARY

VOLUME 274

NEW QUESTS IN STELLAR ASTROPHYSICS: THE LINK BETWEEN STARS AND COSMOLOGY

Proceedings of the International Conference held in Puerto Vallarta, México, 26–30 March 2001

edited by

MIGUEL CHÁVEZ

Instituto Nacional de Astrofísica Óptica y Electrónica, Puebla, México

ALESSANDRO BRESSAN

Osservatorio Astronomico di Padova, Padova, Italy

ALBERTO BUZZONI

Osservatorio Astronomico di Brera, Milan; and Telescopio Nazionale Galileo, Italy

and

DIVAKARA MAYYA

Instituto Nacional de Astrofísica Óptica y Electrónica, Puebla, México

KLUWER ACADEMIC PUBLISHERS

DORDRECHT / BOSTON / LONDON

A C.I.P. Catalogue record for this book is available from the Library of Congress.

ISBN 1-4020-0644-6

Published by Kluwer Academic Publishers,
P.O. Box 17, 3300 AA Dordrecht, The Netherlands.

Sold and distributed in North, Central and South America
by Kluwer Academic Publishers,
101 Philip Drive, Norwell, MA 02061, U.S.A.

In all other countries, sold and distributed
by Kluwer Academic Publishers,
P.O. Box 322, 3300 AH Dordrecht, The Netherlands.

Printed on acid-free paper

Printed in the Netherlands.

Contents

Part VI High-Mass Stellar Products and Cosmology

Part VII Stellar Environments in High-Redshift Galaxies

Preface

The international conference on "The Link between Stars and Cosmology" was held in Puerto Vallarta, Mexico, on March 26-30, 2001. This was the first event in the spirit of a new series of INAOE meetings, that intend to address the "New Quests in Stellar Astrophysics", providing an important opportunity for international scientists to join and contribute to an up-to-date view of the leading problems in this field of research.

A little over 100 participants from 13 countries got together and reviewed, through 57 oral and poster contributions, a variety of topics that spanned from solar physics to the properties of the primeval Universe. The headline of this conference was actually to emphasize the emerging interplay between stellar astrophysics and observational cosmology, in a new (and perhaps unusual) approach to the global problem.

This meeting is number 21 of a recent but yet important score of events organized by the INAOE Astrophysics Department during the last six years of its renewed scientific activity. These years have coincided with a remarkable growth of scientific research in Mexico, where Astrophysics played a leading role both in terms of basic investigation and technological effort (the GTM/LMT, reviewed later in this book, will be the world's largest millimeter radio telescope when it starts operating in 2004 from the Volcano La Negra, close to Puebla).

The throughput of this wide intellectual process has also been palpable in this meeting, where about one third of the participants were graduate and undergraduate students from all over the globe, including 7 different Mexican institutions, who could attend the conference thanks to a generous funding by INAOE, the Universities of Puebla (BUAP) and Guadalajara (UdeG). Even more encouraging, the role of these young scientists has not only been limited to a "silent" and respectful audience, but also most of them actively participated presenting fresh results of their research.

Aside the fruitful discussion and interchange that involved participants all along the meeting, we should especially thank all the invited speakers for spending a relevant effort in their contributions to give a timely and focused picture of the leading problems that link stellar physics and galaxy evolution in the overall context of high-redshift cosmology.

A word of gratitude is also due to INAOE and CONACyT for financial and logistic support, as well as to the University of Guadalajara, through its Centro Universitario de la Costa, for its precious collaboration and organizing facilities. Social events, during the days of the meeting, have been organized with the help of the City Mayor of Puerto Vallarta.
Last but not least, of course, it is a pleasure to acknowledge here the untiring work of all the friends of the Scientific and Local organizing committees, as well as the technical staff of INAOE, who substantially contributed to the successful organization of this meeting. To all of them our warmest appreciation.

The Editors

Scientific Organizing Committee:

Timothy Beers (Michigan State University, USA)
Alessandro Bressan (Osservatorio di Padova, Italy)
Elias Brinks (Univ. of Guanajuato, Mexico)
Alberto Buzzoni (Telescopio Nazionale Galileo, Italy)
Luis Carrasco (INAOE, Mexico)
Miguel Chavez (INAOE, Mexico)
George Coyne (Vatican Observatory)
Wendy Freedman (Carnegie Obs. USA)
Ken Freeman (Mount Stromlo, Australia)
Manuel Peimbert (UNAM, Mexico)
Alfonso Serrano (INAOE, Mexico)
Benjamin Taylor (Brigham Young Univ., USA)
Yuzuru Yoshii (Univ. of Tokyo, Japan)

Local Organizing Committee:

Divakara Mayya (INAOE)
Jose Ramon Valdés (INAOE)
Soledad del Río (INAOE)
Lino Rodríguez (INAOE)
Emanuele Bertone (Oss. di Brera, Italy)

Technical Support:
Rosario Sánchez
Elizabeth Rodríguez
Gorgonio Cerón
Tomás Flores

I

STELLAR ATMOSPHERES

A FEW THINGS WE DO NOT KNOW ABOUT STARS AND MODEL ATMOSPHERES

Robert L. Kurucz
Harvard-Smithsonian Center for Astrophysics. Cambridge, MA 02138, USA

Abstract We list a few things that we do not understand about stars and that most people ignore. These are all hard problems. We can learn more cosmology by working on them to reduce the systematic errors they introduce than by trying to derive cosmological results that are highly uncertain.

1. Optimism and Pessimism

People sometimes complain that I am too pessimistic and that I criticize too much. In fact I am the most optimistic person. I believe that the human race is tremendously improvable and that humans can solve any problem. But the most important step in solving a problem is to realize that the problem exists. When I identify a problem I tell, or try to tell, the people who are capable of doing something about it. I also work on correcting the problem myself, if I am capable.

A pessimist does not believe that problems can be solved so does not question the present and does not search for errors. A pessimist acts so "optimistically" about the present that a pessimist prevents progress. Why worry about basic physics when everything is fine as it is?

The papers below are on my web site kurucz.harvard.edu. Some of them are also on the ASTRO-PH preprint server at Los Alamos.

A Few Things We Do Not Know About the Sun and F Stars and G Stars. I gave part of this talk at The Workshop on Nearby Stars at Ames two years ago. They wanted to know the state of the art in computing model atmospheres and spectra to determine whether they could see small abundance effects in the nearby stars or any spectral signature of planets. (I doubt it.)

Radiatively-Driven Cosmology. Most cosmologists never took a stellar atmospheres course and do not have experience with radiation. They do not realize the power of radiative acceleration compared to gravity.

A Correction to the pp Reaction. What if the pp reaction is a three-body reaction, two protons and an electron?

M. Chávez et al. (eds.),
New Quests in Stellar Astrophysics: The Link Between Stars and Cosmology, 3–14.

Vegan Astrophysics. This is a gedanken experiment to show the importance of basic physics.

2. We do not know how to make realistic model atmospheres; we do not understand convection

Recently I have been preoccupied with convection because the model atmospheres are now good enough to show shortcomings in the convective treatment. Here I will outline what I have learned. I will mainly list the conclusions I have come to from examining individual convective models and from examining grids of convective models as a whole. Eighteen figures illustrating the points made here can be found in Kurucz (1996).

Every observation, measurement, model, and theory has seven characteristic numbers: resolution in space, in time, and in energy, and minimum and maximum energy. Many people never think about these resolutions. A low resolution physics cannot be used to study something in which the physical process of interest occurs at high resolution unless the high resolution effects average out when integrated over the resolution bandpasses.

What does the sun, or any convective atmosphere, actually look like? We do not really know yet. There is a very simplified three-dimensional radiation-hydrodynamics calculation discussed in the review by Chan, Nordlund, Steffen, and Stein (1991). It is consistent with the high spatial and temporal resolution observations shown in the review by Topka and Title (1991). Qualitatively, there is cellular convection with relatively slowly ascending, hot, broad, diverging flows that turn over and merge with their neighbors to form cold, rapidly descending, filamentary flows that diffuse at the bottom. The filling factor for the cold downward flowing elements is small. The structure changes with time. Nordlund and Dravins (1990) discuss four similar stellar models with many figures. Every one-dimensional mixing-length convective model is based on the assumption that the convective structure averages away so that the emergent radiation depends only a one-dimensional temperature distribution.

There is a solar flux atlas (Kurucz, Furenlid, Brault, and Testerman 1984) that Ingemar Furenlid caused to be produced because he wanted to work with the sun as a star for comparison to other stars. The atlas is pieced together from eight Fourier transform spectrograph scans, each of which was integrated for two hours, so the time resolution is two hours for a given scan. The x and y resolutions are the diameter of the sun. The z resolution (from the formation depths of features in the spectrum) is difficult to estimate. It depends on the signal-to-noise and the number of resolution elements. The first is greater than 3000 and the second is more than one million. It may be possible to find enough weak lines in the wings and shoulders of strong lines to map out relative positions to a few kilometers. Today I think it is to a few tens of kilometers.

The resolving power is on the order of 522,000. This is not really good enough for observations made through the atmosphere because it does not resolve the terrestrial lines that must be removed from the spectrum. (In the infrared there are many wavelength regions where the terrestrial absorption is too strong to remove.) The sun itself degrades its own flux spectrum by differential rotation and macroturbulent motions. The energy range of the atlas is from 300 to 1300 nm, essentially the range where the sun radiates most of its energy.

This solar atlas is of higher quality than any stellar spectrum taken thus far but still needs considerable improvement. If we have difficulty interpreting these data, it can only be worse for other stars where the spectra are of lower quality by orders of magnitude.

To analyze this spectrum, or any other spectrum, we need a theory that works at a similar resolution or better. We use a plane parallel, one-dimensional theoretical or empirical model atmosphere that extends in z through the region where the lines and continuum are formed. The one-dimensional model atmosphere represents the space average of the convective structure over the whole stellar disk (taking account of the center-to-limb variation) and the time average over hours. It is usually possible to compute a model that matches the observed energy distribution around the flux maximum. However, to obtain the match it is necessary to adjust a number of free parameters: effective temperature, surface gravity, microturbulent velocity, and the mixing-length-to-scale-height-ratio in the one-dimensional convective treatment. The microturbulent velocity parameter also produces an adjustment to the line opacity to make up for missing lines. Since much of the spectrum is produced near the flux maximum, at depths in the atmosphere where the overall flux is produced, averaging should give good results. The parameters of the fitted model may not be those of the star, but the radiation field should be like that of the star. The sun is the only star where the effective temperature and gravity are accurately known. In computing the detailed spectrum, it is possible to adjust the line parameters to match many features, although not the centers of the strongest lines. These are affected by the chromosphere and by NLTE. Since very few lines have atomic data known accurately enough to constrain the model, a match does not necessarily mean that the model is correct.

From plots of the convective flux and velocity for grids of models I have identified three types of convection in stellar atmospheres:

- normal strong convection where the convection is continuous from the atmosphere down into the underlying envelope. Convection carries more than 90% of the flux. Stars with effective temperatures 6000K and cooler are convective in this way as are stars on the main sequence up to 8000K. At higher temperature the convection carries less of the total flux and eventually disappears starting with the lowest gravity models. Intermediate gravities have intermediate behavior. Abundances have to be uniform through the atmosphere into the envelope. The highly convective models seem to be reasonable representations of real stars, except for the shortcomings cited below.

• atmospheric layer convection where, as convection weakens, the convection zone withdraws completely up from the envelope into the atmosphere. There is zero convection at the bottom of the atmosphere. Abundances in the atmosphere are decoupled from abundances in the envelope. For mixing-length models the convection zone is limited at the top by the Schwarzschild criterion to the vicinity of optical depth 1 or 2. The convection zone is squashed into a thin layer. In a grid, this layer continues to carry significant convective flux for about 500K in effective temperature beyond the strongly convective models. There is no common-sense way in which to have convective motions in a thin layer in an atmosphere. The solution is that the Schwarzschild criterion does not apply to convective atmospheres. The derivatives are defined only in one dimensional models. A real convective element has to decide what to do on the basis of local three-dimensional derivatives, not on means. These thin-layer-convective model atmospheres may not be very realistic.

• plume convection. Once the convective flux drops to the percent range, cellular convection is no longer viable. Either the star becomes completely radiative, or it becomes radiative with convective plumes that cover only a small fraction of the surface in space and time. Warm convective material rises and radiates. The star has rubeola. The plumes dissipate and the whole atmosphere relaxes downward. There are no downward flows. The convective model atmospheres are not very realistic except when the convection is so small as to have negligible effect, i.e. the model is radiative. The best approach may be simply to define a star with less than, say, 1% convection as radiative. The error will probably be less than using mixing-length model atmospheres.

Using a one-dimensional model atmosphere to represent a real convective atmosphere for any property that does not average in space and time to the one-dimensional model predictions produces systematic errors. The Planck function, the Boltzmann factor, and the Saha equation are functions that do not average between hot and cold convective elements. We can automatically conclude that one-dimensional convective models must predict the wrong value for any parameter that has strong exponential temperature dependence from these functions.

Starting with the Planck function, ultraviolet photospheric flux in any convective star must be higher than predicted by a one-dimensional model (Bikmaev 1994). Then, by flux conservation, the flux redward of the flux maximum must be lower. It is fit by a model with lower effective temperature than that of the star. The following qualitative predictions result from the exponential falloff of the flux blueward of the flux maximum:

• the Balmer continuum in all convective stars is higher than predicted by a one-dimensional model;

• in G stars, including the sun, the discrepancy reaches up to about 400nm;
• all ultraviolet photoionization rates at photospheric depths are higher in real stars than computed from one-dimensional models;
• flux from a temperature minimum and a chromospheric temperature rise masks the increased photospheric flux in the ultraviolet;
• the spectrum predicted from a one-dimensional model for the exponential falloff region, and abundances derived therefrom, are systematically in error;
• limb-darkening predicted from a one-dimensional model for the exponential falloff region is systematically in error;
• convective stars produce slightly less infrared flux than do one-dimensio-nal models.

The Boltzmann factor is extremely temperature sensitive for highly excited levels:
• the strong Boltzmann temperature dependence of the second level of hydrogen implies that the Balmer line wings are preferentially formed in the hotter convective elements. A one-dimensional model that matches Balmer line wings has a higher effective temperature than the real star;
• the same is true for all infrared hydrogen lines.

The Saha equation is safe only for the dominant species:
• neutral atoms for an element that is mostly ionized are the most dangerous because (in LTE) they are much more abundant in the cool convective elements. When Fe is mostly ionized the metallicity determination from Fe I can be systematically offset and can result in a systematic error in the assumed evolutionary track and age.
• in the sun convection may account for the remaining uncertainties with Fe I found by Blackwell, Lynas-Gray, and Smith (1995);
• the most striking case is the large systematic error in Li abundance determination in extreme Population II G subdwarfs. The abundance is determined from the Li I D lines which are formed at depths in the highly convective atmosphere where Li is 99.94% ionized (Kurucz 1995b);
• molecules with high dissociation energies such as CO are also much more abundant in the cool convective elements. The CO fundamental line cores in the solar infrared are deeper than any one-dimensional model predicts (Ayres and Testerman 1981) because the cooler convective elements that exist only a short time have more CO than the mean model.

Given all of these difficulties, how should we proceed? One-dimensional model atmospheres can never reproduce real convective atmospheres. The only practical procedure is to compute grids of model atmospheres, then to compute diagnostics for temperature, gravity, abundances, etc., and then to make tables of corrections. Say, for example, in using the Hα wings as a diagnostic of effective temperature in G stars, the models may predict effective temperatures that are 100K too high. So if one uses an Hα temperature scale it has to be corrected by 100K to give the true answer. Every temperature scale by any method has to be corrected in some way. Unfortunately, not only is this tedious, but it is very difficult or impossible because no standards exist. We do not know the energy distribution or the photospheric spectrum of a single star, even the sun. We do not know what spectrum corresponds to a given effective temperature, gravity, or abundances. The uncertainties in solar abundances are greater than 10%, except for hydrogen, and solar abundances are the best known. It is crucial to obtain high resolution, high signal-to-noise observations of the bright stars.

2.1 We do not consider the variation in microturbulent velocity

Microturbulent velocity in the photosphere is just the convective motions. At the bottom of the atmosphere it is approximately the maximum convective velocity. At the temperature minimum it is zero or near zero because the convecting material does not rise that high. There is also microturbulent velocity in the chromosphere increasing outward from the temperature minimum that is produced by waves or other heating mechanisms. In the sun the empirically determined microturbulent velocity is about 0.5 km/s at the temperature minimum and about 1.8 km/s in the deepest layers we can see. In a solar model the maximum convective velocity is 2.3 km/s. The maximum convective velocity is about 0.25 km/s in an M dwarf and increases up the main sequence. The convective velocity increases greatly as the gravity decreases. I suggest that a good way to treat the behavior of microturbulent velocity in the models is to scale the solar empirical distribution as a function of Rosseland optical depth to the maximum convective velocity for each effective temperature and gravity. Why does this matter? Microturbulent velocity increases line width and opacity and produces effects on an atmosphere like those from changing abundances. At present, models, fluxes, colors, spectra, etc are computed with constant microturbulent velocity within a model and from model to model. This introduces systematic errors within a model between high and low depths of formation, and between models with different effective temperatures, and between models with different gravity. Microturbulent velocity varies along an evolutionary track. If microturbulent velocity is produced by convection, microturbulent velocity is zero when there is no convection, and diffusion is possible.

By now I should have computed a model grid with varying microturbulent velocity but I am behind as usual.

2.2 We do not understand spectroscopy; we do not have good spectra of the sun or any star

Very few of the features called "lines" in a spectrum are single lines. Most features consist of blends of many lines from different atoms and molecules. All atomic lines except those of thorium have hyperfine or isotopic components, or both, and are asymmetric (Kurucz 1993). Low resolution, low-signal-to-noise spectra do not contain enough information in themselves to allow interpretation. Spectra cannot be properly interpreted without signal-to-noise and resolution high enough to give us all the information the star is broadcasting about itself. And then we need laboratory data and theoretical calculations as complete as possible. Once we understand high quality spectra we can look at other stars with lower resolution and signal-to-noise and have a chance to make sense of them.

2.3 We do not have energy distributions for the sun or any star

I get requests from people who want to know the solar irradiance spectrum, the spectrum above the atmosphere, that illuminates all solar system bodies. They want to interpret their space telescope observations or work on atmospheric chemistry, or whatever. I say, "Sorry, it has never been observed. NASA and ESA are not interested. I can give you my model predictions but you cannot trust them in detail, only in, say, one wavenumber bins." The situation is pathetic.
I am reducing Brault's FTS solar flux and intensity spectra taken at Kitt Peak for .3 to 5 μm. I am trying to compute the telluric spectrum and ratio it out to determine the flux above the atmosphere but that will not work for regions of very strong absorption. Once that is done the residual flux spectra can be normalized to low resolution calibrations to determine the irradiance spectrum. The missing pieces will have to be filled in by computation. Spectra available in the ultraviolet are much lower resolution, much lower signal-to-noise, and are central intensity or limb intensity, not flux. The details of the available solar atlases can be found in two review papers, Kurucz (1991; 1995a).

2.4 We do not know how to determine abundances; we do not know the abundances of the sun or any star

One of the curiosities of astronomy is the quantity [Fe]. It is the logarithmic abundance of Fe in a galaxy, cluster, star, whatever, relative to the solar abundance of Fe. What makes it peculiar is that we do not yet know the solar abundance of Fe and our guesses change every year. The abundance has varied by a factor of ten since I was a student. Therefore [Fe] is meaningless unless the solar Fe abundance is also given so that [Fe] can be corrected to the current value of Fe.
For an example I use Grevesse and Sauval's (1999) solar Fe abundance determination. I am critical, but, regardless of my criticism, I still use their abundances. There are scores of other abundance analysis papers, including some bearing my name, that I could criticize the same way.

Grevesse and Sauval included 65 Fe I "lines" ranging in strength from 1.4 to 91.0 mÅ and 13 Fe II "lines" ranging from 15.0 to 87.0 mÅ. They found an abundance log Fe/H + 12 = 7.50 ± 0.05.

Another curiosity of astronomy is that Grevesse and Sauval have decided a priori that the solar Fe abundance must equal the meteoritic abundance of 7.50 and that a determination is good if it produces that answer. If the solar abundance is not meteoritic, how could they ever determine it?

There are many "problems" in the analysis. First, almost all the errors are systematic, not statistical. Having many lines in no way decreases the error. In fact, the use of a wide range of lines of varying strengths increases the systematic errors. Ideally a single weak line is all that is required to get an accurate abundance. Weak lines are relatively insensitive to the damping treatment, to microturbulent velocity, and to the model structure. The error is reduced simply by throwing out all lines greater than 30 mÅ. That reduces the number of Fe I lines from 65 to 25 and of Fe II lines from 13 to 5. As we discussed above, the microturbulent velocity varies with depth but Grevesse and Sauval assume that it is constant. This problem is minimized if all the lines are weak.

As we discussed above "lines" do not exist. The lines for which equivalent widths are given are all parts of blended features. As a minimum we have to look at the spectrum of each feature and determine how much of the feature in the "line" under investigation and how much is blending. Rigorously one should do spectrum synthesis of the whole feature. We have solar central intensity spectra and spectrum synthesis programs. For the sun we have the advantage of intensity spectra without rotational broadening. In the flux spectrum of the sun and of other stars there is more blending. The signal-to-noise of the spectra is several thousand and the continuum level can be determined to on the order of 0.1 per cent so the errors from the spectrum are small. With higher signal-to-noise more detail would be visible and the blending would be better understood. Most of the features cannot be computed well with the current line data. None of the features can be computed well without adjusting the line data. Even if the line data were perfect, the wavelengths would still have to be adjusted because of wavelength shifts from convective motions.

Fe has 4 isotopes. The isotopic splitting has not been determined for the lines in the abundance analysis. For weak lines it does not affect the total equivalent width but it does affect the perception of blends.

It is possible to have undetectable blends. There are many Fe I lines with the same wavelengths, including some in this analysis, and many lines of other elements. We hope that these blends are very weak. The systematic error always makes the observed line stronger than it is in reality so they produce an abundance overestimate.

There are systematic errors and random errors in the gf values. With a small number of weak lines on the linear part of the curve of growth it is easy to correct the abundances when the gf values are improved in the future.
We are left with 3 relatively safe lines of Fe I and 1 relatively safe line of Fe II. These have the least uncertainty in determining the blending by my estimation. Grevesse and Sauval found abundances of 7.455, 7.453, and 7.470 for the Fe I lines and 7.457 for the Fe II line. Thus from the same data the Fe abundances is 7.46 instead of 7.50.

2.5 We do not have good atomic and molecular data; one half the lines in the solar spectrum are not identified

It is imperative that laboratory spectrum analyses be improved and extended, and that NASA and ESA pay for it. Some of the analyses currently in use date from the 1930s and produce line positions uncertain by 0.01 or 0.02 Å. New analyses with FTS spectra produce many more energy levels and one or two orders of magnitude better wavelengths. One analysis can affect thousands of features in a stellar spectrum. Also the new data are of such high quality that for some lines the hyperfine or isotopic splitting can be directly measured. Using Pickering (1996) and Pickering and Thorne (1996) I am now able to compute Co I hyperfine transitions and to reproduce the flag patterns and peculiar shapes of Co features in the solar spectrum. Using Litzen, Brault, and Thorne (1993) I am now able to compute the five isotopic transitions for Ni I and to reproduce the Ni features in the solar spectrum. These new analyses also serve as the basis for new semiempirical calculations than can predict the gf values and the lines that have not yet been observed in the lab but that matter in stars. I have begun to compute new line lists for all the elements and I will make them available on my web site, kurucz.harvard.edu.

2.6 Cepheids have convective pulsation but the models do not; we do not have high quality spectra over phase for any Cepheid

Cepheids are convective with velocities the same order of magnitude as the pulsation velocities. The sum of the velocities is supersonic and the difference is order zero. It is completely unphysical to try to compute the convection and the pulsation independently. Convective pulsation is a 3-dimensional radiation-hydrodynamics problem that must be solved as a whole.
If a hot Cepheid has a radiative phase, it becomes convective as it cools. The transition phase has space-time-random outward plumes that become supersonic. The surface is covered with spikes or bumps that cool by radiating toward the side.

All of this physics is displayed in the spectra of nearby Cepheids that are bright enough to be observed at 1 km/s resolution and S/N 3000. It would be perfectly feasible to make an atlas of such high resolution spectra every hour through the phases and then to read out the story, and also to use it to estimate boundary conditions for convective pulsation calculations.

2.7 We do not understand abundance evolution in early type stars

This is a simplified, qualitative outline. Since there is no convection the atmosphere and upper layers mix very slowly. The bulk of the material of the star has approximately scaled solar abundances, $[Fe] \geq 0$. When the star is formed the material in the atmosphere is the last to be accreted. It consists of dregs of the infall material that has been depleted of elements that are able to condense into grains. A young star has low metal abundances in the atmosphere and so appears to have $[Fe] << 0$. As the star ages heavy elements with many lines are levitated into the atmosphere by radiative acceleration. Some elements, such as He, settle inward from gravity. The abundances become closer to solar, $[Fe] \leq 0$. The star grows older and the abundances continue to increase in the atmosphere so that the star becomes a metallic line star with $[Fe] > 0$. If the star has strong magnetic spots, the abundances can be selectively enhanced by many orders of magnitude in the spots. The star is called "peculiar". A radiative wind selectively reduces abundances in the atmosphere because radiative acceleration affects some elements more than others. The only safe way to investigate early type stars is to obtain high quality spectra and spectrophotometry and to compute models and spectra for each star individually. Colors integrate away too many details. Astroseismology may be able to show abundance variation with depth.

From an evolutionary point of view, all main sequence early-type stars in our galaxy have slightly over solar abundances.

2.8 Many early type stars are oblate fast rotators

Early-type stars that are not in binaries are generally fast rotators. They are oblate because of the reduced gravity at the equator. The temperature can be several thousand degrees hotter at the poles than at the equator. Plane-parallel models like mine can be found that represent some average behavior but rigorously one must compute three dimensional rotating models. The real star has more ultraviolet flux from the poles and more infrared flux from the equator than the plane-parallel models so the ionizing radiation field around an early-type star is prolate. It is probably not safe to use any unary early-type star as a photometric standard for calibrating theoretical photometry.

References

Ayres, T. R. & Testerman, L. 1981, ApJ, 245, 1124.

Bikmaev, I. 1994, personal communication

Blackwell, D. E., Lynas-Gray, A. E., & Smith, G. 1995, A&A, 296, 217.

Chan, K. L., Nordlund, Å, Steffen, M., & Stein, R. F. 1991, in Solar Interior and Atmosphere, eds. A. N. Cox, W. C. Livingston, & M. Matthews (Tucson: Univ. of Arizona Press), p. 223.

Grevesse, N. & Sauval, A. J. 1999, A&A, 347, 348.

Kurucz, R. L. 1991, in The Solar Interior and Atmosphere, eds. A. N. Cox, W. C. Livingston, & M. Matthews (Tucson: Univ. of Arizona Press), p. 663.

Kurucz, R. L. 1993, Physica Scripta, T47, 110.

Kurucz, R.L. 1995a, in ASP Conf. Series 81, Laboratory and Astronomical High Resolution Spectra, eds. A. J. Sauval, R. Blomme, & N. Grevesse, 17.

Kurucz, R. L. 1995b, ApJ, 452, 102.

Kurucz, R. L. 1996, in ASP Conf. Series 108, Model Atmospheres and Stellar Spectra, eds. S. Adelman, F. Kupka, & W. W. Weiss, (San Francisco: ASP), p. 2.

Kurucz, R. L., Furenlid, I., Brault, J., & Testerman, L. 1984, Solar Flux Atlas from 296 to 1300 nm, (Sunspot, N. M.: National Solar Observatory).

Litzen, U., Brault, J. W., & Thorne, A. P. 1993, Physica Scripta, 47, 628.

Nordlund, Å, & Dravins, D. 1990, A&A, 228, 155.

Pickering, J. C. 1996, ApJS, 107, 811.

Pickering, J. C. & Thorne, A. P. 1996, ApJS, 107, 761.

Topka, K. P. & Title, A. M. 1991, in Solar Interior and Atmosphere, eds. A. N. Cox, W. C. Livingston, & M. Matthews (Tucson: U. of Arizona Press), p. 727.

COOL STELLAR ATMOSPHERES

Peter H. Hauschildt, Andreas Schweitzer
Dept. of Physics and Astronomy & Center for Simulational Physics,
University of Georgia, Athens, GA 30602-2451, USA

F. Allard
C.R.A.L (UML 5574) Ecole Normale Superieure, 69364 Lyon Cedex 7, France

J. W. Ferguson, D. Alexander
Dept. of Physics, Wichita State University, Wichita, KS 67260-0032, USA

and

E. Baron
Dept. of Physics and Astronomy, University of Oklahoma,
440 W. Brooks, Rm 131, Norman, OK 73019-0225, USA

Abstract We give an overview about the state-of-the-art in cool stellar (and sub-stellar) atmosphere simulations. Recent developments in numerical methods and parallel supercomputers, as well as in the quality of input data such as atomic and molecular line lists have led to substantial improvements in the quality of synthetic spectra when compared to multi-wavelength observations. A wide range of objects from M dwarfs and giants down to substellar objects is considered. We discuss effects such as atomic and molecular NLTE (and) line blanketing, external irradiation, and formation and opacities of dust particles and clouds; each of which affects the structure of the atmospheres and their spectra. Current models can simultaneously fit many of the observed features of a given star with a single model atmosphere, however, a number of problems remain unsolved and will have to be addressed in the future, in particular for very low mass stars and substellar objects.

Introduction

Stellar atmosphere modeling has experienced a renaissance in the past decade with the advent of better algorithms and faster computers. This has allowed

M. Chávez et al. (eds.),
New Quests in Stellar Astrophysics: The Link Between Stars and Cosmology, 15–26.

research groups to remove or relax many of the "standard" assumptions that were made in the 70s and 80s and that had become accepted wisdom over the years. Surprisingly (or not) the new calculations show that many of these assumptions are actually quite bad and can lead to spurious results or incorrect interpretations of observed spectra. The intricate connection between geometry (plane parallel or spherical), line blanketing (atomic and/or molecular) and non-LTE effects (using small to extremely large model atoms and molecules) began to emerge slowly as crucial ingredients for physically correct and meaningful interpretations and analyses of stellar spectra. Unfortunately, easy and simple solutions do not really work for stellar atmospheres (although everybody likes the easy way out and some of them are useful for teaching purposes) and have actually hindered progress and reduced the reliability of results.

New observational techniques opened and continue to open up new areas of stellar atmosphere research. The most important advance has been detailed observations of very low mass stars and after decades of searching, brown dwarfs and extrasolar giant planets. Modeling these objects requires sophisticated stellar atmosphere type modeling with complex equations of state and 100's of millions of molecular spectral lines in order to even approximately reproduce the observed spectra. These new observations have prompted further evolution of stellar atmosphere modeling and helped rejuvenate the field in general.

In the following we will briefly introduce the numerical methods that modern stellar atmosphere research employs (there are also plenty of legacy applications and codes that are still widely used) and then discuss some results that are of interest in the context of this meeting.

1. Methods and Models

For our model calculations, we use our multi-purpose stellar atmosphere code `PHOENIX` (version 10.9; Hauschildt, Baron, & Allard 1997; Baron & Hauschildt 1998; Hauschildt, Allard, & Baron 1999; Hauschildt et al. 1999; Hauschildt & Baron 1999). Details of the numerical methods are given in the above references, so we do not repeat the description here.

1.1 Line opacities

One of the most important recent improvements of cool stellar atmosphere models is the treatment of molecular line opacity. Our combined molecular line list includes about 550 million molecular lines. The lines are selected for every model from the master line list at the beginning of each model iteration to account for changes in the model structure (see below). Both atomic and molecular lines are treated with a direct opacity sampling method (dOS). We do *not* use pre-computed opacity sampling tables, but instead dynamically select the relevant LTE background lines from master line lists at the beginning of each iteration for every model and sum the contribution of every line within a search window to compute the total line opacity at *arbitrary* wavelength points. The latter feature is crucial in NLTE calculations in which the wavelength grid is both irregular and variable from iteration to iteration due to changes in the physical conditions. This approach also allows detailed and depth dependent line profiles to be used during the iterations. Although the direct line treatment seems at first glance computationally prohibitive, it leads to more accurate models. This is due to the fact that the line forming regions in cool stars and planets span a huge range in pressure and temperature so that the line wings form in very different layers than the line cores. Therefore, the physics of the line formation is best modeled by an approach that treats the variation of the line profile and the level excitation as accurately as possible. To make this method computationally more efficient, we employ modern numerical techniques, e.g., vectorized and parallelized block algorithms with high data locality (Hauschildt, Baron, & Allard 1997) and we use high-end workstations or parallel supercomputers for the model calculations.

In the calculations presented in this contribution, we have included a constant statistical velocity field, $\xi = 2\mathrm{km\ s}^{-1}$, which is treated like a microturbulence. The choice of lines is dictated by whether they are stronger than a threshold $\Gamma \equiv \chi_l/\kappa_c = 10^{-4}$, where χ_l is the extinction coefficient of the line at the line center and κ_c is the local b-f absorption coefficient (see Hauschildt, Allard, & Baron 1999, for details of the line selection process). This typically leads to about 10–250 million lines which are selected from master line lists, depending on model parameters. The profiles of these lines are assumed to be depth-dependent Voigt or Doppler profiles (for very weak lines). Details of the computation of the damping constants and the line profiles are given in Schweitzer et al. (1996). We have determined by test calculations that the details of the line profiles and the threshold Γ do not have a significant effect on either the model structure or the synthetic spectra. In addition, we include about 2000 photo-ionization cross sections for atoms and ions (Mathisen 1984; Verner & Yakovlev 1995).

1.2 Equation of State

The equation of state (EOS) is an enlarged and enhanced version of the EOS used in Allard et al. (1997). We include about 500 species (atoms, ions and molecules) in the EOS. This set of EOS species was determined in test calculations. The EOS calculations themselves follow the method discussed in AH95. For effective temperatures, $T_{\mathrm{eff}} < 2500\,\mathrm{K}$, the formation of dust particles has to be considered in the EOS. In our models we allow for the formation (and dissolution) of a variety of grain species.

1.3 Non-LTE

The NLTE treatment of large model atoms or molecules such as H_2O and TiO which have several million transitions is a formidable problem which requires an efficient method for the numerical solution of the multi-level NLTE radiative transfer problem. Classical techniques, such as the complete linearization or the Equivalent Two Level Atom method, are computationally prohibitive for large model atoms and molecules. Currently, the operator splitting or approximate Λ-operator iteration (ALI) method (e.g., Cannon 1973; Rybicki 1972, 1984; Scharmer 1984) seems to be the most effective way of treating complex NLTE radiative transfer and rate equation problems. Variants of the ALI method have been developed to handle complex model atoms, e.g., Anderson's multi-group scheme (Anderson 1987, 1989) or extensions of the opacity distribution function method (Hubeny & Lanz 1995). However, these methods have problems if line overlaps are complex or if the line opacity changes rapidly with optical depth, a situation which occurs in cool stellar atmospheres. The ALI rate operator formalism (Hauschildt 1993; Hauschildt & Baron 1999), on the other hand, has been used successfully to treat very large model atoms such as Fe directly and efficiently (Hauschildt & Baron 1995; Hauschildt et al. 1996; Baron et al. 1996; Hauschildt & Baron 1999).

2. Results

In the following sections we will give a few representative results that highlight important new developments in stellar atmospheres.

2.1 Line blanketing

The number of molecular lines that are important in M dwarf (and later) atmospheres is quite large. About 215 million molecular lines are selected (see above) for a typical giant model with $T_{\rm eff} \approx 3000$ K whereas about 130 million molecular lines have to be considered for a dwarf model with the same effective temperature. The large "density" (in wavelength space) of molecular lines causes large line blanketing effects as illustrated in Fig. 1 for a very simple case. The nearly complete coverage of the optical spectrum by TiO lines and of the near IR spectrum by H_2O lines effectively locks the peak of the spectral energy distribution in place at around 1.1 μm even for substantially different $T_{\rm eff}$, in stark contrast to the behavior expected for blackbodies. Line blanketing also produces a strong metallicity effect on the spectra as illustrated in Fig. 2. Lowering the metal abundances reduces both the TiO and H_2O opacities by roughly the same amount. However, the H_2O opacity in the near IR is replaced by increasingly (with lower metallicity) stronger collision induced opacities (due to the larger pressures in the spectrum forming regions). Therefore, the spec-

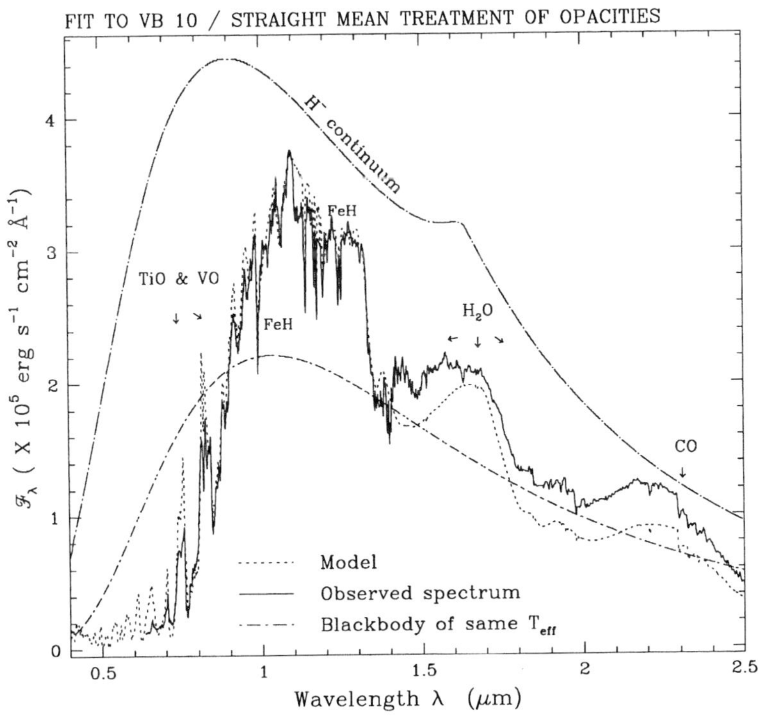

Figure 1. Best fit of Allard & Hauschildt (1995b) to the spectrum of the dM8e star VB 10 (Allard & Hauschildt 1995a). The corresponding H^- continuum obtained by neglecting molecular opacities only in the radiative transfer (long dot-dashed) reveals the magnitude of these opacities in a typical late-type M dwarf. The Planck distribution of the same $T_{\rm eff}$ is also shown for comparison. From Allard et al. (1997).

trum gets *bluer* with lower metallicities, even for comparatively low effective temperatures.

2.2 NLTE effects

Due to their very low electron temperatures, the electron density is extremely low in M stars; absolute electron densities are even lower than found in low density atmospheres, such as those of novae and SNe. Collisional rates due to collisions with electrons, which tend to restore LTE, are thus very small in cool stars. This in turn could significantly increase the importance of NLTE effects in M stars when compared to, e.g., solar type stars with much higher electron densities and temperatures. Collisions with molecular hydrogen and helium will at least partly compensate for the diminished electron collisions, but cross-sections for these processes are not very well known. Therefore, the

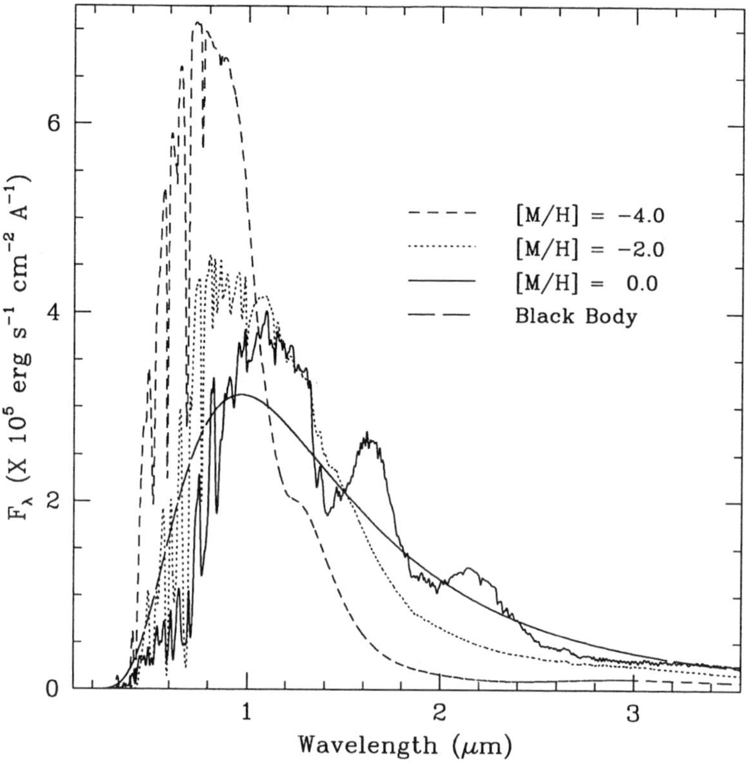

Figure 2. Spectral distributions of emerging fluxes at the stellar surface for 3,000 K models with metallicities corresponding roughly to the solar neighborhood ([M/H] = 0.0), halo ([M/H] = −2.0), and Population III ([M/H] = −4.0) stars. A black-body of the same effective temperature (smooth curve) is shown for comparison. From Allard et al. (1997).

assumption of LTE for atoms and molecules in cool stars is by no means certain and needs to be verified for each species individually. We have performed test calculations to place an upper limit for the importance of atomic NLTE effects in cool stars by only considering electron collisions.

We have calculated a small number of NLTE models in order to investigate the importance of NLTE effects on the structure of the model atmospheres. The results for cooler models were discussed in Hauschildt et al. (1997) and are not repeated here. Figure 3 shows an overview of selected NLTE species for models with $T_{\rm eff} = 4000\,{\rm K}$, $\log(g) = 0.0$ and solar abundances. The total number of NLTE levels in each model is 4532 with a total of 47993 primary NLTE lines (see Hauschildt & Baron 1999; Hauschildt et al. 1999 and references therein for details). For most of the species, the departure coefficients are always close to

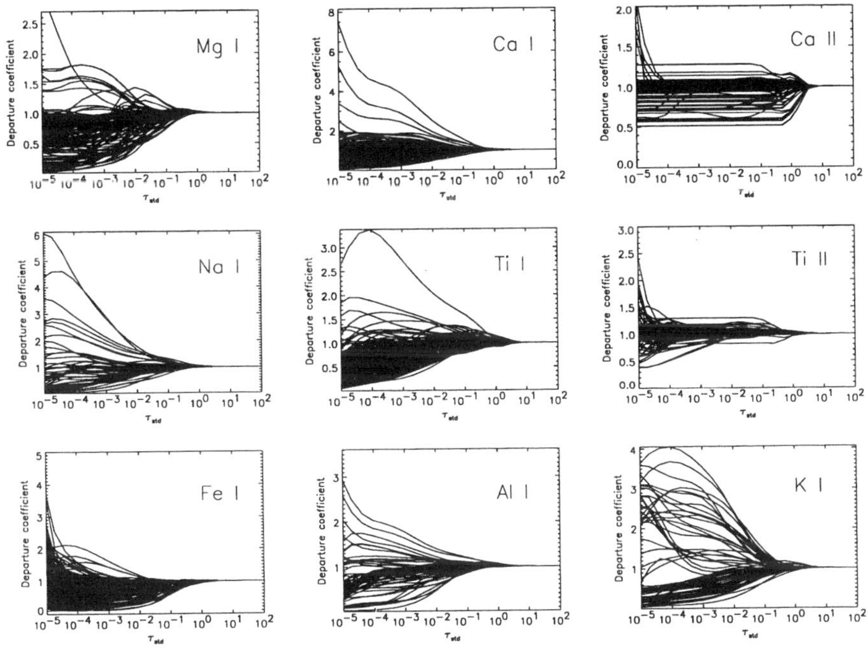

Figure 3. Overview over selected departure coefficients for a NLTE model with $T_{\rm eff} = 4000\,{\rm K}$, $\log(g) = 0.0$, and solar abundances.

unity, in particular for species with resonance lines and photoionization edges in the UV part of the spectrum. The species shown in Figure 3 are species with the most pronounced departures from LTE. The departures are generally too small to significantly affect the structure of the atmospheres. Results for NLTE calculations for the CO molecule show that the high cross-sections of H_2 and He collisions restore LTE very successfully in the case of dwarf stars (Schweitzer, Hauschildt, & Baron 2000).

2.3 Dust and cloud formation

The effects of dust formation on the atmosphere are mainly (a) the removal of important opacity sources (e.g., TiO, VO) from the gas phase and a corresponding weakening of their spectral lines; and (b) the presence of additional opacities produced by the grains themselves, cf. Fig. 4. The latter depends on the behavior of the macroscopic dust particles:

a) They might remain as dust clouds in the layers where the dust originally formed and thus cause strong optical and IR opacities due to these clouds ("Dusty" models).

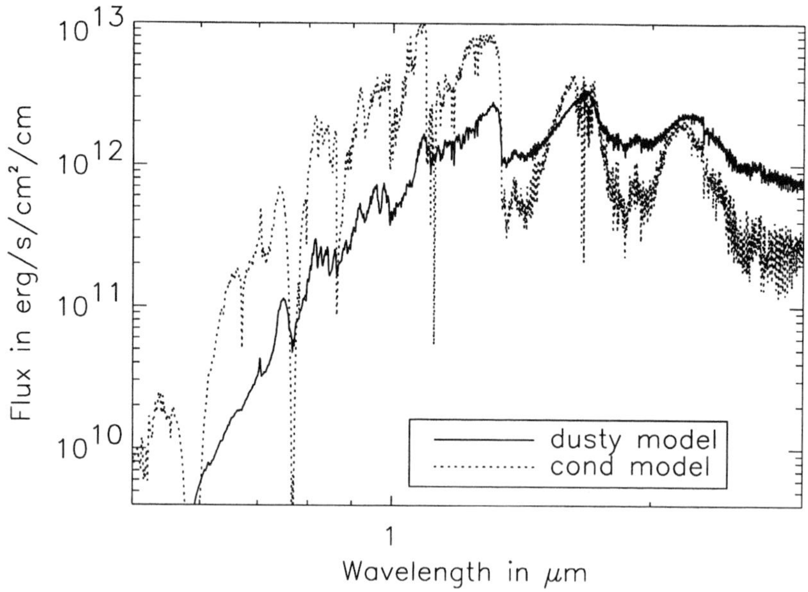

Figure 4. This plot shows the difference between synthetic spectra calculated for a model atmospheres assuming complete settling of all formed dust particles below the layers where spectrum forms ("cond" model) and for a model that assumes that the dust particles remain close to the layer in which they formed ("dusty" model) for $T_{\text{eff}} = 1700$ K, $\log(g) = 4.5$ and solar abundances.

b) They could rain out and settle below the line and continuum forming regions, resulting in no grain opacities detectable in the spectrum ("Cond" models).

c) They can form depleted clouds in the atmosphere so that dust opacities would only be present in the cloud layers but not necessarily in all the layers where the dust had originally formed ("Settle" models).

Observational evidence (see below) suggests that case (a) is realized for $T_{\text{eff}} >$ 1800 K (late M dwarfs to early L dwarfs) whereas for giant planets and extreme T dwarfs case (b) appears to be more appropriate. In the intermediate regime, partial clouds appear to form and case (c) must be investigated. At the present time, this sequence is still very tentative and the physical models of dust and cloud formation have to be refined to obtain a truly physical picture of brown dwarfs and extrasolar giant planets.

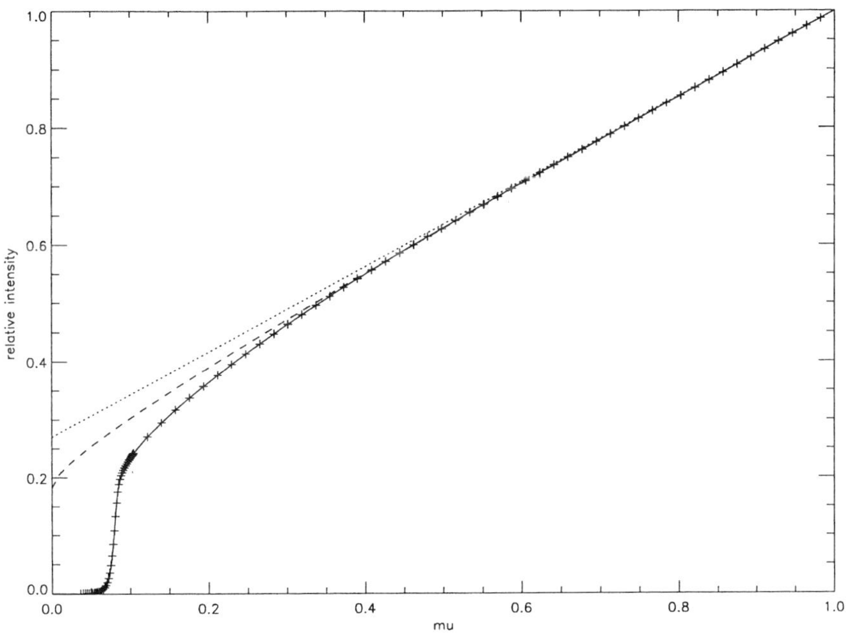

Figure 5. Example of limb darkening for a subgiant model (calculated using spherical geometry and spherically symmetric radiative transfer) with $T_{\rm eff} = 5000\,$K, $\log(g) = 2.5$ and solar abundances. The computed (normalized to the center of the stellar disk) mono-chromatic intensities at ≈ 5000Å are shown (+ symbols) and compared to linear (dotted line) and square root (dashed line) limb darkening laws.

2.4 Microlensing and Stars

Limb darkening is an important tool for the investigation of the structure of stellar atmospheres, in particular for subgiants and giants. It is now possible to use fortuitous microlensing events to study the detailed spectral variations of a microlensed giant, allowing us to more strictly test model atmospheres. In Fig. 5 we show the limb darkening in the optical (~ 5000Å) for a cool giant with $T_{\rm eff} = 5000\,$K and $\log(g) = 2.5$ (solar abundances). For comparison, linear and square-root limb darkening laws are also shown. The form of the limb darkening and the deviations of the computed intensities from the simple limb darkening laws changes substantially with wavelength and model parameters (see Orosz & Hauschildt 2000 also). This is caused by the effects of spherical radiation transport, which tends to concentrate the emitted intensities closer to the center of the stellar disk than the plane parallel approximation does. This is due to the fact that curvature close to the rim of the stellar disk reduces the total emissivity and optical depth and thus produces less emitted radiation than plane parallel models in which the pathlength and optical depth close to the rim actually approaches infinity. This illustrates that in order to obtain reliable and useful stellar atmosphere information from microlensing observations we need to use rather sophisticated modeling, simple approximations will in most cases not work. Using detailed limb darkening models as shown here has helped Orosz & Hauschildt (2000) to address a number of problems regarding lightcurves of eclipsing binaries.

3. Summary and Conclusions

In this paper we have discussed a few new results of stellar atmosphere modeling that have helped to resolve some outstanding problems understanding and interpreting observed stellar spectra. During the last decade, progress was made by breakthroughs in both methodology and computer technology, which has lead to substantially improved models and synthetic spectra. In many cases, even our currently "best effort" models cannot reproduce observed spectra satisfactorily, this is in particular the case for L and T dwarfs. However, this is due to physical effects that we "know" but we cannot currently describe well enough (e.g., incomplete line lists for key molecules or dust and cloud formation processes). Another area that requires much more work is our detailed understanding of winds from both hot and cool stars. There is currently a lot of effort being put into the solution of these key problems, although it is clear that once they are solved, others will pop up in unexpected places.

Acknowledgments

This work was supported in part by the CNRS, INSU and by NSF grants AST-0086246 and AST-9720704, NASA ATP grant NAG 5-8425, NASA ADP grant NAG5-9222, and LTSA grant NAG 5-3619, as well as NASA/JPL grant 961582, NASA grant AST-0086246, as well as grants GO-0-1013C & HST-GO-08671.09-A to the University of Georgia, NASA LTSA grant NAG5-3435 and NASA EPSCoR grant NCCS-168 to Wichita State University. PHH was supported in part by the Pôle Scientifique de Modélisation Numérique at ENS-Lyon. Some of the calculations presented in this paper were performed on the CNUSC IBM SP2, the IBM SP2 and SGI Origin 2000 of the UGA UCNS, on the IBM SP "Blue Horizon" of the San Diego Supercomputer Center (SDSC), with support from the National Science Foundation, and on the IBM SP and the Cray T3E of the NERSC with support from the DoE. We thank all these institutions for a generous allocation of computer time.

References

Allard, F., & Hauschildt, P. H. 1995a, in ESO Astrophys. Symp., The Bottom of the Main Sequence and Beyond, ed. C. Tinney (Heidelberg: Springer-Verlag), p. 32.

Allard, F., & Hauschildt, P. H. 1995b, ApJ, 445, 433.

Allard, F., Hauschildt, P. H., Alexander, D. R., & Starrfield, S. 1997, ARA&A, 35, 137.

Anderson, L. S. 1987, in Numerical Radiative Transfer, ed. W. Kalkofen (Cambridge: University Press), p. 163.

Anderson, L. S. 1989, ApJ, 339, 559.

Baron, E., & Hauschildt, P. H. 1998, ApJ, 495, 370.

Baron, E., Hauschildt, P. H., Nugent, P., & Branch, D. 1996, MNRAS, 283, 297.

Cannon, C. J. 1973, JQSRT, 13, 627.

Hauschildt, P., Baron, E., Starrfield, S., & Allard, F. 1996, ApJ, 462, 386.

Hauschildt, P., Allard, F., Ferguson, J., Baron, E., & Alexander, D. R. 1999, ApJ, 525, 871.

Hauschildt, P. H. 1993, JQSRT, 50, 301.

Hauschildt, P. H., & Baron, E. 1995, JQSRT, 54, 987.

Hauschildt, P. H., & Baron, E. 1999, Journal of Computational and Applied Mathematics, 102, 41.

Hauschildt, P. H., Allard, F., Alexander, D. R., & Baron, E. 1997, ApJ, 488, 428.

Hauschildt, P. H., Baron, E., & Allard, F. 1997, ApJ, 483, 390.

Hauschildt, P. H., Allard, F., & Baron, E. 1999, ApJ, 512, 377.

Hubeny, I., & Lanz, T. 1995, ApJ, 439, 875.

Mathisen, R. 1984, Photo cross-sections for stellar atmosphere calculations — compilation of references and data, Inst. of Theoret. Astrophys. Univ. of Oslo, Publ. Series No. 1.

Orosz, J. A., & Hauschildt, P. H. 2000, A&A, in press.

Rybicki, G. B. 1972, in Line Formation in the Presence of Magnetic Fields, eds. G. Athay, L. L. House, & G. Newkirk Jr. (Boulder: High Altitude Observatory), p. 145.

Rybicki, G. B. 1984, in Methods in Radiative Transfer, ed. W. Kalkofen (Cambridge: Univ. Press), p. 21.

Scharmer, G. B. 1984, in Methods in Radiative Transfer, ed. W. Kalkofen (Cambridge: Univ. Press), p. 173.

Schweitzer, A., Hauschildt, P. H., Allard, F., & Basri, G. 1996, MNRAS, 283, 821.

Schweitzer, S., Hauschildt, P. H., & Baron, E. 2000, ApJ, 541, 1004.

Verner, D. A., & Yakovlev, D. G. 1995, A&AS, 109, 125.

A CRITICAL APPRAISAL OF ATLAS9 AND NEXTGEN5 MODEL ATMOSPHERES

Emanuele Bertone

Osservatorio Astronomico di Brera, Italy and Università Statale di Milano, Italy

Alberto Buzzoni

Osservatorio Astronomico di Brera, Italy and Telescopio Nazionale Galileo, Spain

Miguel Chávez, and Lino H. Rodríguez-Merino

INAOE, Mexico.

Abstract The fitting atmosphere parameters ($T_{\rm eff}$, g, and [Fe/H]) for over 300 stars in the Gunn & Striker (1983) and Jacoby et al. (1984) catalogs have been obtained relying on the Kurucz (1992) ATLAS9 and Hauschildt et al. (1999) NextGen5 synthesis models. The output results are compared, and a critical appraisal of both theoretical codes is performed.

As a major improvement over the standard ATLAS9 code for model atmospheres by Kurucz (1992), the new NextGen5 synthesis code of Hauschildt et al. (1999) adopts a more refined treatment of molecular opacity, and includes spherical symmetry in the atmosphere layers for low-gravity models. In order to assess the main differences between the two theoretical codes, it is of special interest to investigate the theoretical temperature calibration for stars of different spectral type. In this sense, we devised a procedure to determine the fundamental parameters of a star ($T_{\rm eff}$, $\log g$, [M/H]) by comparing its observed spectral energy distribution with a grid of synthetic spectra. The fiducial best model is identified by minimizing the Δ log(flux) standard deviation, $\sigma(f)$, over the full wavelength range of the observations. An application to the Gunn & Stryker (1983, hereafter "GS") and Jacoby *et al.* (1984, "JHC") atlases provided the fitting parameters for over 300 stars by matching with the solar metallicity model grids of ATLAS9 ($3500 \leq T_{\rm eff} \leq 50\,000$ K, $0.0 \leq \log g \leq 5.0$) and NextGen5 ($2000 \leq T_{\rm eff} \leq 10\,000$ K, $0.0 \leq \log g \leq 5.5$). As shown in Figure 1, when comparing the fiducial temperature for GS and JHC class V stars, the NextGen5 fit results in a significantly warmer value of $T_{\rm eff}$ with respect to ATLAS9, especially at the extreme edges of the $T_{\rm eff}$ range.

M. Chávez et al. (eds.),
New Quests in Stellar Astrophysics: The Link Between Stars and Cosmology, 27–28.

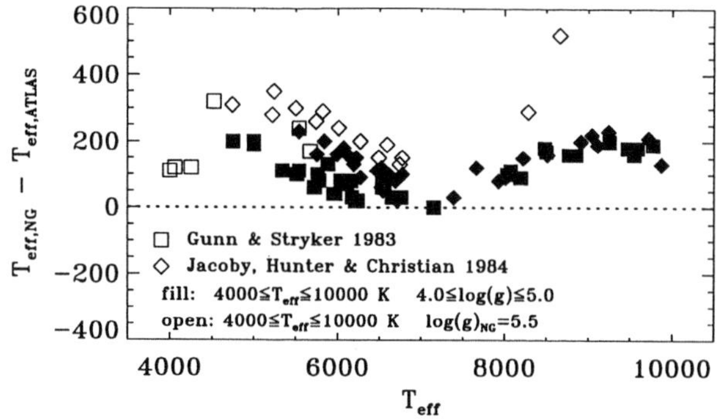

Figure 1. The difference in the fitting value of T_{eff} for the GS and JHC stars as derived from the ATLAS9 and Nextgen5 best-fit models.

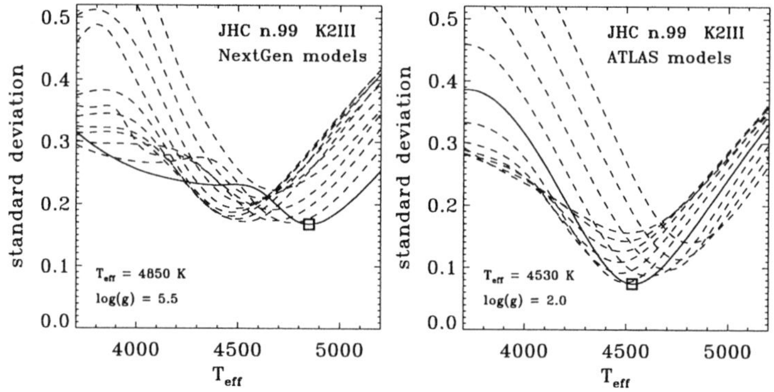

Figure 2. An example of the fitting procedure for star no. 99 in the JHC sample. For ATLAS9 and Nextgen5 models, the open square marks the optimum fit among the different $\sigma(f)$ equi-gravity envelope curves. A solar metallicity is assumed.

Figure 2 is an example of our best-fit procedure for a star in the JHC atlas comparing with both ATLAS9 and NextGen5 models. Equi-gravity envelope curves for the standard deviation $\sigma(f)$ across the temperature range are reported searching for the absolute minimum that marks the fitting T_{eff} and $\log g$. A more univocal and "sharper" solution is in general reached by the ATLAS9 fit, while NextGen5 models display a more entangled trend for the $\sigma(f)$ function, especially as far as cool stars (Sp. type K and M) are concerned.

References

Hauschildt, P. H., Allard, F., & Baron, E. 1999, ApJ, 512, 377.

Kurucz, R. L. 1992 in The Stellar Populations of Galaxies, eds. B. Barbuy & A. Renzini, (Dordrecht: Kluwer), p. 225.

Gunn, J. E., & Stryker, L. L. 1983, ApJS, 52, 121.

Jacoby, G. H., Hunter, D. A., & Christian, C. A. 1984, ApJS, 56, 257.

A PRECISE NEW METHOD TO CORRECTING TEMPERATURE IN STELLAR ATMOSPHERES

O. Cardona
Instituto Nacional de Astrofísica, Optica y Electrónica
A.P. 51 y 216, 72000 Puebla, Mexico

L. Crivellari
Osservatorio Astronomico di Trieste, Italy, and
Instituto de Astrofísica de Canarias, Spain

and

E. Simonneau
Institut d'Astrophysique de Paris, CNRS, France

Abstract

Calculations of energy balance in model stellar atmospheres for cold and moderately cold stars show that the overall energy gained or lost at the Lyman frequencies (lines plus continuum) can be up to 20 orders of magnitude higher than in the rest of the spectrum. Due to the great optical depth at those frequencies, however, monochromatic mean intensity of the radiation and the corresponding source function are numerically equal with many significant figures, and give therefore a vanishing net contribution to the energy balance. As all methods of temperature correction compute both integrals independently, this can lead to a biased temperature correction to the model output. We present here a method to calculate the correction of the temperature in stellar atmospheres that overcomes the aforesaid difficulty.

Introduction

In the present work we discuss a physical problem of great numerical importance for the computation of stellar atmosphere models: the determination of the temperature at each point of the atmosphere from the constraint of energy conservation, in particular from the radiative energy balance.

M. Chávez et al. (eds.),
New Quests in Stellar Astrophysics: The Link Between Stars and Cosmology, 29–32.

The radiative terms that enter the energy balance of a stellar atmosphere are the gain due to the absorption of radiation: $\int \chi(\nu)J(\nu)d\nu$, and the loss brought about by emission: $\int \chi(\nu)S(\nu)d\nu$. It is well known that within each one of the two above integrals the part corresponding to spectral ranges whose opacity $\chi(\nu)$ is very large can overcome by many orders of magnitude the part that corresponds to the remaining frequencies. This is the case of the Lyman continuum and lines in cool (or even mildly cool) stars. (For solar-type stars the energy gains and losses in the spectral lines can be greater by 10 or 15 orders of magnitude than the amount of energy absorbed and, above all, that emitted in the remaining part of the spectrum.) On the other hand, at those frequencies where the opacity $\chi(\nu)$ is very large, the mean intensity $J(\nu)$ of the radiation field shall be equal, up to many significant digits, to the source function $S(\nu)$, and consequently to the Planck function $B(\nu, T)$. Then their net contribution to the energy balance shall be null, albeit separately their contribution to the gain and loss integrals are the most important numerically.
Thus the spectral range whose physical contribution to the overall balance is null will dominate numerically both sides of the energy balance constraint, and consequently the errors on the determination of $J(\nu)$ and $S(\nu)$ at these frequencies will falsify the balance. We shall stress that, because of the numerical approximations employed for the computation of the stellar atmosphere model, the precision of the results for $J(\nu)$ and $S(\nu)$ is of the order of 1%.
In view of that, it results very convenient for this kind of analysis to compute directly the variable

$$\Phi(\nu) \equiv J(\nu) - S(\nu)\,, \tag{1}$$

independently of the computation of $J(\nu)$ and $S(\nu)$. This will enable us to neglect the term $\chi(\nu)[J(\nu) - S(\nu)]$ from the energy balance whenever we consider that the value is small enough, and to take into account the terms $\chi(\nu)J(\nu)$ and $\chi(\nu)S(\nu)$ only for those frequencies that collaborate to the determination of the temperature.
In Fig. 1 we show, for a solar-like model ($T_{eff} = 5770\ K$; $log\ g = 4.5$), the difference $\chi(\nu)[J(\nu) - S(\nu)]$ between absorption and emission as a function of frequency, at a point that corresponds to unity on the Rosseland optical depth scale. At those frequencies where the opacity $\chi(\nu)$ is large (hydrogen free-free for $\nu < 3.10^{14} Hz$; Lyman continuum and metals for $\nu > 2.10^{15} Hz$), the monochromatic energy balance is null, compared with the values corresponding to the other frequencies, at which either emission prevails over absorption (*i.e.* radiative loss) or vice versa (*i.e.* radiative gain). Due to that, to take into account the high opacity frequencies could introduce large numerical errors that would falsify the true energy balance, which is established only by the frequencies in the IR, visible and near-UV spectral range.

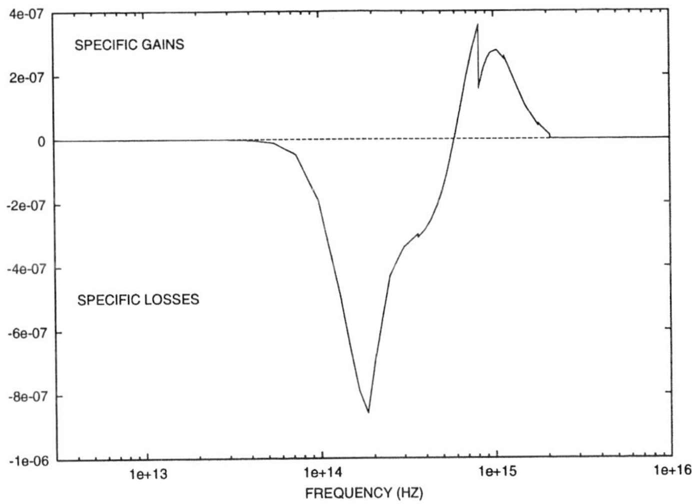

Figure 1. The monochromatic energy balance, $\chi(\nu) = [J(\nu) - S(\nu)]$, for a solar-like model at $\tau_{Rosseland} = 1$. (See text for details.))

1. The Method

The quantity $\Phi(\nu)$ is the integral over directions of the outgoing "non-local intensities" $y^+(\nu,\mu) \equiv I^+(\nu,\mu) - S(\nu)$ and the incoming "non-local intensities" $y^-(\nu,\mu) \equiv I^-(\nu,\mu) - S(\nu)$, where $I^+(\nu,\mu)$ and $I^-(\nu,\mu)$ are the specific outgoing and incoming specific intensities. Thus it holds that

$$\Phi(\nu) \equiv \int_0^1 \frac{y^+(\nu,\mu) + y^-(\nu,\mu)}{2} \, d\mu \, . \tag{2}$$

Then, starting from the standard form of the radiative transfer equation, the equation for the new non-local specific intensities shall be

$$\mu \frac{dy^\pm(\nu,\mu)}{d\tau(\nu)} = \pm y^\pm(\nu,\mu) - \frac{dS(\nu)}{d\tau(\nu)} \, , \tag{3}$$

with the initial conditions

$$y^+(\nu,\mu) = \mu \frac{dS(\nu)}{d\tau(\nu)} \tag{4}$$

for $\tau(\nu) \to \infty$, and

$$y^-(\nu,\mu) = -S(\nu) \tag{5}$$

for $\tau(\nu) = 0$.

Whenever the source function $S(\nu)$ is a linear function of the mean intensity $J(\nu)$, as it happens in those radiative transfer problems that admit a direct solution (or, also, within each step of an iterative procedure employed for solving those that do not admit a direct solution), it results from the definition of $\Phi(\nu)$, as given by eq. (1), that $S(\nu)$ shall be a linear function of $\Phi(\nu)$, too. Therefore we will be able to solve any problem formulated in terms of the non-local intensities $y^{\pm}(\nu, \mu)$ by using the same methods developed in order to solve the problem in terms of the variables $I^{\pm}(\nu, \mu)$.

In order to compute stellar atmosphere models, we must solve a radiative transfer problem, in which the source function appears in the form

$$S(\nu) \;=\; \epsilon(\nu)\, B(\nu, T_0) \;+\; [1 \;-\; \epsilon(\nu)]\; J(\nu)\,. \tag{6}$$

The parameter $\epsilon(\nu)$ is defined as

$$\epsilon(\nu) \;\equiv\; \frac{a(\nu)}{\chi(\nu)} \;=\; \frac{a(\nu)}{a(\nu) \;+\; \sigma(\nu)}\,, \tag{7}$$

where $a(\nu)$ is the absorption coefficient and $\chi(\nu)$ the total opacity. Here, the ratio $\epsilon(\nu)$ and the Planck function $B(\nu, T_0)$ are known functions, while $J(\nu)$, which couples the source function with the specific intensities that are the protagonist of radiative transfer, is the unknown. Of course, $\epsilon(\nu)$ and $B(\nu, T_0)$ are assumed to be known at the beginning of each step of the iteration procedure, during which we have to solve the radiative equations in order to get all the monochromatic mean intensities and the relevant Iteration Factors. The latter will enable us to up-to-date the temperature distribution $T(r)$ in the second part of each step of iterations. (See Simonneau & Crivellari, 1988.) Thus, at each step of iterations we face a set of problems akin to those above discussed. The previous analysis holds valid for their solution.

2. Conclusions

The advantages of the new formulation are self-evident. The gradient of the sources shows up as the protagonist of radiative transfer. The consequence is that we can compute directly the effects of these gradients, hence the flux. The densities (*i.e.* mean intensity and source function) are computed "a posteriori" by adding the local values of the sources. In such a way the numerical values of the source functions, sometimes very large with respect to that of the corresponding gradients, will not blur the computed numerical values of the flux.

Acknowledgments

O.C. acknowledges partial support from Mexican CONACyT, through project 34564-E.

References

Simonneau, E. & Crivellari, L. 1988, ApJ, 330, 415.

NEW PHENOMENA DISCOVERED IN THE ATMOSPHERES OF NORMAL LATE-B TYPE STARS

Swetlana Hubrig
European Southern Observatory, Casilla 19001, Santiago 19, Chile

Glenn M. Wahlgren
University of Lund, Atomic Spectroscopy, Department of Physics, Box 118, 22100 Lund, Sweden

and

Fiorella Castelli
CNR-Gruppo Nazionale Astronomia and Osservatorio Astronomico di Trieste, Via G.B. Tiepolo 11, 34131-Trieste, Italy

Abstract

We report detections of weak emission lines in main sequence B star spectra observed at high spectral resolution and high signal-to-noise ratio. The emission lines originate from highly excited states of singly-ionized species. The mechanism for populating these excited states is currently under investigation and the results may have a significant bearing upon our understanding of the outermost atmospheric regions of B type stars. Another intriguing result of our study is that stars in our sample show evidence for magnetic intensification.

Introduction

There are several perspectives from which to consider high-resolution spectra of main-sequence early-type stars. First, we are interested in the nature of their atmospheres. High-resolution, high signal-to-noise spectra are used to derive the temperature and pressure variation with depth, the chemical composition, the structure and velocity fields on the surface, and the magnetic fields. A second perspective is that since main-sequence early-type stars are uncontaminated by the products of reactions, their chemical composition should reflect that of the interstellar material from which the stars formed. In addition, their

M. Chávez et al. (eds.),
New Quests in Stellar Astrophysics: The Link Between Stars and Cosmology, 33–36.

Table 1. Journal of observations.

HD	HR	Other id.	V	Sp. type	S/N
179761	7287	21 Aql	5.1	B8 III	370
186568	7512		6.1	B8 III	420
196426	7878		6.2	B8 III	460
209459	8404	21 Peg	5.8	B9.5 V	380
219927	8873		6.3	B8 III	320

short lifetimes imply that they will not have moved far from where they were formed. Hence observations of bright nearby early-type stars should provide contemporary cosmic abundance estimates for the solar neighborhood.

1. Observations

The observations were performed on 1999 August 29 at the CFHT with the GECKO spectrograph at a resolving power of R = 120000 and grating settings corresponding to the wavelength intervals 6105–6190 Å. Programme stars are listed in Table 1, where we give the HD number and other identifiers in Cols. 1, 2, and 3, the V magnitude and the spectral type. The achieved signal-to-noise ratios in the continuum are given in Col. 6. The selected five normal late-B type stars have $v \sin i$ values between 4 and 18 km s^{-1}.

2. Detection of optical emission lines

Emission lines in the spectra of normal B type stars have long been sought in connection with ascertaining whether chromospheric conditions exist for them. The mid-B to mid-A type main sequence stars appeared to reside within an emission gap of the H-R diagram between stars cooler than spectral type A7 that display diagnostics of cool star chromospheres and hot B stars, for which emission lines arise from the photoionization of stellar winds induced by radiation pressure. Our search for weak emission lines has been undertaken by comparing observed with synthesized spectra. For each star a LTE model atmosphere and synthetic spectra were generated using the ATLAS 9 and SYNTHE programs (Kurucz 1993), respectively. The stellar parameters effective temperature and gravity were determined from the uvbyβ photometric colors of Hauck & Mermilliod (1998) and a computer code provided by Napiwotzki (1999, private communication) based on the formalisms of Moon & Dworetsky (1985). Overall, we have identified weak emission lines, as characterized by equivalent widths of a few mÅ, associated with some high-excitation levels in Mn II and Cr II. Table 2 presents equivalent width measurements for the typically strongest

Table 2. Stellar Characteristics and Equivalent Width Data

HD	Sp.Type	T_{eff} (K)	log g	$v \sin i$ (km s^{-1})	W_λ(6182.3) Cr II	W_λ(6122.4) Mn II
179761	B8 III	13030	3.43	16	(+0.5)/+2.1	−1.6/+10.3
186568	B8 III	11700	3.50	18	(−0.1)/+1.5	−0.9/+2.5
196426	B8 III	13010	3.84	5.5	−0.6/+1.8	−1.4/+3.5
209459	B9.5 V	10350	3.48	4	−1.0/+0.9	+1.4/+4.7
219927	B8 III	12970	3.63	4.5	−1.2/+2.0	−1.2/+1.8

emission line of Mn II, and Cr II in our data. Each W_λ entry has the form x/y, where x is the observed equivalent width (positive below the continuum, negative above), and y is the equivalent width calculated by spectrum synthesis. Further studies based on high resolution and high signal-to-noise ratio spectra are needed to state whether all main sequence B star atmospheres produce emission lines. The emission lines originate from highly excited states of singly-ionized species. The mechanism for populating these excited states is currently under investigation and the results may have a significant bearing upon our understanding of the outermost atmospheric regions of B type stars.

3. Magnetic fields in normal late-B type stars?

Another intriguing result of our study is that stars in our sample show evidence for magnetic intensification. The anomalous strength of the Fe II λ 6147.7 line relative to Fe II λ 6149.2 in stars with magnetic fields has been used to detect magnetic fields (Hubrig & Castelli 2001). The observed relative differences between the equivalent widths of the two Fe II lines were compared with those derived from synthetic spectra computed neglecting magnetic field effects. All the stars in our sample show evidence for magnetic intensification. No large-scale magnetic fields were ever detected for normal upper-main-sequence stars. Especially striking is the outcome of measurements in the star HD 186568, which has the largest value of $v \sin i$. The difference in equivalent width of the line Fe II 6147.7 Å relative to the line Fe II 6149.2 Å is expected to be 0.72 mÅ, whereas we have measured 2.57 mÅ. Other independent approaches to study magnetic fields in normal late B stars would be the moment technique Mathys (1995), Mathys & Hubrig (1997), or the multi-line Stenflo-Lindegren (Stenflo & Lindegren 1977) technique, both of which look for possible differential broadening of spectral lines having different magnetic sensitivities and which can be very powerful if applied to a suitable sample of spectral lines. Magnetic field detections might also be valuably attempted through the observation of linear polarization in spectral lines. To our knowledge, such observations have never been done for normal late-B type stars.

References

Hauck, B., & Mermilliod, M. 1998, A&ASS, 129, 431.
Hubrig, S., & Castelli, F. 2001, A&A, submitted.
Kurucz, R. L. 1993, CD-ROM 18.
Mathys, G. S. 1995, A&A, 293, 733.
Mathys, G. S., & Hubrig S. 1997, A&ASS, 124, 475.
Moon, T. T., & Dworetsky, M. M. 1985, MNRAS, 217, 305.
Stenflo, J. O., & Lindegren, L. 1977, A&A, 59, 367.

HEATING THE QUIET CORONA BY EXPLOSIVE EVENTS IN UV: RESULTS FROM "SUMER" ON BOARD SOHO

Juan Pablo Torres-Papaqui, and Eduardo Mendoza

Instituto Nacional de Astrofísica, Optica y Electrónica, México

Abstract

We present SUMER observations of UV explosive events (EE) in a range which contains lines from about 750 to 790 Å. The strongest spectral lines of this range have temperatures of maximum abundances in the range $0.8\ 10^5 - 6.3\ 10^5$ K and therefore are expected to be formed in the Transition Region (TR). The aim of this work is to investigate whether the EE originate in lines of a limited range of temperatures or indistinctly at lines of any temperature. We identified the times when the flux of a line attained the 3 and 5-σ levels. Moreover a considerably number of the events attained first a flux of 3 and 5-σ levels at a temperature of $1.5\ 10^5$ K. We interpret the above results as an indication that most of the EE could be originated in the intermediate layers of the TR.

Introduction

The UV explosive events (EE) have been observed at spectral lines at temperatures from about $2\ 10^4$ to 10^6 K Brueckner and Bartoe (1983). They represent a way in which the high magnetic energy of the lower layers of the solar atmosphere may be transformed to heat the Corona. However, up to now it has not been established whether the EE originate in the lower or in upper layers of the TR. Chae et al. (1998) and Brekke et al. (1997) found that the largest red shift take place for lines at about $1.5\ 10^5$ K. In this work we attempt to estimate whether the onset of explosive events has a preferential altitude of the TR or they begin at any altitude. With this purpose we analyzed EE at lines formed at different TR temperatures.

1. Data

The data were acquired with SUMER/SoHO Wilhelm et al. (1997) during 1996 November 14, 15 and 16. The solar radius, from the SoHO location, was 980"

M. Chávez et al. (eds.),
New Quests in Stellar Astrophysics: The Link Between Stars and Cosmology, 37–38.

so one arcsecond corresponded to 715 Km. The spectral resolution was 0.045 Å and the spectral range contains lines from about 750 Å to 790 Å.
At each day the slit was directed towards different locations. In each of them it was not adjusted to the solar rotation. A region was observed first with a slit of 4" width and 300" length in a 30 seconds cadence. After that the same region was observed with a 1" width and 300" length slit in a 570 seconds cadence. At the end of this last set of observations another target was observed, first with the 4" slit (with a 30 s cadence) and then with the 1" slit (in a 570 s cadence) and so on.
Several EE were either partially or fully recorded. From them we selected those observed from the beginning. The lines that showed the most considerably flux enhancement during each event were selected.

2. Results and Discussion

We found that the 3-σ flux was predominantly first attained at the 1.5 10^5 K lines indicates that the EE preferentially begin at mid altitudes of the TR. An energy source located around the altitude with this temperature could be the responsible for the flux enhancement at these lines faster than at the other lines. The possibility that the origin of EE is around the layer with a temperature of 1.5 10^5 K does not considerably disagrees with Teriaca et al. (1999) assumption that magnetic reconnection, and therefore the energy release, during EE take place around 3 10^5 K.
These results seem to indicate that between the height responsible for the emission at the 1.5x10^5 K line and those for the neighboring lines the physical conditions could considerably differ. Probably a sharp thermal/turbulent velocity gradient around these lines is the responsible for this difference. We consider that these results support the possibility that the region of energy release is around the layer with 1.5x10^5 K which is in the intermediate layers of the TR.

References

Brekke, P., Hassler, D.M., & Wilhelm, K. 1997, Solar Phys., 175, 349.
Brueckner, G.E., & Bartoe, J.-D.F. 1983, ApJ, 272, 329.
Chae, J., Yun, H.S., & Poland, A.I. 1998, ApJS, 114, 151.
Teriaca, L., Doyle, J.G., Erdélyi, R., & Sarro, L.M. 1999, A&A, 352, L99.
Wilhelm, K., et al. 1997, Solar Phys., 170, 75.

A GRID OF SYNTHETIC STELLAR UV FLUXES

Lino H. Rodríguez-Merino, Miguel Chávez
Instituto Nacional de Astrofísica, Optica y Electrónica, Mexico

Alberto Buzzoni
Telescopio Nazionale Galileo, Spain, and Osservatorio Astronomico di Brera, Italy

and

Emanuele Bertone
Osservatorio Astronomico di Brera, Italy

Abstract

We present preliminary results of a large project aimed at creating an extended theoretical and observational database of stellar spectra in the ultraviolet wavelength range. This library will consist of IUE spectra at low and high resolution, and a set of LTE and NLTE theoretical fluxes. A first grid of 50 model fluxes with solar metallicity, in the wavelength interval 1000–4400 Å, is reported here. Calculations are based on the Kurucz (1993) SYNTHE code. The models span effective temperatures between 10 000 K and 50 000 K, and a surface gravity in the range $2.5 \leq \log g \leq 5.0$ dex.

Introduction

Over the past two decades, most of the work dealing with the study of stellar aggregates by using synthesis techniques has concentrated in the optical and near IR wavelength interval. Detailed analyses have been carried out to investigate the usefulness of many absorption features as indicators of age and/or global chemical composition of stellar populations Buzzoni (1995). Although in the UV there are several comprehensive works Fanelli et al. (1987); Brown et al. (1996), this spectral window has not yet been fully exploited and represents a promising tool for complementing our current knowledge. Moreover, at intermediate and high redshift the UV spectra of galaxies are detected in the optical and therefore a characterization of UV spectral features is mandatory for the interpretation of data collected with large telescopes.

M. Chávez et al. (eds.),
New Quests in Stellar Astrophysics: The Link Between Stars and Cosmology, 39–40.

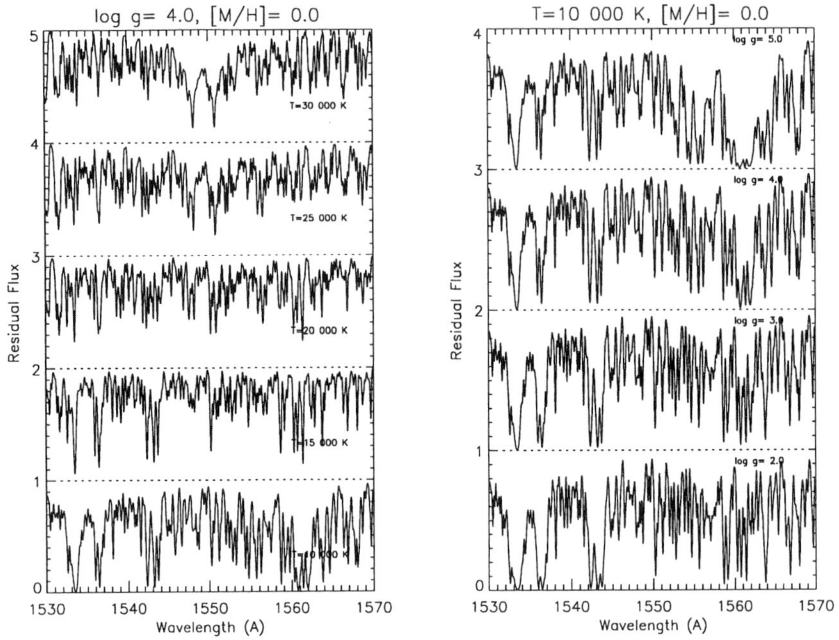

Figure 1. Intensity of prominent UV absorption features vs T_{eff} and log g in the 1530–1570 Å wavelength range. Theoretical spectra computed with Kurucz (1993) synthesis code are shown at the IUE resolution of 0.2 Å FWHM.

1. Synthetic Spectra

Relying on the Kurucz (1993) SYNTHE code for model atmospheres, we computed a preliminary grid of 50 high-resolution ($\lambda/\Delta\lambda$ =50 000) synthetic spectra in the wavelength interval from 1000 Å to 4400 Å. Spectra cover an effective temperature range between 10 000 K and 50 000 K at step of 5000 K, a surface gravity from log g = 2.5 to 5.0 at step of 0.5 dex, and adopt a solar chemical composition. In Fig. 1 we show the behavior of metallic absorption features (mainly Fe) in the interval 1530–1570 Å vs. effective temperature and gravity. We plan to define a set of spectroscopic indices that measure the intensity of absorption lines and calibrate them according to the atmosphere fundamental parameters. This will eventually result in a set of fitting functions for use in population synthesis studies.

References

Brown, T. M., Ferguson, H. C., & Davidsen, A. F. 1996 ApJ, 472, 327.

Buzzoni, A. 1995, ApJS, 98, 69.

Fanelli M., O'Connell R., & Thuan T. 1987, ApJ, 321, 768.

Kurucz, R. L. 1993, CD-Rom 13 and CD-Rom 18, and later upgrades.

A GRID OF LTE ZERO-METALLICITY STELLAR FLUXES

M. Chávez, and O. Cardona
Instituto Nacional de Astrofísica, Optica y Electrónica,
A.P. 51 y 216, 72000 Puebla, Mexico

Abstract

We present a grid of LTE theoretical spectral energy distributions (SEDs) of models composed of only H and He. The calculations were based on the codes developed by L. Auer and considered 71 atmospheric layers in hydrostatic and radiative equilibria. The SEDs consist on 46 wavelength points extending from the extreme ultraviolet to the infrared and cover the parameter space from 10 000 to 140 000 K in effective temperature and 4.0–6.0 dex in surface gravity. The coarse coverage in wavelength makes the grid suitable for continuum studies since the Lyman and Balmer breaks at 912 and 3647 Å respectively are among the most prominent features for the parameter space covered by the grid.

Introduction

Model atmospheres and theoretical fluxes computed from them are a fundamental building block in many astrophysical contexts. They are widely used in diverse methods for determining stellar leading atmospheric parameters when compared to stellar observations (Malagnini et al. 2000) or have been implemented in population synthesis codes to model the spectrophotometric characteristics of stellar systems e.g. Bressan et al. (1994). Another important application is related to the analysis of properties of the interstellar medium such as HII regions or, at cosmological time scales, the investigation of the role of primordial stars on ionizing the intergalactic medium. It is in this later phenomenon that the grid presented in this paper may find direct application. Despite the wide variety of definitions given to the so-called Population III stars there seems to be concensus on their chemical composition. The first generation of stars should have been composed of zero or nearly zero ([M/H] < -6.0) metals. Massive metal-free stars are of much higher effective temperature (T_{eff}) than their population I counterparts. The lack of metal opacity prevents the outer atmospheric layers to blow-off by radiative pressure, allowing the

M. Chávez et al. (eds.),
New Quests in Stellar Astrophysics: The Link Between Stars and Cosmology, 41–44.

formation of extremely massive objects of relatively small radii. Evolutionary tracks of such objects have been computed by several authors. Recently Marigo et al. (2000) presented a grid of evolutionary models showing that 100 $M_\odot$ stars can reach up to 100 000 K or twice as much if mass-loss is included.
Existing grids of model atmospheres are generally limited in the parameter space. Either they do not cover the very high temperature regime or the lowest metallicity available is not zero. As we shall later illustrate, for the current grids to be applicable for the analysis of population III stars, we require the metallicities to be [M/H] < 2.5. Higher metallicities than this threshold include important effects at the extreme and far ultraviolet wavelength intervals (200–1200 Å) because of the presence of numerous lines of highly ionized metals.
In this paper we present a grid of $Z = 0$ LTE stellar fluxes computed from classical LTE model atmospheres based on the codes developed by Auer and Heasley (1971). We also illustrate a qualitative comparison of some of our grid entries with Kurucz's grid of theoretical fluxes (Kurucz 1993).

1. The Auer Model Atmospheres

The calculations were carried out using a program coded by Auer and Heasley (1971) and the method described by Mihalas (1978). The models are plane parallel atmospheres with the approximations of hydrostatic and radiative equilibria. The continuum opacities considered are those of hydrogen and helium including six edges for each ionization stage. For the computation we considered 71 depth points ranging from $-7.0 < \tau_{ROSS} < 2.4$. and a set of 46 wavelength points extending from 113 Å to about 4.5 μ.
Using the code described above we computed a grid of nearly 90 fluxes covering the parameter space between 10 000 K and 140 000 K and surface gravities between 4.0 and 6.0 dex with a step of 0.5 dex. In Fig. 1 we illustrate a subset of fluxes for $\log g = 5.0$ and temperatures from 50 000 to 100 000 K. At this $T_{\rm eff}$ regime the main features are the Balmer and Lyman breaks at 912 Å and 3647 Å respectively, and the helium ionization edge at 504 Å. For temperatures below 50 000 K the coverage is more coarse since, as illustrated in the next section, the more extensive Kurucz grid conveniently covers that range at an increased resolution.

2. Comparison with the Kurucz grid

In this section we show the comparison between the theoretical fluxes described above and those computed from Kurucz model atmospheres using ATLAS9 code (Kurucz 1993). We first illustrate in Fig. 2a the effects of varying chemical composition in a model of Teff = 50 000 and $\log g = 5.0$. Note that for $\lambda >$ 912 Å the effect produces a difference in flux of 10% in almost five decades in [M/H], while for $\lambda < 912$ this difference is at least 25%. In order to determine

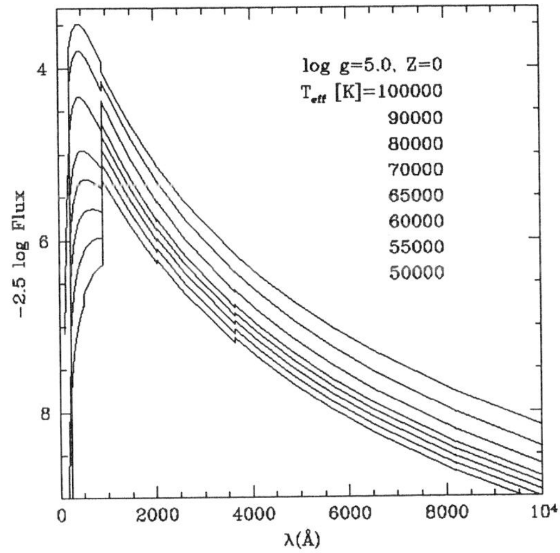

Figure 1. Sequence of energy distributions computed with Auer's code for surface gravity, log $g = 5.0$.

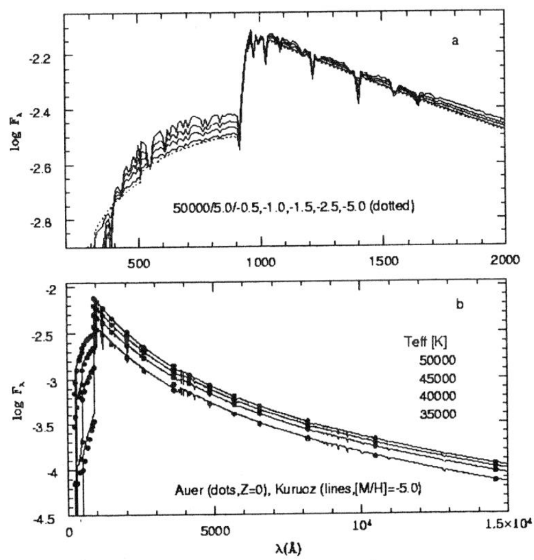

Figure 2. (a) Effects of the chemical composition on the energy distribution around the Lyman break at 912 Å for a model flux of $T_{eff} = 50\,000$ K and log $g = 5.0$. The lowest metallicity [M/H] = −5.0 is represented with a dotted line. (b) Comparison with the Kurucz grid for $T_{eff} \leq 50\,000$ K and gravity log $g = 5.0$

the usefulness of Kurucz grid we have compared a subset of model fluxes for T_{eff}=50 000 K. Fig. 2b displays a sequence of spectra for four models of gravity log g=5.0. Dots correspond to results using Auer's code and solid lines display the energy distribution extracted from Kurucz's grid for the lowest metallicity available, [Fe/H] = -5. There is remarkable good agreement between both sets of fluxes with perhaps minor deviations below 1000 Å. This can be explained by the slight differences in the helium content. The discussion of the effects of helium content on the SEDs is presented in Chavez & Cardona (in preparation).

Acknowledgments

M.C. would like to thank CONACyT for financial support through grant 28506-E. He also acknowledges financial support from INAOE for the organization of this conference.

References

Auer, L, & Heasley, J. 1971, Unpublished Yale University Observatory Report.

Bressan, A., Fagotto, F., & Chiosi, C. 1994, ApJS, 94, 64.

Kurucz, R. L. 1993, ATLAS9 model atmosphere programs and 2 km/s grid, CD-Rom 13 and later updates.

Malagnini, M. L., Morossi, C., Buzzoni, A., & Chavez, M. 2000, PASP, 112, 1455

Marigo, P., Chiosi, C., Girardi, L., & Sarrubi, T. 2000, in MPA/ESO workshop, The First Stars, eds. A. Weiss, T. Abel, & V. Hill (Dordrecht: Kluwer), p. 119.

Mihalas, D. 1978, in Stellar Atmospheres, 2nd edition (New York: W. H. Freeman & Company)

II

POPULATION III AND ZERO-METAL STARS

THE EVOLUTIONARY PROPERTIES OF ZERO METAL STARS IN THE MASS RANGE BETWEEN 4 AND 100 $M_\odot$

Alessandro Chieffi
Istituto di Astrofisica Spaziale del CNR, Roma, Italy

Marco Limongi
Osservatorio Astronomico di Roma a Monteporzio Catone, Italy

Inma Domínguez
Dpto. de Física Teórica y del Cosmos, Universidad de Granada, Spain

and

Oscar Straniero
Osservatorio Astronomico di Collurania, Teramo, Italy

Abstract In this review we will discuss the main evolutionary properties of the zero metallicity stars in the mass interval which extends from the intermediate mass to the massive stars together with their final yields.

Introduction

Current Big Bang models predict that essentially no metals were produced by the Big Bang. If we couple this result to the current belief that metals are synthesized within the stars, we cannot escape the conclusion that zero metal stars certainly formed and that they were able to pollute the interstellar medium with freshly synthesized nuclei. Unfortunately the mass function of this first generation of stars is still unknown. In the past decades the most accepted idea was that only massive stars could be formed by a mixture of pure H and He (due to the presumed lack of a good cooler able to reduce the Jeans mass) while in this last years it has become increasingly convincing that molecular

M. Chávez et al. (eds.),
New Quests in Stellar Astrophysics: The Link Between Stars and Cosmology, 47–54.

hydrogen may have been an efficient cooler able to reduce the Jeans mass down to $\simeq 1.4 M_{\odot}$. At present there are a variety of proposed IMF which range from the formation of just Very Massive Object (Ober, El Eid and Fricke 1983) to the formation of a doubly peaked distribution (one around 10-100 $M_{\odot}$ and one around 2 $M_{\odot}$), (Nakamura and Umemura 1999).

Though the interest in this first stellar generation is largely connected to the yields they have produced, only very few papers analyzed the behavior of these stars up to their final fate (cooling phase or explosion): in particular, the AGB phase of primordial intermediate mass stars has been addressed by Chieffi and Tornambé (1984), Fujmoto et al. (1984) and Chieffi et al. (2001), while the yields produced by massive zero metal stars have been computed, up to now, by Woosley and Weaver (1995) and Limongi, Straniero & Chieffi (2000).

In the following we will briefly review the yields produced by zero metal stars in the range 4-100 $M_{\odot}$. For a review of the evolutionary properties of low mass stars (below 1 $M_{\odot}$) see Schlattl, this volume. The intermediate mass stars will be discussed in section 2 while the massive stars are presented in section three together with their explosive yields.

1. The intermediate mass stars

Intermediate mass stars are usually defined as those stars which ignite He in a not degenerate He core and that form a degenerate CO core after the central He burning. The formation of a degenerate core after the central He burning, coupled to the existence of an efficient shell burning(s), pushes these stars towards the Hayashy track first and towards larger luminosity later, when the penetration of the convective envelope in the deep interior favors the energy transmission through the H rich layers. As the He burning shell advances progressively in mass, the overlying layers expand and cool down so that, first, the H burning shell powers down and then the convective envelope penetrates within the He core (for masses larger than a threshold value which depends on the initial chemical composition) competing for a while with the He shell in reducing the He rich layers. As the He burning shell approaches the H-He discontinuity, it weakens progressively while the overlying layers, no more sustained by the He shell, contract again and heat: the following H-shell re-ignition marks the passage from the previous E(arly)-AGB phase to the T(hermally)P(ulsing)-AGB phase. This phase is characterized by a complex interplay between the H and He burning shells in which the two shells alternate as energy producers (the Thermal Pulses): the expansion of the mantle of the star induced by the He burning at each thermal pulse favors the penetration of the convective envelope within the H-exhausted core and eventually the dredge-up to the surface of freshly synthesized material. A further modification of the surface chemical composition may also come from the partial efficiency of the H burning (Hot

Bottom Burning) at the base of the convective envelope (which would mainly lead to a significant production of ^{14}N). By the way, note that while the third dredge-up is directly connected to the existence of the thermal pulses, the Hot Bottom Burning is, in principle, independent on the existence of the thermal pulses. The final fate of these stars is to loose all the envelope and to end their life as white dwarfs. The general behavior of the AGB stars sketched above holds, in general, in a wide metallicity interval which ranges from that of most metal poor galactic globular clusters up to the solar metallicity. In spite of this well understood qualitative behavior, a quantitative comprehension of even the two key properties of the stars in this phase is still missing, i.e. the efficiency of the third dredge up (which means the amount of material dredged to the surface per thermal pulse) and the mass loss rate(s) during a full (H to He to H) cycle. It is therefore clear that the total yields produced by these stars is still subject to a large degree of uncertainty.
The behavior of primordial metal free intermediate mass stars in the AGB phase has been discussed up to now by Chieffi and Tornambé (1984) and Fujmoto et al. (1984) in the middle of the 80's and recently by Chieffi et al. (2001). The main result of the Chieffi and Tornambé (1984) paper was that a primordial $5M_\odot$ star did not experience any thermal pulse but, on the contrary, that the two burning shells where able to advance in mass at the same rate and hence to be simultaneously active. The obvious consequence of that finding was that the third dredge-up could not occur and hence that it was not possible to bring to the surface (and hence in the interstellar medium) the products of the He burning (mainly C and O) and to convert part of the C in N. In such a scenario the primordial intermediate mass zero metal stars would not contribute to the initial chemical enrichment of C, N and O. On the other hand, the lack of the thermally pulsing phase could have prevented in these stars the arising of a strong mass loss phase and hence favor the possibility for these stars to reach the Chandrasekhar mass and explode as type $1_{1/2}$ Supernovae (thermonuclear supernovae). It goes without saying that this kind of scenario would have important consequences on the first phase of chemical enrichment of the matter which emerged from the Big Bang since it would naturally allow the very early existence of thermonuclear supernovae. The continuous improvement of the input physics included in a stellar evolution code, led Chieffi et al. (2001) to recompute, almost 15 years after that first pioneering work, a set of primordial intermediate mass stars. These updated computations have shown that, at odds with the old findings, primordial intermediate mass stars actually evolve like the more metal rich ones after a first phase of readjustment. For more details see Chieffi et al. (2001). Here it suffices to say that current models of primordial intermediate mass stars predict an (almost) standard thermally pulsing phase and hence a rather canonical production of yields of C, N and O. Table 1 shows the yields which may be derived by our AGB models for two different values

Table 1. Yields from Z=0 stars

M	η	N_{TP}	He	Li	C	N	O
4	1	219	1.3	$2.9 \cdot 10^{-9}$	$3.6 \cdot 10^{-2}$	$6.3 \cdot 10^{-2}$	$9.0 \cdot 10^{-3}$
4	3	90	1.3	$6.9 \cdot 10^{-9}$	$2.1 \cdot 10^{-2}$	$1.2 \cdot 10^{-2}$	$3.2 \cdot 10^{-3}$
4	6	48	1.3	$5.3 \cdot 10^{-9}$	$1.4 \cdot 10^{-2}$	$6.6 \cdot 10^{-4}$	$1.4 \cdot 10^{-3}$
7	1	348	2.2	$2.9 \cdot 10^{-9}$	$3.8 \cdot 10^{-2}$	$5.0 \cdot 10^{-2}$	$6.9 \cdot 10^{-3}$
7	3	196	2.2	$5.2 \cdot 10^{-9}$	$2.1 \cdot 10^{-2}$	$2.8 \cdot 10^{-2}$	$3.8 \cdot 10^{-3}$
7	6	131	2.2	$7.9 \cdot 10^{-9}$	$1.4 \cdot 10^{-2}$	$1.8 \cdot 10^{-2}$	$2.5 \cdot 10^{-3}$

of the parameter (η) which enters in the Reimers mass loss formula and one value (0.01) of the free parameter (f) we adopt to determine the behavior of the base of the convective envelope and hence the efficiency of the third dredge-up Chieffi et al. (2001).

2. The massive stars

The H-burning phase of the massive ($15 \leq M/M_{\odot} \leq 100$) zero metal stars closely resembles that of non zero metallicity stars because these stars produce locally the amount of Carbon which is necessary to fully activate the CNO cycle. The reason is that they are massive enough that the PP-chain can not produce the required amount of energy by keeping the central temperature below ($T \simeq 10^8 K$) and hence they reach the threshold value for the activation of the 3α process. Therefore, since these stars are fully supported by the CNO cycle, the size of the convective core (and hence the He core mass) closely resembles that of stars having the same mass and a low metallicity (e.g. $Z = 10^{-3}$). Since the evolutionary phases beyond the H-burning depend primarily on the size of the He core which forms at the end of the H-burning, also all the advanced burning phases will not be much different from those experienced by metal poor massive stars. We refer the reader to the paper by Limongi, Chieffi and Straniero (2001) for a detailed discussion of the dependence of the advanced burnings on the initial metallicity and also for a discussion of the technique we adopt to simulate the explosion of these stars. Here we just want to briefly discuss the dependence of the yields on the initial mass.
For the sake of simplicity we divide the elements from C to Ni in three groups. The first one includes all the elements which do not depend significantly on the final mass cut, i.e. those which are almost unaffected by the explosion (C and O), those which are produced by the C and Ne burnings and modified later by the C and Ne explosive burnings (Ne, Na, Mg, Al, P and Cl) and those which are produced by the explosive O burning and incomplete explosive Si burning (Si, S, Ar, Ca and K). The second group includes the three elements which are mainly produced by the incomplete explosive Si burning (V, Cr and Mn) while

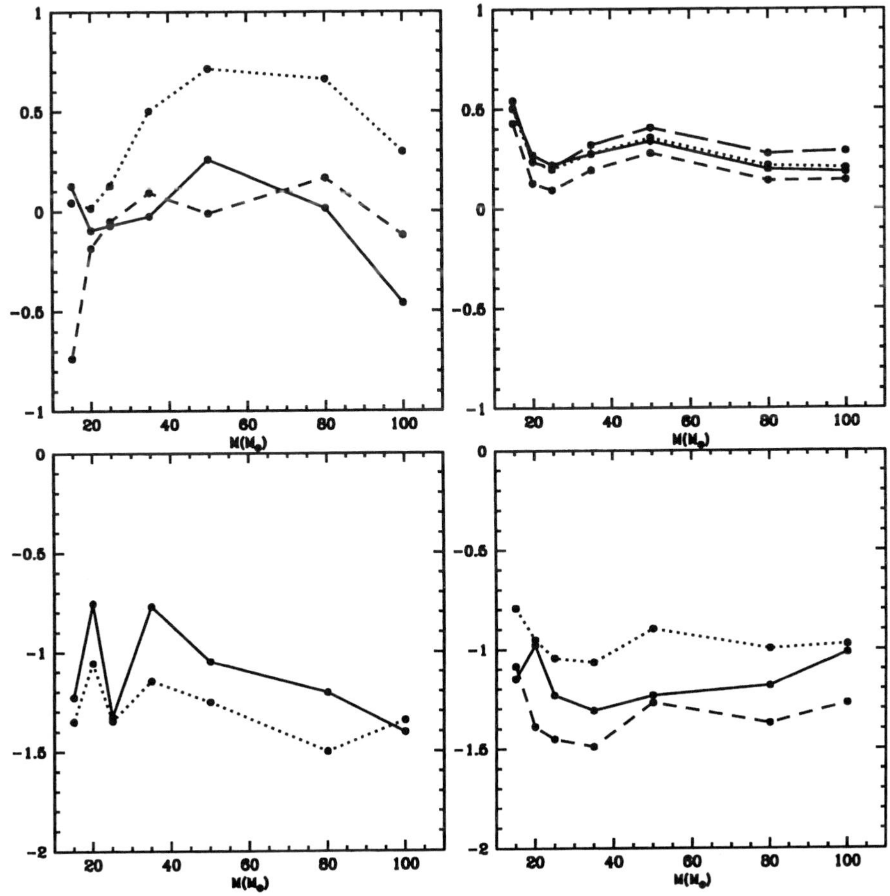

Figure 1. Theoretical ratios [x/Mg] $\equiv Log(x/Mg) - Log(x_{\odot}/Mg_{\odot}$ vs. stellar initial mass. Upper left panel: C (solid line), O (dotted) and Ne (short dashed) Upper right panel: Si (solid), S (dotted), Ar (short dashed) and Ca (long dashed) Lower left panel: Na (solid) and Al (dotted) Lower right panel: P (solid), Cl (dotted) and K (short dashed).

the third one includes the elements whose yields come from the complete and incomplete explosive Si burnings (Sc, Ti, Co and Ni).

Let us consider the first group. Figure 1 shows the trend of the ratios [x/Mg] ($\equiv Log(x/Mg) - Log(x_{\odot}/Mg_{\odot})$ with the initial mass. C (solid), O (dotted) and Ne (short dashed) are shown in the upper left panel; Si (solid), S(dotted), Ar(short dashed) and Ca (long dashed) are shown in the upper right panel; Na (solid) and Al (dotted) are in the lower left panel while P (solid), Cl (dotted) and K (short dashed) are in the lower right panel. We prefer to show the [x/Mg] rather than the [x/Fe] because the [x/Mg] ratios of these elements are largely independent on the mass cut. The trend of C, O and Ne with the mass shows that

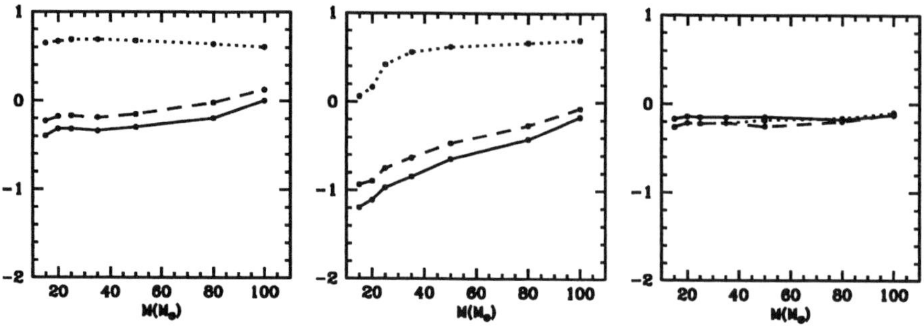

Figure 2. Theoretical abundance ratios for [V/Fe] (solid line), [Cr/Fe] (dotted) and [Mn/Fe] (short-dashed) according to different ^{56}Ni mass ejecta. See text for details.

there is not a monotonic behavior. However, at least in the range $15 - 40\ M_\odot$, it may be seen that it exist a clear monotonic dependence of [O/Mg] and [Ne/Mg] on the initial mass, the first of the two ranging between 0 and 0.7 dex and the second between –0.7 and 0. [C/Mg], on the contrary, remains relatively flat and close to zero up to a mass of 80 $M_\odot$. [Si/Mg], [S/Mg], [Ar/Mg] and [Ca/Mg] behave similarly to each other and appear to be substantially independent on the initial mass and more or less clustered around 0.2-0.3 dex above the solar value. [Na/Mg] and [Al/Mg] are substantially under solar and show a rather scattered behavior which reflects the differences in the presupernova structures at the moment of the core collapse. Also [P/Mg], [Cl/Mg] and [K/Mg] do not show a monotonic dependence on the initial mass even if each of them oscillates within 0.3 dex over the full mass range. Also in these case we predict values which are well below the solar ratios and which range between –1 and –1.5 dex. The trend with the mass of the three elements produced mainly by the incomplete explosive Si burning is shown in Fig. 2: since they depend on the mass cut, we prefer in this case to show the [x/Fe] ratios for two different choices of the amount of Fe ejected by the explosion. The left panel in Fig. 2 shows the trends we obtain by assuming that each star in the mass range $15 - 100\ M_\odot$ ejects 0.01 $M_\odot$ of ^{56}Ni while the central one shows the case in which one assumes that all stars eject 0.1 $M_\odot$ of ^{56}Ni. The solid, dotted and short dashed lines in these two panels refer, respectively, to [V/Fe], [Cr/Fe] and [Mn/Fe]. A comparison between these two panels in Fig. 2 shows that the trend with the mass of these three elements depends significantly on the adopted mass cut. If a low amount of ^{56}Ni is ejected, both V and Mn show a very mild dependence on the initial mass while they show a rather steep gradient with the initial mass if a rather large amount of ^{56}Ni is ejected. Of course these trends must be seen simply as examples of what can be obtained by changing the mass cut. It must be noted, however, that V and Mn show a very similar trend with the mass for both mass cuts so that their internal ratio [V/Mn] is almost independent on both

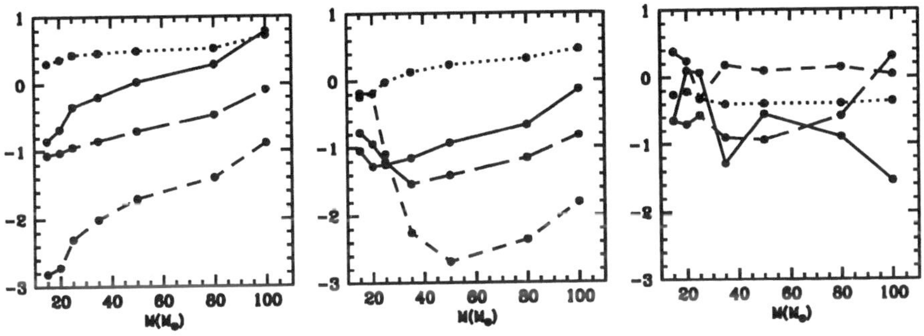

Figure 3. Same as in Fig. 2, but for [Sc/Fe] (solid line), [Ti/Fe] (dotted), [Co/Fe] (short dashed) and [Ni/Fe] (long dashed). See text for further details.

the initial mass and the mass cut: this is shown in the right panel of Fig. 2 where the trend of the ratio [V/Mn] with the mass is shown for three different choices of the ^{56}Ni ejected, i.e. 0.01 $M_{\odot}$ (solid line), 0.1 $M_{\odot}$ (dotted line) and 0.2 $M_{\odot}$ (short dashed line). In other words we predict that the most metal poor stars have $[V/Mn] \simeq -0.2$ and that this value does not depend on the initial mass function.

The behavior of the elements pertaining to the third group is shown in Fig. 3. Once again we show [x/Fe] because the yields of these elements totally depend on the mass cut. The three panels show the trends one obtains for three different choices of the ^{56}Ni ejected. The left and central panels refer, respectively, to the case in which 0.01 $M_{\odot}$ and 0.1 $M_{\odot}$ of ^{56}Ni is ejected; the right panel refers to the case in which the mass cut is fixed at the border of the Fe core. It goes without saying that in this last case the amount of ^{56}Ni ejected depends on the initial mass. The four lines in this figure refer to [Sc/Fe] (solid), [Ti/Fe] (dotted), [Co/Fe] (short dashed) and [Ni/Fe] (long dashed). These three panels clearly demonstrate how both the trends with the initial mass and the absolute values of these ratios largely depend on the adopted mass cut. Only a comparison between observational data and theoretical prediction could help in constraining the mass cut for different masses. Just as an example let us mention that [Ti/Fe] behaves, in the observed metal poor stars (Ryan, Norris and Beers 1996), like the α elements Si, S, Ar and Ca. Such a behavior is not obvious a priori because this element is produced more internally than the other α elements and hence it should not necessarily follow their behavior. Figure 3 shows that [Ti/Fe] depends mildly on the initial mass and largely on the mass cut. Hence, by requiring that [Ti/Fe] $\simeq 0.3$ for [Fe/H] ≤ -1 (i.e. that it behaves like Si, S, Ar and Ca), one can put strong constraints on the range of permitted mass cuts. In particular, Fig. 3 shows that the quoted value ($\simeq 0.3$) is compatible only with an amount of ^{56}Ni ejected not larger than $\simeq 0.1$ $M_{\odot}$ for all the mass interval examined so far.

A deeper comparison between the chemical composition of the extremely metal poor stars and the theoretical predictions will be presented in a forthcoming paper.

References

Chieffi, A., Dominguez, I, Limongi, M., & Straniero, O. 2001, ApJ (in press).

Chieffi, A., & Tornambé, A. 1984, ApJ, 287, 745.

Fujmoto, M.Y., Iben, I. Jr., Chieffi, A., & Tornambé, A. 1984, ApJ, 287, 749.

Limongi, M., Chieffi, A., & Straniero, O. 2001 (in preparation).

Limongi, M., Straniero, O., & Chieffi, A. 2000, ApJS, 129, 625.

Nakamura, F., & Umemura M. 1999, ApJ, 515, 239.

Ober, W. W., El Eid, M. F., & Fricke, K. J. 1983, A&A, 119, 61.

Ryan, S. G., Norris, J. E., & Beers, T. C. 1983, ApJ, 471, 254.

Woosley, S. E., & Weaver, T. A. 1995, ApJS, 101, 181.

THE PECULIAR EVOLUTION OF LOW-MASS POP III STARS

H. Schlattl
Max-Planck-Institut für Astrophysik, Karl-Schwarzschild-Str. 1, D-85741 Garching, Germany, and Astrophysics Research Institute, Liverpool John Moores University, Twelve Quays House, Egerton Wharf, Birkenhead CH41 1LD, UK

S. Cassisi
Osservatorio Astronomico di Collurania, Via M. Maggini, I-64000 Teramo, Italy

A. Weiss
Max-Planck-Institut für Astrophysik, Germany

and

M. Salaris
Astrophysics Research Institute, Liverpool John Moores University, UK

Abstract We have followed the evolution of low-mass, initially metal-free stars from the zero-age main sequence to the late AGB under the assumption of pollution with external, metal-rich material and in the presence of diffusion. We identify conditions under which helium-flash induced mixing between the envelope and carbon-rich layers occurs and compare the resulting surface abundances and post-flash stellar parameters with those of observed extremely metal-poor stars.

Introduction

According to the Big Bang Theory the first stars in the universe should have been formed almost purely of hydrogen and helium. However, up to now solely stars with iron content [Fe/H] ≥ -4 (see, e.g., Beers et al. 1992, the HK-survey) have been observed. We follow the idea that the surfaces of still living stars of the first generation have been polluted by interstellar material which had been enriched with metals, e.g., from nearby supernovae. We are therefore computing the evolution of low-mass initially metal-free stars until the late AGB stage, including external pollution and gravitational settling of heavy elements.

M. Chávez et al. (eds.),
New Quests in Stellar Astrophysics: The Link Between Stars and Cosmology, 55–58.

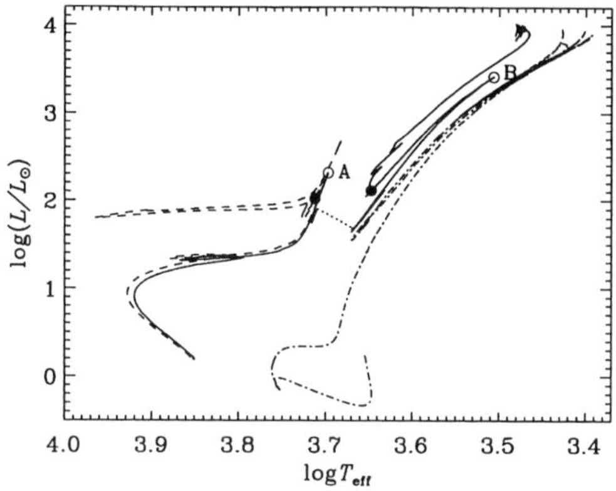

Figure 1. Evolutionary tracks of metal-free stars of 1 $M_\odot$ with $Y_i = 0.23$ (solid) and $Y_i = 0.24$ (dashed line), respectively. The open circles indicate the position of the major ("A") and second He-flash ("B") of the model with $Y_i = 0.23$. For reference the evolution of a 1 $M_\odot$ Pop. I star is shown by the dash-dotted line.

1. The He core flash of low-mass metal-free stars

In Fig. 1 the evolution of two metal-free stellar models of 1 $M_\odot$ with different initial helium content of $Y_i = 0.23$ and $Y_i = 0.24$, respectively, are shown. Note, that for these sequences diffusion and pollution have been neglected. Until the core He flash both models evolve almost identically and — correcting for their lack of metals — as expected similar to metal-rich stars (dash-dotted line in Fig. 1). However, while after the core He flash the $Y_i = 0.24$ model continues its "usual" evolution to the horizontal branch and subsequent asymptotic giant branch, the $Y_i = 0.23$ model shows a completely different behavior: a jump in the HR-diagram at $\log(L/L_\odot) \approx 1.7$ to lower temperatures caused by a dredge-up of large amounts of carbon and nitrogen to the surface.
The origin of this new evolutionary branch is an additional mixing event during the core He flash (see also Fujimoto et al. (1990) or Schlattl et al. (2001)): the expanding He-flash driven convective zone is penetrating into H-rich regions transporting H into deep, very hot regions, where the fresh protons are almost immediately consumed up in the $^{12}\mathrm{C}(\mathrm{p},\gamma)^{13}\mathrm{N}(\beta^+\nu_\mathrm{e})^{12}\mathrm{C}(\mathrm{p},\gamma)^{14}\mathrm{N}$ chain producing a large amount of energy ("H flash", Fig. 2). Note, that in order to follow this phase as good as possible, we treated the convective mixing as a fast diffusive process and computed it in a common numerical scheme simultaneously with the nuclear burning. In the subsequent evolution He burning ceases and the envelope convective zone deepens and merges with the previous He-flash driven convective zone. After the dredge-up the envelope of the new RGB star consists of about 50% of He (mass ratio), up to 7% carbon, 7% nitrogen, and a negligible amount of oxygen. Note, that presently no other scenario exists

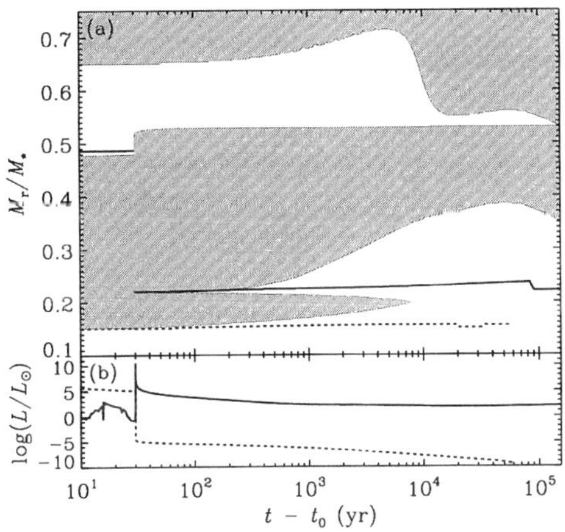

Figure 2. (a) The development of the different convective zones – indicated by shaded areas – as a function of time in our standard model. The solid and dashed lines represent the mass shells of maximum energy release by H- resp. He-burning. (b) The evolution of L_H (solid) and L_{He} (dashed line) of the same model.

which produces such abundance pattern, and thus stars in these phases can be identified unambiguously. The further evolution of the model is very similar to a Pop I star, but due to the large He content the new RGB phase only lasts about 50 million years.

We analyzed which physical property determines, whether the model shows either the H-flash event or the usual HB evolution, and found, that the strength of the degeneracy in the He core is a crucial parameter. Actually, the 1 $M_\odot$ star is a limiting case, which is very sensitive to small changes in the He-core degeneracy, which may result, e.g., from the inclusion of diffusion, the usage of a different equation of state, or the choice of a different initial helium content (for more details see Schlattl et al. 2001). With lower mass stars having a higher degenerate He core, they always experience the H flash, while stars with higher masses follow the "usual" evolution.

2. Comparison with observations

Among the ultra metal-poor stars about 10% have been observed which have a strongly carbon and nitrogen enhanced surface. To compare our models with these stars we computed the evolution of a 0.82 $M_\odot$ Pop III star including microscopic diffusion and external pollution. The amount of material polluting the surface of the initially metal-free object has been adjusted to fit the [Fe/H] value of two observed stars, CS22892-052 and CS22957-027.

In Table 1 the resulting values of [C/Fe] and [N/Fe] of the 0.82 $M_\odot$ sequence after the H flash have been summarized. Although we are able to find mod-

Table 1. Observational properties in two evolutionary stages of the 0.82 $M_\odot$ model and for two extremely metal-poor stars

	Age [Gyr]	$T_{\rm eff}$ [K]	$\log(L/L_\odot)$	[C/Fe]	[N/Fe]	[Fe/H]
CS22892-052	15.6 ± 4.6 (1)	4760 (2)	2.5[a]	1.1 (3)	1.0 (3)	–3.0 (3)
CS22957-027	...	4840 (5)	1.7[b]	2.2 (4)	2.0 (4)	–3.4 (4)
Model	13.7	4500	1.71	3.9	4.1	–3.3
	13.7	4700	2.4	4.0	4.2	–3.3

[a]Derived from $T_{\rm eff}$ and $\log g = 1.3$ of Ref. (2) by assuming a mass of 0.82 $M_\odot$.
[b]Derived like for CS22892-052 from $\log g = 2.25$ of Ref. (5).

REFERENCES. — (1) Cowan et al. (1997); (2) Sneden et al. (1996); (3) Norris et al. (1997a), (4) Norris et al. (1997b); (5) Bonifacio et al. (1998)

els at the correct luminosity and effective temperature, carbon and nitrogen is strongly overproduced in the surface. We conclude that our improved mixing-burning scheme is not yet sufficient to calculate the H-flash phase correctly. In the future, we are going to investigate whether the observed abundances of s-process elements can be reproduced by our scenario. In addition, we will verify, whether an improved description of convection diminishes the discrepancy between model and observation, and whether computing this phase in a 1D-code is sufficient at all. Nevertheless, the explanation for the mysterious carbon-enhanced ultra metal-poor stars to be post He-flash Pop III stars appears to be a promising approach, and moreover, this scenario can explain the almost equal surface enhancement of carbon and nitrogen without additionally producing oxygen.

Acknowledgments

H. Schlattl wants to thank the "Sonderforschungbereich 375-95 für Astro-Teilchenphysik" of the Deutsche Forschungsgemeinschaft for its financial support.

References

Beers, T. C., Preston, G. W., & Shectman, A. 1992, AJ, 103, 1987.

Bonifacio, P., Molaro, P., Beers, T. C., & Vladilo, G. 1998, A&A, 332, 672.

Cowan, J. J., Pfeiffer, S. G., Kratz, K. L., Tielemann, F. K., Sneden, C., Burles, S., Tytler, D., & Beers, T. C. 1997, ApJ, 521, 194.

Fujimoto, M. Y., Iben, I. Jr., & Hollowell, D. 1990, ApJ, 349, 580.

Norris, J. E., Ryan, S. G., & Beers, T. C. 1997a, ApJ, 488, 350.

Norris, J. E., Ryan, S. G., & Beers, T. C. 1997b, ApJ, 489, L169.

Schlattl, H., Cassisi, S., Weiss, A., & Salaris, M. 2001, ApJ (submitted).

Sneden, C., McWilliam, A., Preston, G. W., Cowan, J. J., Burris, D. L., & Armosky, B. J. 1996, ApJ, 467, 819.

ZERO-METAL STELLAR YIELDS, THE IMF FOR POPULATION III AND NUCLEOSYNTHETIC CONSTRAINTS

C. Abia, I. Domínguez
Dpto. de Física Teórica y del Cosmos, Universidad de Granada, Spain

O. Straniero
Osservatorio di Collurania, Teramo, Italy

M. Limongi
Osservatorio di Roma a Monteporzio Catone, Italy

A. Chieffi
Istituto di Astrofisica Spaziale del CNR, Frascati, Italy

and

J. Isern
Institut d'Estudis Espacials de Catalunya, Barcelona, Spain

Abstract We study the nucleosynthesis constraints that the recent detection of metals in different components of the high redshift universe and the abundance ratios measured in the extremely metal-poor stars of our Galaxy, imposes on an early generation of stars (Population III).

Introduction

In current cold dark matter (CDM) hierarchic cosmological models, shortly after the Big Bang (redshifts $z \sim 10 - 30$) the primordial gas cooled rapidly and fragmented. From the first structures, with typical masses between 10^5 and 10^8 $M_\odot$, it is believed that a very early generation of stars, Population III, formed. Since the chemical content of these first structures was primordial (H,

M. Chávez et al. (eds.),
New Quests in Stellar Astrophysics: The Link Between Stars and Cosmology, 59–62.

^{4}He and traces of D, ^{3}He and ^{7}Li), this first generation of stars must have been formed essentially free of metals ($Z \sim 10^{-9}$).
The possible existence of Population III was originally postulated because of the presence of metal overabundances in Population II stars and especially due to the so-called G-dwarf problem. The idea of a pre-galactic metal enrichment has become increasingly important in recent years because of the discovery of intergalactic metals in Ly-α forest clouds ($N(HI) \leq 10^{17}$ cm^{-2}) at $z = 3 - 3.5$ showing metal abundances of $Z \sim 10^{-3} - 10^{-2} Z_{\odot}$ (e.g. Cowie and Songalia 1998). The carbon and silicon abundances measured are [C/H]~ -2.5 and [Si/H]~ -2.3, although with an important dispersion within the forest. Whether the metals found have a local origin associated with the Ly-α lines observed, or are due to an early episode of star formation is still unclear. The answer to this question might come from metal abundance studies in very low density clouds (N(H I)$\leq 10^{14}$ cm^{-2}).
An indirect way to infer the possible existence of Population III and the nature of its stellar mass function is to search for the imprints that could have been left in the IGM and/or in the subsequent generation of stars formed from the material ejected, that is the *chemical* imprint. Here we reexamine the issue of the chemical pollution that could have originated Population III. To do so, we make use of chemical yields from strictly zero metal ($Z = 0$) stellar models in the mass range $3 \leq M/M_{\odot} \leq 200$ through a simple chemical evolution approach.

1. Observational Constraints and Input Material

The main observational constraint concerns the detection of metals in the Lyman-α forest spectra of distant QSOs. We assume as a *working hypothesis* that the metal contamination (see above) observed in the Ly-α forest clouds represents typically the maximum level of metal pollution in the IGM due to the Population III. Our second constraint are the abundance ratios found in the most extremely metal-poor stars in our galaxy. The abundance patterns in these stars might well be the key to understanding the nucleosynthetic processes in the very early Galaxy before these extremely metal-deficient stars formed. For instance, one of the most striking results of these studies is that $\sim 25\%$ of the stars with [Fe/H]≤ -2.5 exhibit large carbon and nitrogen abundances relative to iron, [C,N/Fe]≥ 1 (Rossi, Beers and Sneden 1999).
Concerning the IMF for Pop III, since metals are the most important coolants in present-day star formation, many studies have emphasized the importance of the radiative cooling by H-based molecules because the primordial gas was deficient in heavy elements. These studies agree in several facts: *i)* there are physical arguments for assuming that the slope of the present IMF has changed in time, *ii)* the first star-forming clouds typically fragmented into massive clumps of

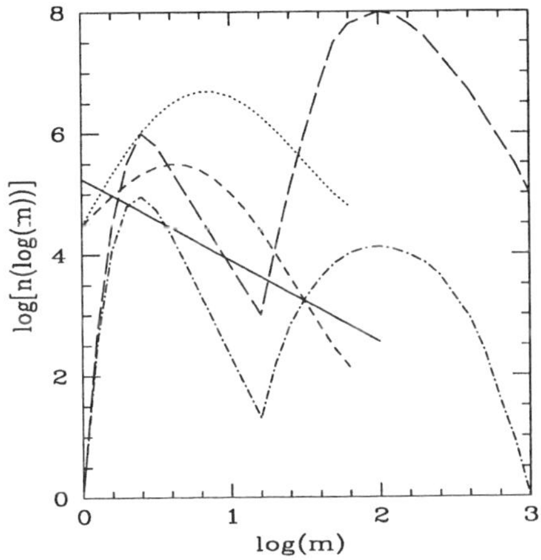

Figure 1. Comparison of the IMFs used in this work over the stellar mass range $1 \leq M/M_{\odot} \leq 10^3$. Solid line: Salpeter mass spectrum. Dotted line: Yoshii and Saio (1986) adopting $L \propto M^{1.5}$ relation for the protostars (YSa). Short dashed line: the same adopting $L \propto M^3$ (YSb). Long dashed line: Nakamura and Umemura (2001) for 1σ density fluctuations in the primordial gas density (NUa). Dashed-dotted line: Nakamura and Umemura (2001) for 3σ fluctuations. See the above references for details.

$10^2 - 10^3$ $M_{\odot}$, *iii)* there exists the possibility of further fragmentation into smaller masses, producing an IMF peaking at about the intermediate stellar mass range. Here we adopt and test IMFs for Pop III that basically recover these ideas. Figure 1 shows the IMFs used in the mass range 1-10^3 $M_{\odot}$, and clearly demonstrates how these IMFs differ.
According to recent theoretical analyses of the evolution of Population III stars we have adopted the following scheme for the fate of metal-free stars: (1) Stars with $M \geq 250$ $M_{\odot}$ collapse to a black hole (BH) with no metal ejection. (2) Very massive objects (VMO), $100 \leq M/M_{\odot} \leq 250$, produce a type II supernova-like event and eject metals leaving no compact remnant. (3) Stars in the range $40 \leq M/M_{\odot} \leq 100$ probably collapse into a BH without metal contribution. (4) Stars in the range $10 \leq M/M_{\odot} \leq 40$ give a SNII and finally, (5) stars $1 \leq M/M_{\odot} \leq 8$ will pass through the AGB phase leaving a white dwarf as the stellar remnant. We use here $Z = 0$ stellar yields from Ober et al. (1993) for VMO, Limongi et al. (2001) for SNII and Chieffi et al. (2001) for low and intermediate mass stars.

2. The Starburst Model

We apply the computed yields and the IMFs proposed for the zero-metal stars to study the early chemical enrichment of the IGM by means of the simple starburst model approach. Within this model it is straightforward to show that the chemical enrichment in a given isotope $\Delta X_i = \rho_i/\rho_{gas} - X_i^o$ is given by

$$\Delta X_i = \frac{p}{[1-(1-R)p]}(y_i - X_i^o) \tag{1}$$

where y_i is the stellar yield, R is the fraction of the mass returned to the IGM by the dying stars, X_i^o is the initial mass fraction in the gas of the isotope i and $p \equiv \rho_\star/\rho_{gas}$ is the fraction of gas mass which goes into stars (i.e. the star formation efficiency) which is considered as a free parameter. On the other hand, the stellar remnant density (Ω_{sr}) can be also constrained for a given baryon density ($\Omega_b \sim 0.019$) and a choice of the star formation efficiency p: $\Omega_{sr} = (1-R)\,p\,\Omega_b$. Table 1 shows the main results of the starburst model for different IMFs.

Table 1. Results from the starburst model.

IMF	$Z/Z_\odot$	p	Ω_{sr}	[C/H]	[N/H]	[Si/H]	[Fe/H]
Salpeter	0.6	7×10^{-3}	4×10^{-5}	–2.5	–1.8	–3.1	–3.4
YSa	1.5	3×10^{-3}	8×10^{-6}	–2.3	–2.0	–2.4	–2.7
YSb	0.5	9×10^{-3}	3×10^{-5}	–2.2	–1.6	–3.0	–3.3
NUa	8	5×10^{-4}	1×10^{-6}	–3.2	–3.7	–2.2	–2.2
NUb	12	3×10^{-4}	3×10^{-6}	–3.1	–5.6	–2.1	–2.1

From Table 1 it is apparent that very low values of the star formation efficiency are required to not exceed the abundances measured in the Ly-α forest systems. This contrasts with the current estimates of this value in spiral galaxies ($\geq$ 1%). Note, also the resulting low Ω_{sr} values, i.e.: the stellar remnants of the hypothetical Population III stars would not contribute to the baryonic dark matter. Large [C,N/Fe] ratios are only obtained with an IMF heavily weighted in the $4-8$ M$_\odot$ (YSa & YSb IMFs), while small [Si/Fe] enhancements are obtained, as observed in the damped Lyman-α systems, almost for any IMF choice. A more complete discussion of the nucleosynthetic constraints imposed to the Population III stars can be found in Limongi et al. (this conference).

References

Chieffi, A., Dominguez, I., Straniero, O., & Limongi, M. 2001, ApJ, (in press).

Cowie, L. L., & Songaila, A. 1998, Nature, 394, 44.

Limongi, M., Chieffi, S., & Straniero, O. 2001, ApJ, (submitted).

Nakamura, F., & Umemura, M. 2001, ApJ, 548, 19.

Ober, W. W., El Eid, M. F., & Fricke, K. J. 1983, A&A, 119, 61.

Rossi, S., Beers, T., & Sneden, C. 1999, in ASP Conf. Series 165, The Galactic Halo, eds. B. K. Gibson, T. S. Axelrod, & M. E. Putman (San Francisco: ASP), p. 67.

Yoshii, Y., & Saio, H. 1986, ApJ, 301, 587.

THE STELLAR IMF IN VERY METAL-DEFICIENT GAS

Fumitaka Nakamura
Niigata University, Japan, and UC Berkeley, USA

and

Masayuki Umemura
University of Tsukuba, Japan

Abstract In the context of the star formation through the fragmentation of an extremely metal-deficient protogalactic cloud, the gravitational collapse of filamentary gas clouds is explored with H_2 and HD chemistry. It is found by 1D hydrodynamical simulations that the cloud evolution is prescribed mainly by the initial density (n_0) and H_2 abundance ($x_{H_2,0}$). In particular, it turns out that the evolution of low-density filaments ($n_0 \lesssim 10^5$ cm^{-3}) bifurcates at a critical H_2 abundance of $x_{H_2,cr} \simeq 3 \times 10^{-3}$, beyond which HD cooling overwhelms H_2 cooling. The numerical results indicate that the stellar IMF is likely to be double-peaked and deficient in sub-solar mass stars, where the high mass peak of the IMF is around $10 M_\odot$ or $10^2 M_\odot$, dependently on the initial density and H_2 abundance. If the gas in protogalactic clouds is photoionized by UV radiation or shock-heated, the H_2 abundance could exceed $x_{H_2,cr} \simeq 3 \times 10^{-3}$ by H^- reactions. Then, the high mass peak would be $O(10) M_\odot$.

Introduction

Star formation in the early universe should have proceeded in very metal-deficient gas. When the metallicity is lower than $Z \lesssim 10^{-2} Z_\odot$, the thermal property of such metal-deficient gas is essentially the same as that of the metal-free gas because cooling by heavy metals is less effective than primordial molecular cooling. Among the primordial molecules, the importance of H_2 on the star formation in the early universe have been emphasized by many authors because they are the most abundant molecules in metal-free gas. The H_2 molecules provide significant cooling through rotational and vibrational transitions which can lower the gas temperature down to a few hundred K and accordingly reduce the Jeans masses to stellar mass scales.

M. Chávez et al. (eds.),
New Quests in Stellar Astrophysics: The Link Between Stars and Cosmology, 63–68.

HD molecules are considered to be the second most abundant molecules in primordial gas during the post-recombination era. In spite of its low abundance ($[HD/H_2] \sim 10^{-3}$), HD can provide significant cooling at low temperature gas because HD has higher radiative transition probabilities and lower excitation temperatures than H_2. Therefore, HD cooling can lower the gas temperature down to $T \lesssim 100$K and accordingly, the Jeans mass is also reduced.
In this paper, we examine the effects of the HD cooling on the formation of stars in very metal-deficient gas, and elucidate the role of HD molecules for the stellar initial mass function (IMF) there.

1. Numerical Model and Methods

Our numerical model and method are the same as those of Nakamura and Umemura (2001a). We consider an infinitely long cylindrical gas cloud. We deal with the following 14 species: e, H, H^+, H^-, H_2, H_2^+, He, He^+, He^{++}, D, D^+, D^-, HD, and HD^+.
The density is assumed to be uniform along the cylinder axis and the radial distribution is expressed as $\rho = \rho_0 \left(1 + r^2/R_0^2\right)^{-2}$, where $R_0 = \sqrt{2fkT_0/(\pi G \rho_0 \mu)}$ is the effective radius, ρ_0 is the central mass density, T_0 is the initial gas temperature, μ is the mean molecular weight, and f is the ratio of the gravitational force to the pressure force. Our model is specified by four parameters: n_0, T_0, f, and the H_2 abundance $x_{H_2,0}$. It should be noted that a higher $x_{H_2,0}$ can be translated to a higher electron abundance of a parent cloud. This is because when the ionization degree of the parent cloud is higher, the H_2 molecules form more efficiently via radiative reactions of $H + e \rightarrow H^- + h\nu$ and $H + H^- \rightarrow H_2 + e$.
We take into account the following thermal processes: (1) H cooling by radiative recombination, collisional ionization, and collisional excitation, (2) H_2 line cooling, (3) cooling by H_2 collisional dissociation, (4) heating by H_2 formation, and (5) HD line cooling.

2. Numerical Results

In this section, we examine the collapse of the filaments, taking into account the H_2 and HD cooling. As shown below, there is a critical initial H_2 abundance, above which HD cooling predominantly regulates the cloud evolution. Also, it is found that the HD cooling does not play an important role for high-density gas. Thus, the evolution of the primordial filaments is classified into three cases, depending upon the initial density and initial H_2 abundance; (1) low-density filaments with high H_2 abundance, (2) low-density filaments with low H_2 abundance, and (3) high-density filaments.
For low density filaments with $x_{H_2,0} \gtrsim 3 \times 10^{-3}$, HD cooling overwhelms H_2 cooling during the contraction, and the temperature descends down to $T \sim 50$ K accordingly. When the density reaches a critical density of HD ($n_{cr} \sim$

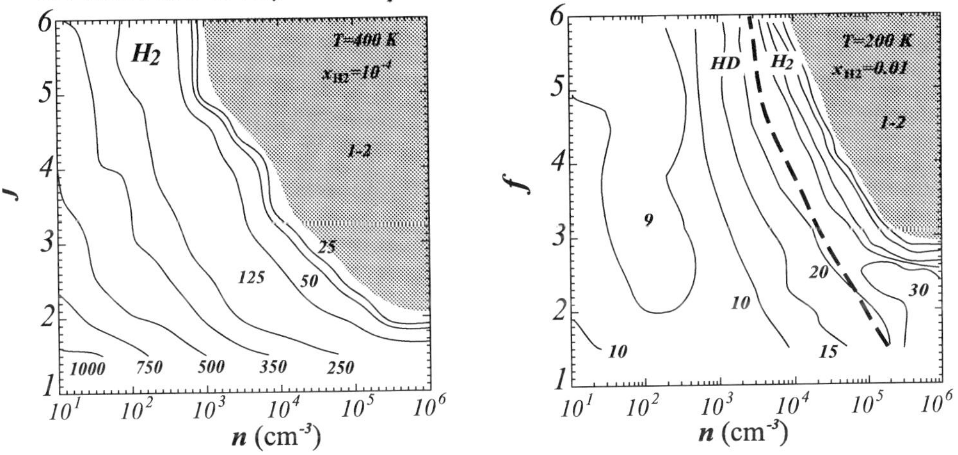

Figure 1. Mass distributions of the fragment derived from 1D simulations for the models with (a) $x_{H_2,0} = 1 \times 10^{-4}$ and (b) 1×10^{-2}, respectively. The abscissa and ordinate denote the initial central density and the parameter f, respectively. The solid lines denote the contours of the fragment mass which are labeled with adjacent numbers. The dashed lines show the lines at which the HD cooling rate is equal to H_2 cooling rate at the epoch of fragmentation. In the left regions of the dashed lines, HD cooling is more efficient than H_2 cooling.

10^{4-5} cm^{-3}), the cloud contraction tends to become quasi-static. Thus, the fragmentation will take place after that stage. For low density filaments with $x_{H_2,0} \lesssim 3 \times 10^{-3}$, because of the low x_{H_2}, the temperature stays a relatively high value at a few hundred K, and HD cooling does not play a role during the contraction. The contraction proceeds quasi-statically after the density reaches the critical density of H_2 ($n > n_{H_2,cr} = 10^{3-4}$ cm^{-3}). As for high density filaments with $n_0 \gtrsim 10^5$ cm^{-3}, the temperature stays at a relatively high value because both H_2 and HD cooling is saturated, and therefore HD does not play a significant role for the thermal evolution. The contraction of such a high-density filament is accelerated by the effective three-body H_2 formation when the density reaches 10^{8-9} cm^{-3}. Then, the fragmentation is not expected to take place until the cloud becomes opaque to H_2 lines at the density of 10^{12-13} cm^{-3}.

Figures 1a and 1b show the distributions of the fragment mass derived from the 1D simulations (see Nakamura and Umemura (2001b) for more detail). For the models with $x_{H_2,0} < x_{H_2,cr} \approx 3 \times 10^{-3}$ (Figures 1a), the mass distribution of the fragments is quite similar to the case without HD (Paper II), because HD cooling does not play an important role in the thermal evolution of the filaments. There is a steep boundary at $n \sim 10^4 - 10^5$ cm^{-3} in the mass distribution of the fragments for $f \gtrsim 3$. For the models with $n_0 \gtrsim 10^5$ cm^{-3}, the fragment masses take their minima at $1 \sim 2$ M$_\odot$, whereas, for the models with $n_0 \lesssim 10^5$ cm^{-3}, they are greater than $\sim 10^2$ M$_\odot$. [For the models with low densities ($n_0 \lesssim 10^5$ cm^{-3}), the contraction becomes quasi-static when the density reaches the critical density of H_2 and then linear density fluctuations can grow nonlinearly before the three-body H_2 formation becomes dominant ($n_0 \gtrsim 10^{8-9}$ cm^{-3}). In contrast, for the models with high densities ($n_0 \gtrsim 10^{5-6}$ cm^{-3}), the contraction time does not exceed the fragmentation time until the H_2 lines become optically thick at $n \sim 10^{12} - 10^{13}$ cm^{-3}.] This sensitivity in the fragment mass comes from the rapid increase in x_{H_2} due to the three-body reactions.

On the other hand, when $x_{H_2,0} > 3 \times 10^{-3}$, HD cooling is more effective than H_2 cooling for low-density filaments. Thus, the maximum mass of low-density region reduces to a few tens $M_\odot$. The minimum mass does not change because HD is not a dominant coolant in the evolution of dense filaments. Therefore, similarly to the models with low $x_{H_2,0}$ the dependence of the fragment mass on n_0 exhibits a steep boundary around $n_0 = 10^4 - 10^5$ cm^{-3} for $f \gtrsim 3$.

The existence of a steep boundary in the mass distribution implies that the IMF is likely to be bimodal if both low-density and high-density filaments bear stars. The low-mass peak is around a few $M_\odot$, which is not sensitive to the abundance of H_2 formed in a parent cloud. The high-mass peak is $\approx 10^2 M_\odot$ if $x_{H_2,0} \lesssim 3 \times 10^{-3}$, while it is $\approx 10 M_\odot$ if $x_{H_2,0} \gtrsim 3 \times 10^{-3}$. In the next section, we discuss some implications of such bimodal IMF in the very metal-deficient gas.

3. Implications for Galaxy Formation

Recently, Susa and Umemura (2000) investigated the pancake collapse of pregalactic clouds under UV background radiation. They found that once the pancaking disk is shielded against external UV radiation, the H_2 molecules form efficiently via H^- reaction with abundant free electrons produced by UV background, and the resultant abundance reaches $x_{H_2} \approx 3 \times 10^{-3}$. The pancake disks probably fragment into filaments in which stars would form. In this case, HD cooling is expected to become efficient in low-density filaments and then, the high mass peak of the IMF would go down to ~ 10 $M_\odot$.

The time-decreasing UV background radiation is also likely to influence star formation in dwarf galaxies which may be related to excess number count of faint blue objects observed in the Hubble Deep Field. Corbelli, Galli, and Palla (1997) (Corbelli, Galli, and Palla (1997)) studied the effects of the declined UV background on the thermal evolution of the protogalaxies. They found that there is a critical redshift of $z \sim 1 - 2$, below which the declined UV radiation is shielded by the gas disks where the H_2 abundance reaches 10^{-2} owing to high ionization degree by the UV radiation. The enhanced H_2 formation promotes a rapid transition toward the cold H I phase with $\sim 10^2$ K. Thus, in such dwarf galaxies, high mass peak of the IMF would decrease to $\sim 10\ M_\odot$ owing to HD cooling.

References

Corbelli, E., Galli, D., & Palla, F. 1997, ApJ, 487, L53.
Nakamura, F., & Umemura, M. 2001a, ApJ, 548, 19.
Nakamura, F., & Umemura, M. 2001b, ApJ (submitted).
Susa, H., & Umemura, M. 2000, ApJ, 537, 578.

COSMOLOGICAL ORIGIN OF THE LOWEST METALLICITY HALO STARS

X. Hernández
Instituto de Astronomía, UNAM, Mexico

and

A. Ferrara
Osservatorio di Arcetri, Italy

Abstract

We explore the predictions of the standard hierarchical clustering scenario, regarding the numbers and metallicities of PopIII stars found within our Galaxy today. An analytic, extended Press-Schechter formalism is used to get the mass functions of halos which will host PopIII stars at a given redshift, and which will end up in the Milky Way (MW) today. Each of these is modeled as a mini galaxy, with a detailed treatment of the dark halo structure, angular momentum distribution, final gas temperature and disk instabilities, all of which determine the fraction of the baryons which are subject to star formation. Use of new primordial metallicity stellar evolutionary models allows us to trace the history of the stars formed and give accurate estimates of their expected numbers today. A first comparison with observational data suggests that the IMF of the first stars was increasingly high mass weighted towards high redshifts, leveling off at $z \approx 9$ at a characteristic stellar mass scale $m_s = 10 - 15 M_{\odot}$. Details can be found in Hernandez & Ferrara (2001).

Introduction

We use the extended Press-Schechter formalism to calculate the conditional probability that a virialized halo of mass M_1 at z_1 will become part of a more massive halo of mass M_0 at a later time, z_0. This allows us to calculate the mass functions of fragments which will end up as part of the MW today. In all calculations we used a ΛCDM power spectrum normalized to $\sigma_8 = 0.91$, in a $\Omega_\Lambda = 0.7$ flat universe ($\Omega_M = 0.3$), with $H_0 = 65$ km s^{-1}Mpc^{-1}. Taking a baryonic fraction of 0.05 and $M_{MW} = 2.0 \times 10^{12} M_{\odot}$ for the MW mass.

M. Chávez et al. (eds.),
New Quests in Stellar Astrophysics: The Link Between Stars and Cosmology, 69–72.

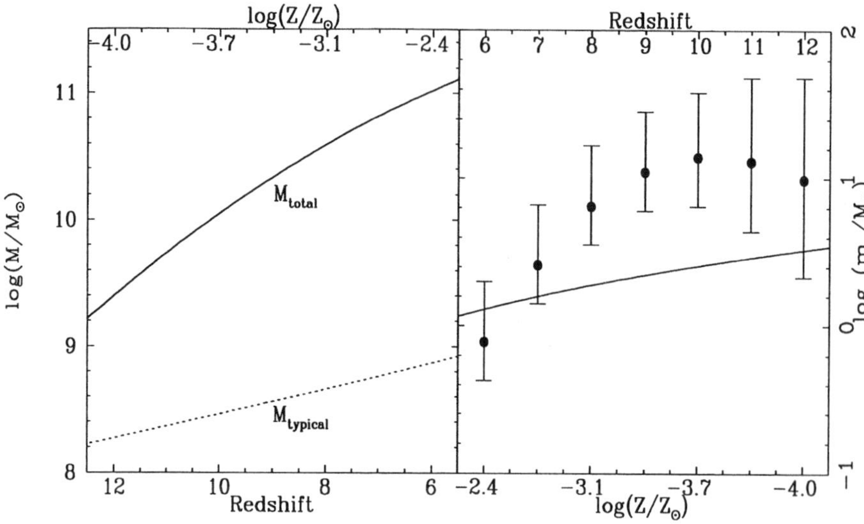

Figure 1. a) Total mass in PopIII halos that ends up in a MW sized galaxy today, as a function of redshift (solid line). The dotted curve shows the typical mass of these PopIII halos. The upper axis shows the average metallicity of the IGM at that redshift. b)Inferred values of m_s as a function of redshift (points with error bars). The solid curve gives the redshift evolution of the Jeans mass of cold star forming clouds, identified with m_s, resulting from the temperature evolution of the cosmic microwave background (CMB)

We associate a maximum metallicity to the stars formed within these halos through the mean mass weighted metallicity of the universe at each redshift, as calculated by Gnedin & Ostriker (1997). A lower mass limit taken from Tegmark et al. (1997) excludes systems having virial temperatures such that they cannot cool in a Hubble time. An upper limit corresponding to the mass at which supernova can not expel the baryonic component excludes systems where self enrichment can continue and stars of metallicities higher than the limit taken will be formed.

Fig. 1a shows the integral of the PopIII halos mass functions between the selected limits as a function of redshift. The upper axis shows the mean mass weighted metallicity corresponding to each redshift. Fig. 1a also gives the typical mass of a single PopIII halo, $M_{typical}$, showing that each present day MW sized system will have had between 10 and 100 PopIII halos as progenitors, contributing stars with metallicities in the $-4.0 < [Z/Z_{\odot}] < -2.4$ range. These two curves intersect at $z \approx 15$, corresponding to a metallicity of $[Z/Z_{\odot}] = -4.3$. At this limit many present day MW sized systems will have had only one PopIII halo and many will have had none at all. We thus expect that the numbers of stars in the Milky Way with metallicities below $[Z/Z_{\odot}] = -4.3$ should be minimal, indeed, no such stars have been found.

1. Forming the First Stars

We model each of the PopIII halos as a small galactic system. The gas forms a disk in centrifugal equilibrium with an exponential surface density profile fixed by the total baryonic mass and angular momentum. This last is given by the λ parameter, which is chosen at random from a cosmological distribution. The final temperature of the gas in the disk is taken to be $300K$, as the main cooling mechanism is H_2 (Ferrara 1998). This determines c_s, the sound speed within the disk, which together with κ, the epicycle frequency obtained from the final rotation curve, yields Toomre's stability for the disk: $Q = c_s\kappa/\pi G\Sigma$. In the regions where $Q > 1$ the disk is locally stable and not subject to star formation. Finally, a gas to star conversion efficiency $f_\star = 0.05$ is used to turn the total mass within the instability region into a final mass of stars.
Once the total gas mass turned into stars as a function of the metallicity has been determined, we use the IMF of Larson (1998) to turn this into a total number of stars. This function offers a convenient parameterization of the IMF:

$$dN/dlog\, m \propto (1 + m/m_s)^{-1.35}. \quad (1)$$

In the above equation m_s is a characteristic mass scale, of order $0.35M_\odot$ for a present day solar neighborhood IMF. This scale mass can be increased to explore the consequences of a top heavy IMF applying to stars of metallicities much lower than solar, as the ones we are treating.
We combine this sampling of the IMF using $m_s = 0.35M_\odot$ with recent Padova stellar models of primordial metallicities (Girardi et al. 2000) to exclude those stars which due to their age are no longer visible. Sampling the IMF from a fixed lower luminosity up to the tip of the RGB of the theoretical isochrones allows us to obtain the number of stars visible today upwards of a certain limiting luminosity, as a function of metallicity.

2. Comparisons with Data and Conclusions

To compare our results with observations we take the sample of Ryan & Norris (1991) of low metallicity halo stars in the local disk. They include a magnitude limited sample, (their NLTT sample) which is complete down to $0.6M_\odot$ and $V = 13.0$, the same limit we use to sample our IMF. From the theoretical isochrones, this corresponds to a distance of 290 pc, which we take as the radius of the sample.
From N-body simulations White & Springel (2000) have argued that most of the lowest metallicity stars in a present day galaxy will be found in the stellar halo, and not in the bulge. We therefore make the assumption that the lowest metallicity stars share the isothermal distribution of the dark halo.
Taking the Ryan & Norris (1991) data we derive the total number of stars with $[Z/Z_\odot] < -3.0$ in the Galaxy as 4.43×10^6. To obtain numbers at other

metallicities we use the relative metallicity distribution function of the more recent Beers et al. (1998) work, which combines several new samples.
For $m_s = 0.35M_\odot$ our results fall somewhat above the observational measurements, implying that our assumption of a constant $m_s = 0.35M_\odot$, essentially a present day solar neighborhood IMF, is not valid. Reconciling our estimates with the observations requires not only the assumption of a higher m_s in the past, but throughout the redshift range studied, an increasing trend for this value. Note that our derived numbers of PopIII stars scale linearly with $f_\star$
In Fig. 1b we show the values of m_s which we infer for the IMF of PopIII objects, by imposing that model and observations agree. The solid curve indicates the redshift evolution of the Jeans mass in star forming clouds, which we identify with m_s. Although for $z < 9$ our results are consistent with the increase in m_s coming solely from that in the Jeans mass due to the CMB, the trend for $6 < z < 9$ is clearly for a rise in m_s beyond what this effect can account for. In fact, as Larson (1998) points out, this is only the *minimum* increase to be expected, as other heating sources, e.g. massive stars themselves or a UV background, were likely to be present. Furthermore, a reduced opacity consistent with these low metal abundances would have the effect of skewing the IMF towards larger masses, thus operating in the same direction.
For $z > 9$, the values of the characteristic mass appear to stabilize around the value $m_s = 10 - 15M_\odot$, indicative of a very top heavy IMF at these early epochs, although the increase of our error bars makes determining trends in this redshift range harder.
We stress that observational limits on the numbers and metallicity distribution of very low metallicity stars in the MW can place sharp restrictions on the IMF of the first stars. This through modeling the full range of phenomena giving rise to such stars, establishing a connection between the formation scenario of the Milky Way and present day observations of low metallicity stars.

References

Beers, T. C., Rossi, S., Norris, J. E., Ryan, S. G., Molaro, P., & Rebolo, R. 1998, SSRv, 84, 139.
Ferrara, A. 1998, ApJL, 499, 17.
Girardi, L., Bressan, A., Bertelli, G., & Chiosi, C., 2000, A&AS 141, 371.
Gnedin, N. Y., & Ostriker, J. P. 1997, ApJ, 486, 581.
Hernandez, X., & Ferrara, A. 2001, MNRAS in press.
Larson, R. B. 1998, MNRAS, 301, 569.
Ryan, S. G., & Norris, J. E. 1991, AJ, 101, 1865.
Tegmark, M. et al. 1997, ApJ, 474, 1.
White, S. D. M., & Springel, V. 2000, in The First Stars, eds. A. Weiss, T. G. Abel, & V. Hill (Heidelberg: Springer-Verlag), p. 327.

COSMOLOGICAL CHEMISTRY AND THE FORMATION OF THE FIRST OBJECTS

A. Lipovka, J. Saucedo, J. Campos, and I. Soto
Centro de Investigación en Fisica, Universidad de Sonora, Mexico

Abstract The present report discusses the role of primordial molecules in the evolution of the Universe at epochs $10 < Z < 300$. The recent results of "BOOMERANG" and "MAXIMA" allow for the possibility of non-standard big bang nucleosynthesis scenarios, so it becomes important to consider not only primordial molecules formed by H, D, He, and Li, but also by primordial C, N, and O. Molecular abundances in proto-objects under different conditions are estimated and possible changes in their evolution are discussed. It is stressed that under plausible assumptions of the relative abundance of oxygen and carbon = $10^{-9} - 10^{-10}$, one can obtain observable optical depths for the lines of primordial molecules due to the luminescence and scattering of CMBR by molecules of proto-objects moving with peculiar velocities at high redshift. By taking into account the formation of the first stars at redshift $10 < Z < 100$ it is possible to observe their evolution in the molecular lines.

Introduction

The investigation of the chemistry of primordial molecules in the post-recombination epoch was first recognized in the papers by Saslaw and Zipov (1967), Lepp and Shull (1984), Dalgarno and Lepp (1987), and Black (1988). In the last few years there has been a lot of progress in the investigation of primordial chemistry in different contexts and applications, especially for proto-objects. The possibility of a direct observation of remote proto-objects at epochs $z > 20$ in molecular lines has been recently discussed by Dubrovich (1994), de Bernardis et al. (1993), Signore, et al. (1994). The interest in this problem is due to the possibility of determining primordial abundances of such key species as Li and D and consequently, new constraints to primeval nucleosynthesis models. Furthermore, this interest is stimulated by the importance of obtaining data at early epochs in the evolution of the Universe, which has lead to the observable distribution of matter at the present time. In this case, the influence of primary molecules on the cooling of proto-objects and, hence, on their evolution, becomes fundamental. As a rule, within the Standard Big Bang Nucleosyn-

M. Chávez et al. (eds.),
New Quests in Stellar Astrophysics: The Link Between Stars and Cosmology, 73–76.

thesis model (SBBN), the following primordial molecules can be considered: H_2, H_2^+, H_3^+, HD, HD^+, H_2D^+, HeH^+, LiH, and LiH^+. Possible observational tests of these molecules have been discussed by several authors (e.g. Dubrovich 1997 and Signore et al. 1994). Several attempts have been made to detect emission by rotational lines of LiH from proto-objects at large redshifts (de Bernardis et al. 1993). Here, we consider new physical effects based on the absence of equilibrium in the rest-frame of the moving proto-object and the fluorescence of non-equilibrium radiation by molecules in their ro-vibrational structure (Dubrovich 1994, 1996). Furthermore, recent CMBR measurements by BOOMERANG and MAXIMA (de Bernardis et al. 2000, Jaffe et al. 2001) argue in favor of a new value for the baryon density Ω_b. These new results leave an open possibility for Nonstandard Big Bang Nucleosynthesis models (NBBN) (Burles et al. (2000)). Some of the NBBN models lead to considerable abundances of primordial C, N, and O. For this reason it is important to investigate in proto-objects, the possibility of detecting molecular line emission by molecules such as CH, H_2O, and OH, to be able to constrain NBBN models.

1. Proto-objects

The interest in small objects with masses $M < 10^{10} M_{\odot}$is due to their role in the formation of the present structure of matter distribution, as well as by the peculiarities of their evolution. Objects with these masses (protoclusters, protogalaxies and protostars), are thought to be responsible for the formation of the large scale structure observed today. The spectrum of masses of these objects is interesting because it was formed during the early Universe, hence, it can provide information about fundamental parameters of the overdensity of matter. After recombination the Jeans mass decreases to about $10^5 M_{\odot}$ and the collapse of proto-objects of small masses becomes possible. This is essentially reflected on the amplitude size of the Spectrum of Scale Fluctuations (SSF). The most common spectrum, among the variety of primordial fluctuations $\Delta\rho$ at small scales is

$$\Delta\rho/\rho = (M/M_{\star})^{(-1/2+n/6)}(1+z)^{-1}, \tag{1}$$

where n is the parameter of the model, M is the mass of the protocloud and $M_{\star} = 10^{15} M_{\odot}$. A model with n=1 was adopted in the mass spectrum calculations of proto-objects by Puy and Signore (1996). As it was shown by them, according to the standard cosmological scenario, the first proto-objects began to form at epochs $50 < Z < 500$. The masses of these objects depended on the redshift at the epoch when they became gravitationally bound. So, according to the theory, objects characterized by mass values $M = 10^6 M_{\odot}$ stopped their expansion at $Z = 560$, and $M = 10^7 M_{\odot}$ at $Z = 260$.

According to the Saha equation, the creation of molecules started at approximately the same time ($Z < 400$). The relative abundances of the standard set of molecules formed by H, D, He, and Li, has been discussed by several authors (e.g. Puy et al. (1993), Latter & Black (1991), Galli & Palla (1998) and references therein). However, all these calculations have been done considering a homogeneous medium under equilibrium conditions, whereas the relatively high concentrations and temperatures in proto-objects, should increase the formation rate of the molecules.

As it was mentioned above, in the case of NBBN, the presence of rather high abundances of primordial N, C and O might be possible (up to 10^{-9} as suggested by Rauscher et al. 1994). This possibility is due to the presence of micro-inhomogeneities in the initial distribution of matter at $z = 10^9$. In this case molecules such as CH, OH, H_2O, OH^+ could be created in significant amounts at epochs with redshift $Z < 300$. For this reason the search of these molecules in proto-objects is of particular importance. If they are detected or at least, upper limits of their abundances are found, we could not only restrict the variety of the NBBN models, but also estimate the mass spectrum of the first objects, their physical parameters, and find out their formation epoch.

The calculations of relative abundances for some molecules, presented in this section, were made for proto-objects located at $Z = 100$ and for different kinetic temperatures. The relative abundance of O and C was taken $[O]/[H] = [C]/[H] = 10^{-10}$ and a concentration of gas in the cloud of approximately 1 - 10 cm^{-3}. The relative abundances for OH, H_2O, and CH at T_k =300K are 10^{-18}, 10^{-22}, and 10^{-14}. Whereas at T_k = 1000K they are 10^{-13}, 10^{-16} and 10^{-17}, respectively.

As it was mentioned above, the emission we are discussing can be formed due to the interaction of the lines of primordial molecules with the CMBR. The intensity of the emission is determined by the degree of deviation of the CMBR spectrum from that of the black-body in the restframe of the protocloud, and by the molecular abundances. It would be interesting to know if the observed lines are amplified due to fluorescence, because in this case the value $\Delta T/T$ will increase 3 - 5 orders of magnitude with respect to the simple scattering of the line (Dubrovich 1996).

In the case of NBBN, fluorescence can lead to significant emission by proto-objects in the molecular lines of CH, OH, H_2O, etc. For example, for $[C]/[H] = 10^{-14}$, $\Delta T/T = 10^{-5}$ and a linewidth $\Delta\lambda/\lambda = 10^{-5}$. The wavelength for the first vibrational transition of the CH molecule at the present time is $\lambda = 0.034(1+z)$cm, and for OH $\lambda = 0.027(1+z)$cm. One can see that for $Z = 100 - 200$ these lines will appear in the cm region of the spectrum.

2. First stars

At latter epochs ($Z = 50 - 20$), the physical conditions in the gas were no longer in quasi-equilibrium and homogeneous, because by then the first stars were formed. It is clear that the primordial molecules were able to re-emit not only the peculiarity of the CMBR, but also the radiation of the first stars. On the other hand, the evolution of the first stars and the re-ionization of the Universe lead to active formation of the said molecules (CH, OH, H_2O, ...). Even assuming a very small amount of matter injection by the first stars ($\sim 10^{-5}$ of the mass) to the ISM, the relative abundances of oxygen and carbon were of the order of 10^{-9}, hence it is possible that the optical depth of the ro-vibrational lines is $\tau \approx 1$.

However, even if the enrichment of matter by the heavy elements due to the first supernovae is insignificant, its effect on re-ionizing most of the primordial hydrogen turned out to be very important. In this case the abundance of molecules such as H_2^+, HeH^+, will increase and the gas temperature can reach values higher than 10^3K. The molecule HeH^+ has a large dipole moment, so the first two rotational transitions will produce a SSF at wavelengths $\lambda_1 = 0.149(1+z)$mm., and $\lambda_2 = 0.075(1+z)$mm. These SSF will reflect the structure of large-scale inhomogeneity during the non-linear stage of evolution. The profiles of lines of HeH^+ in this case will be defined by the distribution of the gas velocity in the protogalaxy.

References

Black J. 1988, Molecular Astrophysics (Cambridge: Univ. Press).

Burles S., Nollet, K., & Turner, M. 2000, (astro-ph/0008495).

Dalgarno, A., & Lepp, S. 1987, in IAU Symp. 118, Astrochemistry, (Dordrecht: Reidel), p. 109.

de Bernardis et al. (the ARGO project) 1993, A&A, 271, 683.

de Bernardis et al. 2000, Nature, 404, 955.

Dubrovich, V. K. 1994, A&A Transactions, 5, 7.

Dubrovich, V. K. 1997, A&A, 324, 27.

Dubrovich, V. K., & Lipovka, A. A. 1995, A&A, 296, 301.

Galli D., & Palla F. 1998, A&A, 335, 403.

Jaffe A. et al. 2001, Phys. Rev. Letter, 86, 3475.

Latter, W., & Black, J. 1991, ApJ, 371, 161.

Lepp, S., & Shull, M. 1984, ApJ, 280, 465.

Puy, D., & Signore, M. 1996, A&A, 305, 371.

Puy D., Alecian G., Lebourlot J., Leorat J., Pineau des Forets 1993, A&A, 267, 337.

Rauscher, T., Applegate, J. H., Cowan J. J., Thielemann F. K., Wiescher H. 1994, ApJ, 429, 499.

Saslaw, W., & Zipov, D. 1967, Nature, 216, 967.

Signore, M. et al. 1994, ApJS, 92, 535.

THE PRIMORDIAL LI: NEW OBSERVATIONS

Piercarlo Bonifacio
Osservatorio Astronomico di Trieste, Italy

Abstract In this contribution I make use of new data to investigate the *Spite* plateau and deduce an updated value for the primordial Li abundance. A preliminary result on globular cluster NGC 6397, based on recent VLT+UVES observations, suggests that its Li abundance is the same as that in field stars.

Introduction

In 1982 F. Spite and M. Spite (Spite & Spite 1982) discovered that the Li abundance in a sample of metal–poor stars is constant, regardless the effective temperature and the metallicity. This is referred to as the *Spite plateau* and is interpreted as evidence that the Li observed in these stars is primordial. This interpretation requires that the Li in the atmospheres of the *plateau* stars has not been depleted or diluted and the Galactic Li production has not significantly enriched stars at these low metallicities. There are three main arguments which support the absence of significant depletion in metal–poor turn-off stars, two observational and one theoretical: 1) the *plateau* is characterized by a very low dispersion ; 2) the fragile ^{6}Li isotope has been detected in three *plateau* stars (HD 84937, HD 338529 and G271-62) ; 3) no significant Li depletion is predicted by *standard* models of stellar evolution. On the other hand the flatness of the the *plateau* with [Fe/H] testifies against significant Li production at early times in the Galaxy.

The flatness of the *Spite plateau* has been challenged by several papers (Thorburn 1994; Ryan et al. 1996, 1999, 2001) in which slopes with both [Fe/H] and T_{eff} have been claimed. Claims against the reality of both slopes have been made by Molaro et al. (1995) , Spite et al. (1996) and Bonifacio & Molaro (1997) , thus the matter is still open to debate and only new and more accurate observations may settle the controversy. During the last four years I have been working in collaboration with A. Alonso and P. Molaro in order to collect IR photometry, from which determine accurate effective temperatures, and high resolution, high S/N spectra, from which we can measure accurate equivalent widths.

M. Chávez et al. (eds.),
New Quests in Stellar Astrophysics: The Link Between Stars and Cosmology, 77–80.

1. Results

The observational data shall be described in detail elsewhere, suffice it to say here that IR photometry has been collected at the TCS 1.5m telescope at Tenerife and spectra both at La Silla with NTT+EMMI and Paranal with VLT+UVES. Our sample now consists of 98 metal-poor stars of which 73 satisfy the conditions $T_{\rm eff} > 5700$ K and $[{\rm Fe/H}] \leq -1.5$, which define the *plateau*, as suggested by Bonifacio & Molaro (1997).

Li abundances have been computed as described in Bonifacio & Molaro (1997), using ATLAS9 model atmospheres with α enhanced opacities and no convective overshooting. The mean lithium abundance for the 73 *plateau* stars is $A({\rm Li}) = 2.22$ with a dispersion of 0.12 dex about the mean. The average error on A(Li) is 0.078 ± 0.020, if we compute the total error by summing under quadrature the errors due to errors in EW and $T_{\rm eff}$. The average absolute error, obtained by summing linearly the errors due to EW and $T_{\rm eff}$, is instead 0.100 ± 0.030. A Kolmogorov-Smirnov test of the hypothesis that the data come from a normal distribution with mean 2.22 and $\sigma = 0.078$ gives a probability of diversity of 96%, thus the hypothesis is rejected, but at a very low level. If the σ is raised only to 0.09 the probability of diversity drops to the 82% and the hypothesis is accepted. From this elementary statistical analysis we may conclude that quite likely the observed dispersion is entirely due to the observational errors; under the most restrictive hypothesis on the errors we may claim *at most* 0.02 dex of dispersion in excess of what due to observational errors. Thus the very low dispersion of the *plateau* is a very strong argument against any Li depletion. On the other hand it poses a very tight constraint on any possible slope which must be less than 0.02 dex over $\sim$ 2 dex in [Fe/H].

Non parametric tests fail to detect any correlation between A(Li) and [Fe/H]. On the other hand a correlation between A(Li) and $T_{\rm eff}$ is detected. Parametric fits with different techniques detect a mild slope of the order of 0.03 dex$\times$100 K. This is only slightly higher than what found by Bonifacio & Molaro (1997) and is quite likely a signature of the mild depletion predicted by *standard* models. In fact if we correct the data for this depletion, *and* for NLTE effects, as done by Bonifacio & Molaro (1997) the slope drops to the range 0.015 – 0.020 dex$\times$100 K. Let us restrict our attention to a *high quality* sample i.e. *plateau* stars with errors in the Li EW less than 1 mÅ, which comprises 22 stars. The slope with $T_{\rm eff}$ drops to $\sim$ 0.01 dex$\times$100 K and to zero once we consider the corrections for standard depletion and NLTE. Needless to say that no slope with [Fe/H] is detected for this high quality sample either.

With the present data the only way to recover a slope with [Fe/H] is to perform a bivariate fit of *both* [Fe/H] and $T_{\rm eff}$ to the *full* sample of stars. The bivariate fit to the sample of 22 stars with accurate EWs, instead, does not recover any slope. This rings a bell! Let us consider Li EWs as a function of $T_{\rm eff}$, there

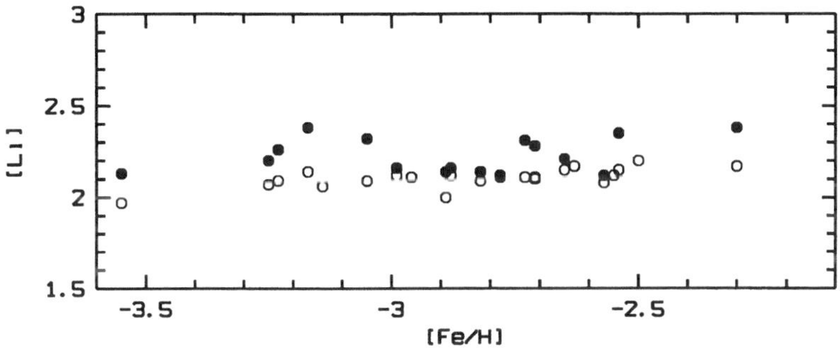

Figure 1. A(Li) versus [Fe/H] for the stars of Ryan et al. (1999) (open symbols) and using IRFM temperatures (filled dots).

is an obvious trend in that the most metal–poor stars have the smallest EWs. This just reflects the observational bias that the most metal–poor stars are also the hottest. However when we measure weak lines (EW $\leq$ 25 mÅ), if the S/N ratio is not adequate we tend to underestimate the width of the line because we place the continuum too low. I suggest that the main driver of the slope with [Fe/H] found by previous researchers is in fact this bias which tends to underestimate Li EWs for metal–poor stars. There are surely, however, other reasons and the main one is the temperature scale adopted. This case is nicely illustrated by looking at the data of Ryan et al. (1999), which is of fairly high quality, although not as good as the present *high quality* sample. We now have IRFM temperatures for 17 out of the 22 stars of Ryan et al. (1999); in Fig. 1 we show the A(Li) derived using the EWs of Ryan et al. (1999) and the IRFM temperatures as a function of the [Fe/H] of Ryan et al. (1999), together with the original data of Ryan et al. (1999). The visual impression that with IRFM temperatures there is no detectable slope is confirmed by both parametric and non-parametric analysis. This demonstrates that, contrary to what claimed by Ryan et al. (1999), the difference between their results and those of Bonifacio & Molaro (1997) lies in the different temperature scales adopted and *not* in the adopted metallicities.

Considering the *high quality* sample the best estimate of the primordial Li abundance, once the effects of standard depletion and NLTE are accounted for, is A(Li) $= 2.317 \pm (0.014)_{1\sigma} \pm 0.05_{sys}$ which corresponds to two possible solutions for the baryonic density $\Omega_b h^2 = 0.0055 \pm 0.0004$ and $\Omega_b h^2 = 0.015 \pm 0.001$. The *high* density solution is consistent with the most recent results obtained from the fluctuations of the cosmic microwave background as observed by BOOMERANG, which imply $\Omega_b h^2 = 0.021 \pm 0.003$ Netterfield et al. (2001).

In the course of an ESO Large Programme devoted to the study of globular clusters 5 TO stars of the cluster NGC 6397 have been observed with UVES at the VLT. Only preliminary results are available, however it is already clear that the five stars have the same effective temperature and the same metallicity (Gratton et al. 2001). This result is in agreement with the early results of Molaro & Pasquini (1994) and Pasquini & Molaro (1996). By taking the atmospheric parameters from Gratton et al. (2001) we determine the Li abundance to be A(Li) $= 2.35 \pm 0.04$ in remarkable agreement with the abundance of the field stars. While this result is preliminary and may change slightly due to a better extraction of the spectra and to a change in the adopted effective temperature, in order to be on the same scale as that adopted for the field stars (the IRFM scale), it is clearly hinting at a common origin for the Li in the stars of NGC 6397 and of the halo field. This strengthens the arguments for the primordial origin of Li in metal–poor stars.

Acknowledgments

I am grateful to A. Alonso and P. Molaro and also to R. Gratton and the whole ESO-LP 165.L-0263 collaboration for allowing me to anticipate some of the results of our common work.

References

Bonifacio P. & Molaro P. 1997, MNRAS, 285, 847.

Gratton R. et al. 2001, A& A, 369, 87.

Molaro P. & Pasquini L. 1994, A&A, 28, L77.

Molaro P., Primas, F., & Bonifacio, P. 1995, A&A, 295, L47.

Netterfield et al. 2001, ApJ, submitted (astro-ph/0104460).

Pasquini L. & Molaro P. 1996, A&A, 307, 761.

Ryan S. G., Beers, T. C., Deliyannis, C. P., & Thorburn, J. A. 1996, ApJ, 458, 543.

Ryan S. G., Norris, J. E. & Beers, T. C. 1999, ApJ, 523, 654.

Ryan S. G., Kajino, T., Beers, T. C., Suzuki, T. K., Romano, D., Matteucci, F., & Rosolankova, K. 2001, ApJ, 549, 55.

Spite F. & Spite M. 1982, A&A, 163, 140.

Spite M., Francois, P., Nissen, P. E., & Spite, F. 1996, A&A 307, 172.

Thorburn, J. A. 1994, ApJ, 421, 318.

III

STARS IN THE MILKY WAY

STAR FORMATION HISTORY IN THE SOLAR NEIGHBORHOOD: THE LINK BETWEEN STARS AND COSMOLOGY

V. Avila-Reese, C. Firmani,* and X. Hernández
Instituto de Astronomía, U.N.A.M., A.P. 70-264, 04510 México, D.F.

Abstract

Using a cosmological galactic evolutionary approach to model the Milky Way, we calculate the star formation history (SFH) of the solar neighborhood. The good agreement we obtain with the observational inferences suggests that our physical model describes accurately the long term/large spatial trends of the local and global Milky Way SFH. In this model, star formation is triggered by disk gravitational instabilities and self-regulated by an energy balance in the ISM. The drivers of the SFH are the cosmological gas infall rate and the gas surface density determined by the primordial spin parameter. A ΛCDM cosmology was used throughout.

Introduction

Galaxies are the "ecosystems" where stars are born and evolve interacting with the ISM. On the other hand galaxies are the structural units of the universe as a whole. Thus, the study of galaxy formation and evolution connects in a natural way stellar and ISM astrophysics with cosmology. The Milky Way (MW) is among the best studied galaxies, in particular at the solar neighborhood. The available observations for the solar neighborhood provide valuable information not only about its present-day properties but also about its past history. For example, the possibility of resolving stars and constructing a color-magnitude diagram (CMD) for the solar neighborhood allows us in principle to recover its star formation history (SFH).

Star formation (SF) is a key ingredient for galaxy formation and evolution. Most models of galaxy formation in a cosmological context extend the hierarchical

* Also Osservatorio Astronomico di Brera, Italy

M. Chávez et al. (eds.),
New Quests in Stellar Astrophysics: The Link Between Stars and Cosmology, 83–90.

assembly of cold dark matter (CDM) halos to luminous galaxies using *ad hoc* and/or empirical recipes for calculating the SF (e.g., White & Frenk 1991; Kauffmann, White, & Guiderdoni 1993; Baugh, Cole, & Frenk 1996). An important improvement in this kind of approaches is to link the SF process to the structure, dynamics and hydrodynamics of modeled galaxies, and to test the model with detailed observations, for example the SFH in the solar neighborhood.

We have developed an approach of disk galaxy formation and evolution in a cosmological context which includes a physically self-consistent scheme to calculate the SF process of the growing disk (Firmani & Avila-Reese 2000; Avila-Reese & Firmani 2000, hereafter FA00 and AF00 respectively). Using this approach and in the light of the available observational information, we will explore the drivers of the local and global SFH in the MW and their potential connection to the cosmological background.

1. Observational inferences

An objective inference of the SFH in the solar neighborhood comes from by comparing synthetic CMDs to the observed one (e.g., Chiosi et al. 1989). These CMD inversion methods have been improved more recently with a more rigorous statistical analysis (Tolstoy & Saha 1996; Hernández, Valls-Gabaud & Gilmore 2000 and more references therein). For example, the advanced Bayesian analysis introduced by Hernández et al., allows the recovery of the underlying SFH without the need of assuming any *a priori* SFH. With the *Hipparcos* catalog the time resolution of this technique is ~ 0.05 Gyr, however, the age range which it allows to explore is small (the last 3 Gyr); this limitation is related to the incompleteness of the catalog. With a standard parametric maximization technique, Bertelli & Nasi (2001) were able to recover the SFH along the whole life of the disk solar neighborhood, at expenses of a low time resolution. Their method is useful to describe only the *general trend of the SFH over the total life of the system*, and not its detailed shape. The robust conclusion we can extract from this study is that the average SF rate (SFR) over the period 10-6 Gyr (in look back time) could not have been larger than that over the 6-0 Gyr period. In Fig. 1a, we plot the surface SFR $\dot{\Sigma}_s$ vs. look back time corresponding to the solution for the model with more degrees of freedom of those used in Bertelli & Nasi (jointed circles). That the solar neighborhood SFR in the past could not have been very different from the present-day one is also suggested by the birthrate parameter b and the differential metallicity distribution of G-dwarfs (e.g., Boissier & Prantzos 1999).

A quantity more accurately determined than the SFH shape, is the look back time at which stars began to form in the solar neighborhood. Using the photometric and kinematic data of the *Hipparcos* satellite, Binney, Dehnen, & Bertelli (2000)

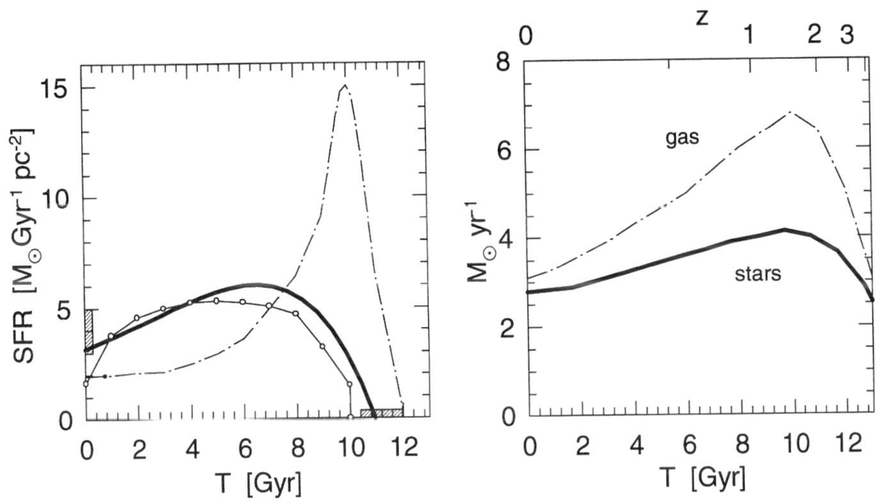

Figure 1. *(a) Left:* Surface SF and gas accretion histories for the solar neighborhood corresponding to the fiducial model (thick solid and dot-dashed lines). The jointed circles give the inferences of Bertelli & Nasi 2001, and the boxes are independent observational estimates for the age of the solar neighborhood and its present-day SFR. *(b) Right:* Global SF and gas accretion histories for the fiducial MW model.

inferred an age for the disk at the solar radius of 11.2 ± 0.75 Gyr (horizontal shaded box in Fig. 1a). Regarding its present-day SFR, observational inferences give values around $3 - 5\ M_{\odot}Gyr^{-1}pc^{-2}$ (vertical shaded box).
Methods alternative to the CMD inversion have been also used to infer the SFH in the solar neighborhood, among them the construction of the stellar age distribution using the empirical correlation of chromospheric activity in G-K dwarfs with their age (e.g., Rocha-Pinto et al. 2000; see references on other methods therein). These methods are subject of strong observational uncertainties and biases. For example, the chromospheric activity-age relation presents a strong scatter of around a factor 2 in age which should be included as a time smoothing kernel on the inferred SFH. Interestingly, if such a smoothing is applied on the results of Rocha-Pinto et al. (2000), a trend similar to that reported in Bertelli & Nasi (2001) appears.

2. A theoretical approach

As it was discussed above, the observational data in the solar neighborhood allow to infer a general trend of its SFH, which due to stellar diffusion it may actually correspond to the average SFH of an annulus of $\sim 2 - 3$ kpc. We will present predictions regarding local and global SFHs for the MW modeled using a deductive disk galaxy evolution approach (FA00; AF00). This approach is based on the hierarchical formation scenario, assuming that disks form by gas

accretion more than by repeated merging of sub-unities which would destroy the disk. A brief description of our approach is as follows:
A disk forms *inside-out* within a growing CDM halo. The evolution of the halo is set by its mass aggregation history (MAH). For a given present-day total virial mass M_v, a statistical set of MAHs is calculated from the initial Gaussian density fluctuation field once the cosmology and power spectrum are fixed (Avila-Reese, Firmani & Hernández 1998). Given the MAH, the halo density profile is calculated with a generalization of the secondary infall model, based on spherical symmetry and adiabatic invariance, but allowing for non-radial orbits (Avila-Reese et al. 1998). The halo density profile depends on the MAH. The average MAH yields density profiles very close to the Navarro, Frenk & White (1997) profile, while the other less probable MAHs yield a variety of density profiles which also agree with what is found in N-body cosmological simulations (Avila-Reese et al. 1998, 1999). A fraction f_d of the mass of each collapsing spherical shell is transferred in a virialization time into a central disk gas layer. Assuming angular momentum conservation, a given gas element of the shell falls into the equatorial plane at the position where it reaches centrifugal equilibrium. The specific angular momentum of each shell is obtained from the derivative of λ, assumed to be constant in time. We calculate the halo contraction due to the gravitational drag of the infalling gas. A nearly exponential disk surface density arises naturally in this inside-out scheme of disk formation. The λ parameter determines the scale length and surface density of the disk.

Star formation. SF at a given radius in the gaseous disk is triggered by gravitational instabilities, i.e. whenever the Toomre parameter $Q_g(r)$ is less than a given threshold, q (numerical and observational studies suggest $q \approx 2$). The SF rate (SFR) is calculated from the equation that relates the energy input rate (mainly due to SNae) to the (turbulent) energy dissipation rate assuming that $Q_g(r)$ is always equal to q at all radii, i.e. we allow only for the stationary solution of a *self-regulated* SF mechanism (Firmani & Tutukov 1994; Firmani, Hernandez, & Gallagher 1996). A key parameter in this scheme is the turbulent dissipation time $t_d(r)$ which we approximate as in Firmani et al. (1996). Estimates of $t_d(r)$ for the solar neighborhood obtained in compressible magneto-hydrodynamic simulations of the ISM (Avila-Reese & Vázquez-Semadeni 2001) are in agreement with the Firmani et al. approximation. The evolution of the stellar populations is followed with a parameterization of simple population synthesis models. A Salpeter IMF with minimal and maximal masses of $0.1M_\odot$ and $100M_\odot$, and solar metallicities were used. The azimuthally averaged dynamics of the evolving gas and stellar disks coupled with the dark halo are treated by solving the corresponding hydrodynamical equations.

Modeling the Milky Way. For a given cosmology (here: $\Omega_m = 0.3, \Omega_\Lambda = 0.7$, $H_0 = 65$ km s^{-1}Mpc^{-1}, $\sigma_8 = 0.9$) a galaxy model is completely determined by the virial mass M_v, the MAH, the spin parameter λ, and the disk mass fraction f_d. These factors and their statistical distribution are related to the cosmological background, and we have shown that they are able to produce disk galaxies which match the Tully-Fisher relation and its scatter, and several correlation across the Hubble sequence (FA00; AF00). In order to calculate a model representative of the MW, we have to chose correctly these input factors. Fortunately, each one of them is tightly related to a present-day MW feature, although with some (small) degeneracy. In Hernandez, Avila-Reese & Firmani (2001, HAF01) we showed that the sensitivity of the local SFH to this degeneracy and to the uncertainty of the observables that constrict the input parameters are small.

The MAH determines mainly the galaxy color index. Unfortunately the MAH cannot be described by one simple parameter and the color index of the MW is not well determined. Therefore, does not make sense to fix a very particular MAH for the MW. We shall use the most probable realization of the MAHs (the average one), but we also will explore other possible MAHs. As a matter of fact, the $B - K$ color estimated for the MW disk (~ 3.3, see Kent et al. 1991 for the $K-$magnitude estimate) is similar to the average color of galaxies of its type, suggesting that the MW is indeed an average galaxy.

The total mass M_v, the spin parameter λ, and the disk mass fraction M_d are fixed in such a way that the modeled MW obeys the observable constraints related to these parameters: the circular velocity at 50 kpc, the disk scale length r_s, and the disk+bulge mass, mainly the mass contained in stars, M_s. For the average MAH, these constraints fix $M_v = 2.8\ 10^{12} M_\odot$, $\lambda = 0.02$, and f_d=0.021 (the fiducial model). For this model the constraints and predictions, and the corresponding observational estimates are given in Table 1 (see HAF01 for the references). The shortcoming of the model is an excessively peaked rotation curve w.r.t. observations. This is associated with the large central concentrations of CDM halos. The predicted SFH does not change significantly if a shallow core is introduced in the MW CDM halo (HAF01).

Predicted local and global star formation histories. In Fig. 1a we show the SF and gas infall histories, $\dot\Sigma_s$ and $\dot\Sigma_g$, of the fiducial model at the radius $R_0 = 8.5$ kpc. There is a remarkable agreement with the observational estimates. The disk at R_0 begins to form at look back time $\sim$ 11 Gyr ($z \approx 2$), the SFR attains a maximum at $8 - 6$ Gyr ($z \approx .9 - .6$) and then it decreases by a factor of ~ 2 towards the present. With these SF and gas infall histories the present-day stellar and gas surface densities are 41.3 and 13.8 $M_\odot$pc^{-2}, respectively.

In HAF01 we have experimented with a wide range of MAHs (the MAH influences strongly the SFH), and we have found that the SFH predicted with the most probable (average) MAH is the best in matching the observational data.

Table 1. Properties of the MW: observations and model results

Observable	Predicted value [a]	Observed value
$V_c(50)$ [b] $[\mathrm{km\ s^{-1}}]$	208	206 ± 10
r_s [kpc]	3.0	3.0 ± 0.5
M_s [c] $[10^{10}\ M_\odot]$	4.4	4-5
M_v $[10^{12}\ M_\odot]$	2.8	1-4
V_{max} [d] $[\mathrm{km\ s^{-1}}]$	235	220 ± 10
L_B $[10^{10} L_{B\odot}]$	1.7	1.8 ± 0.3
r_B [e] [kpc]	4.3	4-5
$\Sigma_{0,K}$ [f] $[L_{K\odot} pc^{-2}]$	810	1000 ± 200
f_g	0.23	$0.15 - 0.20$
SFR $[M_\odot \mathrm{yr}^{-1}]$	2.9	2-6
Solar neighborhood		
Σ_s $[M_\odot \mathrm{pc}^{-2}]$	41.3	43 ± 5 [g]
Σ_g $[M_\odot \mathrm{pc}^{-2}]$	13.6	13 ± 3
$\dot{\Sigma}_s$ $[M_\odot \mathrm{Gyr}^{-1}\ \mathrm{pc}^{-2}]$	3.1	3-5
$B - K$ [mag]	3.15	3.13

$^{(a)}$ The MW model was obtained tuning three input parameters in order to reproduce the first three quantities, which are constraints and not predictions. $^{(b)}$ Circular (asymptotic) velocity at 50 kpc. $^{(c)}$ Stellar (disc+bulge) mass. $^{(d)}$ This maximum rotation velocity does not take into account the nuclear and bulge region and it is for Galactocentric radii smaller than 15 kpc. $^{(e)}$ B–band disc scale length. $^{(f)}$ Disc central surface brightness in the K–band, taking a mass to light ratio of 1.0 for this band. $^{(g)}$ This estimate includes the contribution of stellar remnants.

Using Padova stellar evolutionary models we computed synthetic CMDs for this SFH taking a solar metallicity for the last 2 Gyr, and one third the solar before, as a first approximation to the enrichment history. The obtained $B - K$ color index is 3.15, in excellent agreement with estimates for the solar neighborhood, $B - K = 3.13$ (Binney & Merrifield 1999). In the future we shall compare directly the *Hipparcos* CMD with our theoretical CMD (thanks to A. Bressan for suggesting this to us). Repeating the experiment for SFHs resulting from other MAHs, we obtain $B - K$ colors up to 0.25 mag redder and 0.15 mag bluer than 3.15. Again, the average MAH seems to be the optimal choice for the Galaxy.

The global SF and gas infall histories (in $M_\odot\ \mathrm{yr}^{-1}$) for the fiducial MW model are shown in Fig. 1b. The maximum SFR is attained at $z \approx 1.3$, then decreases by a factor of 1.5 towards $z = 0$. A nearly constant SFH is a common feature in the evolution of our modeled disk galaxy population. Our predicted SF and gas infall histories do not deviate from the constraints of the phenomenological chemo-spectrophotometric models (e.g., Boissier & Prantzos 1999). Therefore, we can expect our model to be also in agreement with the data regarding metallicities and abundance gradients in the solar neighborhood and the Galaxy.

3. Drivers of the SFH and concluding remarks

Within our SF scheme, the two important drivers of the SFH are the gas infall history and the gas disk surface density. The former is related to the cosmological MAH while the latter depends mainly on the spin parameter λ. For a roughly 1-σ range in the MAHs and λ's, most of the predicted global SFHs of disk galaxies present a shallow maximum at $z \approx 1 - 2.5$ and then slightly decrease until $z = 0$ by a factor < 3 (see Fig. 1b, for the case of a MW-like galaxy). The shape of the SFH of the universe inferred from high-redshift observations does not agree with this nearly constant behavior of the SFH of disk galaxies, suggesting that other galaxy populations (starburst dwarfs?, bulges?, ellipticals?) dominated at larger redshifts in the integral SFR of the universe.
The local SFH (e.g. at radius R_0) also is controlled by the (cosmological) gas infall history and the local gas surface density. The stationary self-regulation mechanism combined with the disk gravitational instability triggering criterion lead to a $\dot{\Sigma}_s(r) \propto {\Sigma_g}^n$ law, with $n \approx 2$ in rough agreement with observations. In the light of our inside-out scenario, the formation epoch T of the MW disk at a given radius (e.g. R_0) depends on the angular momentum of the infalling gas, i.e. on λ. On the other hand, λ determines r_s. Interestingly, once we fixed λ to reproduce the observed r_s, the predicted T_0 is in agreement with the observational estimates, suggesting that T_0 and r_s could indeed be related to the only parameter λ. Note that we have used $H_0 = 65$ km s^{-1}Mpc^{-1}; cosmic time is inversely proportional to H_0, so that the predicted T_0 also depends on H_0. A similar connection, although more trivial, appears to be related to the $B - K$ color index of the solar neighborhood and its SFH. According to our model, both of them are related to the MAH, and for the average MAH the predictions agree with the observational inferences.
The overall self-consistency of our cosmological MW model predictions and the agreement found with the inferred SFHs in the solar neighborhood, suggest that the main ingredients of the SF process in the MW were correctly taken into account. Our stationary solution may represent the long term/large spatial trends of the SFH in the solar neighborhood. Complementary mechanisms of SF or SF enhancement could introduce fluctuations of short temporal and spatial character.

Acknowledgments

V.A. and X.H. thank the Organizing Committee for providing them financial support in order to assist to this conference. This work was partially supported by CONACyT grant J33776-E to V.A.

References

Avila-Reese V., & Firmani C., 2000, Rev. Mex. Astron. Astrofis., 36, 23 (AF00).
Avila-Reese V., & Vázquez-Semadeni E., 2000, ApJ, 553, 645
Avila-Reese V., Firmani C., & Hernández X., 1998, ApJ, 505, 37.
Avila-Reese V., Firmani C., Klypin A., Kravtsov A., 1999, MNRAS, 310, 527.
Baugh C.M., Cole S., & Frenk C.S., 1996, MNRAS, 283, 1361.
Bertelli G., & Nasi E., 2001, AJ, 121, 1013.
Binney J., Dehnen W., & Bertelli G. 2000, MNRAS, 318, 658.
Binney J., & Merrifield M. 1998, "Galactic Astronomy" (Princeton Univ. Press), p. 556.
Boissier S., & Prantzos N., 1999, MNRAS, 307, 857.
Chiosi C., Bertelli G., Meylan G., & Ortolani S., 1989, A&A, 219, 167.
Firmani C., & Avila-Reese V. 2000, MNRAS, 315, 457 (FA00).
Firmani C., & Tutukov A.V. 1994, A&A, 288, 713.
Firmani C., Hernández, X., & Gallagher, J. 1996, A&A, 308, 403.
Hernández X., Avila-Reese V., & Firmani C., 2001, MNRAS, 327, 329 (HAF01).
Hernández X., Valls-Gabaud D., & Gilmore G. 2000, MNRAS, 316, 605.
Kauffmann G., White, S.D.M., & Guiderdoni, B. 1993, MNRAS, 264, 201.
Kent S., Dame T., & Fazio G., 1991, ApJ, 378, 131.
Navarro J., Frenk C.S., & White S.D.M., 1997, ApJ, 490, 493.
Rocha-Pinto, H., Scalo, J., Maciel, W.J., & Flynn, C. 2000, ApJ, L115.
Tolstoy E., & Saha A., 1996, ApJ, 462, 672.
White S.D.M, & Frenk C.S. 1991, ApJ, 379, 52.

CARBON AND OXYGEN GALACTIC ABUNDANCE GRADIENTS PREDICTED BY DIFFERENT STELLAR YIELDS

L. Carigi

Instituto de Astronomía, Universidad Nacional Autónoma de México, Mexico

Abstract

Chemical evolution models for the Galactic disk under an inside-out formation scenario are presented for seven sets of stellar yields, all dependent on metallicity. The effects of massive-star yields and low-and-intermediate-mass-star yields on the C/O chemical history of the solar neighborhood and on abundance gradients are discussed.

The increase of C/O with metallicity in the disk of the solar vicinity is due to massive stars sole. Models with yields by Maeder (1992) or yields by Portinari, Chiosi, & Bressan (1998) can reproduce the increase of C/O with metallicity in the solar neighborhood, while models assuming yields by Woosley & Weaver (1995) and Woosley, Langer, & Weaver (1993) can not. Models based on yields by Maeder (1992) are in agreement with the C/O gradient, while those based on yields by Woosley & Weaver (1995) and Woosley et al. (1993) or Portinari et al. (1998) are not. Model results call for: i) yields that account for metal-dependent stellar winds, and ii) a more complicated mass-loss rate law than $Z^{0.5}$ for massive stars.

The early rise of C/O with O/H is due to massive stars and low-and-intermediate-mass stars. C/O abundance ratios predicted with yields by Marigo, Bressan, & Chiosi (1996, 1998) are higher that those by Renzini & Voli (1981), and these last ones higher than those by van den Hoek & Groenewegen (1997).

Introduction

In a chemical evolution model that assumes an infall of H and He gas, without mass loss by galactic winds or radial flows, the predicted C/O abundance ratio depends mainly on the stellar yields and the initial mass function. In this work the initial mass function is fixed and therefore, the C/O value is used to test different sets of stellar yields dependent on metallicity.

This test is performed comparing the model predictions with the observed fact that C/O increases with metallicity in the solar vicinity and with the negative C/O gradients determined from nearby H II regions and B-stars.

M. Chávez et al. (eds.),
New Quests in Stellar Astrophysics: The Link Between Stars and Cosmology, 91–94.

1. Chemical Evolution Models

All the models are built to reproduce two observational constraints: the O/H abundance gradient from H II regions and B-stars (see Figure 1(b)), and the σ_{gas} distribution (Matteucci & Chiappini, 1999) from 4 to 10 kpc. Models are similar to the infall model of Allen, Carigi, & Peimbert (1998), but with some different assumptions explained in detail by Carigi (2000).
Specifically, I have assumed: a) Three sets of metal-dependent stellar yields from massive stars ($8 < m/\mathrm{M}_\odot < 85$): *i)* M yields: Maeder (1992) for $9 \leq m/\mathrm{M}_\odot \leq 85$; *ii)* WWL yields: Woosley & Weaver (1995) for $11 \leq m/\mathrm{M}_\odot \leq 40$, and Woosley et al. (1993) for $m = 60$ $\mathrm{M}_\odot$ and $m = 85$ $\mathrm{M}_\odot$; *iii)* PCB yields: Portinari et al. (1998) for $9 \leq m/\mathrm{M}_\odot \leq 85$. Since $Z_{max} = Z_\odot$ for the M and WWL yields, then yields ($Z > Z_\odot$) = yields($Z_\odot$). b) Three sets of metal-dependent stellar yields for for low-and-intermediate-mass stars (LIMS): *i)* RV yields: Renzini & Voli (1981) for $1 \leq m/\mathrm{M}_\odot \leq 8$. *ii)* HG yields: van den Hoek & Groenewegen (1997) for $0.8 \leq m/\mathrm{M}_\odot \leq 8$. *iii)* MBCP yields: Marigo et al. (1996) for $0.8 \leq m/\mathrm{M}_\odot \leq 3$; Marigo et al. (1998) for 4, 4.5, and 5 $\mathrm{M}_\odot$; Portinari et al. (1998) for 6 and 7 $\mathrm{M}_\odot$.
Models are divided in three groups, assuming M, WWL, or PCB yields. M and WWL yields have been complemented with RV, HG, and MBCP yields. There is just one model for the PCB yields, because the PCB+MBCP yields form a complete and consistent set.

2. Model Results and Discussion

Figure 1 presents the predicted (a) C/O chemical enrichment history for the solar neighborhood ($r_\odot = 8$ kpc) and (b)-(c) O/H and C/O gradients for the Galactic disk, when M, WWL, and PCB yields are assumed. From Figure 1(a), one can see that the C/O abundance always increases with time in the M-models and the PCB-model, in agreement with most of the observations. Furthermore, the results of the WWL-models do not show a recent rise of C/O, in disagreement with observations. From Figure 1(c), it can be seen that the negative C/O gradients predicted by the M-models are in very good agreement with the observations, while the positive gradients predicted by WWL-models or PCB-model are in contradiction with the observations. Moreover, M-models are the sole models that reproduces the high C/O values of M17.
A main difference between these sets of yields resides on the stellar-wind assumptions. Woosley & Weaver (1995) do not consider stellar winds at any stage, then C and O yields are almost independent of Z, this fact implies a saturation in C/O. Therefore, the W-models reach a plateau in C/O that decreases slowly with time and produce almost flat C/O gradients. M and PCB assume a mass-loss rate proportional to $Z^{0.5}$ during the post-main-sequence phases. In M and PCB, the C yields increase and O yields decrease with Z. This fact

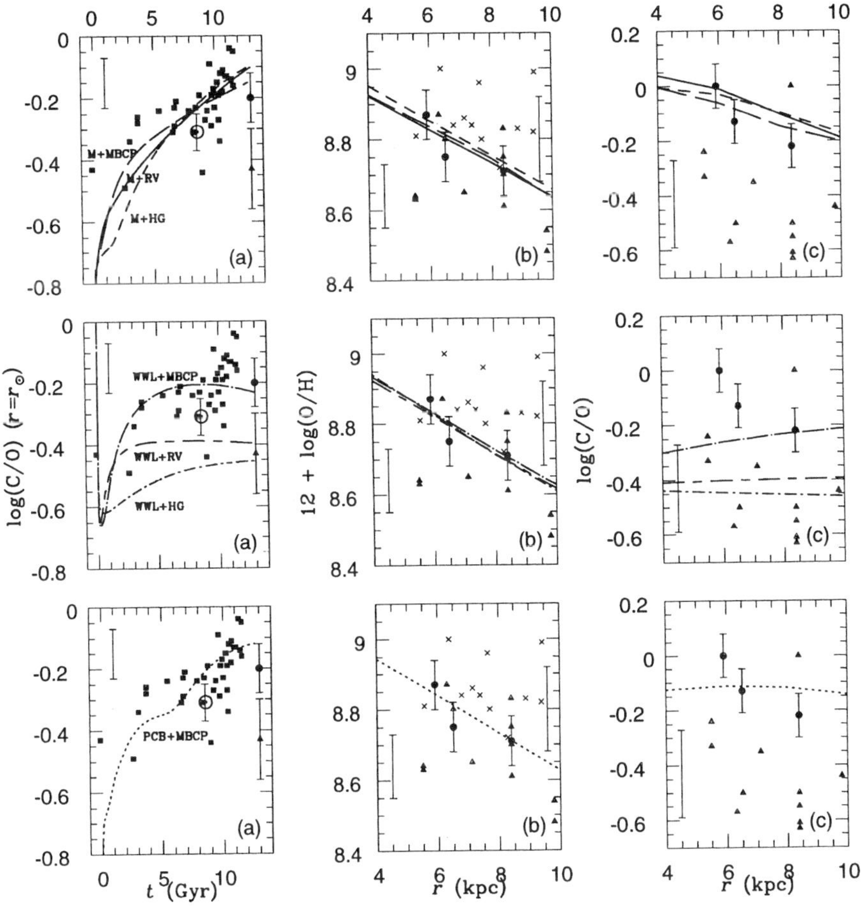

Figure 1. (a) The C/O evolution of the solar vicinity. (b)-(c) O/H and C/O predicted and observed abundance ratios. Models considering yields from massive stars complemented by yields from low-and-intermediate-mass stars are labeled as follow: M = Maeder (1992); WWL= Woosley & Weaver (1995) and Woosley, Langer, & Weaver (1993); PCB = Portinari, Chiosi, & Bressan (1998); RV = Renzini & Voli (1981); HG = van den Hoek & Groenewegen (1997); MBCP = Marigo, Bressan, & Chiosi (1996, 1998) and Portinari et al. (1998). Observational data: filled circles: H II regions from Peimbert (1999); filled triangles: B-stars from Gummersbach et al. (1998); crosses: B-stars from Smartt & Rolleston (1997); filled squares: dwarf stars from Gustafsson et al. (1999); $\odot$: solar value from Grevesse & Sauval (1998). Error bars in the left and the right in (b)-(c) represent the typical errors from data by Gummersbach et al. (1998) and Smartt & Rolleston (1997), respectively.

causes the M-models and the PCB-model to reproduce the rise of the C/O in the solar vicinity.

The increase of C yields and the decrease of O yields with Z can not be extrapolated to $Z > Z_\odot$. According to PCB, the C to O yields ratio decreases from $Z = Z_\odot$ to $Z = 2.5\ Z_\odot$. The supermetallic and massive stars have early and intense winds, which occur before the stars synthesize C and O. Therefore, their C and O yields are similar to those of metal-poor stars, that mainly eject C and O in the SN explosion. This behavior of the C and O yields produce a saturation of the C/O abundance ratio at ~ -0.1 dex, which produces flatter gradients for the PCB-model. Since in M-models I have assumed that the supermetallic stars behave like metal-solar stars, the PCB-model would predict a negative C/O gradient if the mass-loss rate for $Z = 2.5\ Z_\odot$ is similar to, at least, that for $Z = Z_\odot$

The sets of yields for LIMS dependent on Z that exist in the literature (RV, HG, and MBCP) can not explain the recent rise of C/O in the solar vicinity. However, LIMS have an important role in the early evolution of C/O.

3. Conclusions

Based on basic chemical evolutionary models for the Galactic disk, assuming different sets of stellar yields dependent on metallicity, I reach the following conclusions: a) Metal-dependent stellar winds are essential to properly follow the chemical evolution of the Galaxy. b) To improve the agreement with the C/O Galactic abundances and the C/O evolution with metallicity, the present models call for a more complicated mass-loss rate law than $Z^{0.5}$: $\dot{M}_{wind} \propto Z^n$, such that $n \sim 0.5$ if $Z \leq Z_\odot$ and $0.5 < n < 0.7$ if $Z > Z_\odot$.

References

Allen, C., Carigi, L., & Peimbert, M. 1998, ApJ, 494, 247.

Carigi, L. 2000, Rev. Mex. Astron. Astrof., 36, 171.

Grevesse, N., & Sauval, A.J. 1998, Space Sci. Rev., 85, 161.

Gummersbach, C.A., Kaufer, A., Schäfer, D.R., Szeifert, T., & Wolf, B. 1998, A&A, 338, 881.

Gustafsson, B., Karlsson, T., Olsson, E., Edvardsson, B. & Ryde, N. 1999, A&A, 342, 426.

van den Hoek, L.B., & Groenewegen, M.A.T. 1997, A&AS, 123, 305 (HG).

Maeder, A. 1992, A&A, 264, 105 (M).

Marigo, P., Bressan, A., & Chiosi, C. 1996, A&A, 313, 545.

—— 1998, A&A, 331, 580.

Matteucci, F., & Chiappini, C. 1999, in Chemical Evolution from Zero to High Redshift, ed. J. Walsh & M. Rosa, (Garching: ESO), p. 83.

Peimbert, M. 1999, in Chemical Evolution from Zero to High Redshift, ed. J. Walsh & M. Rosa, (Garching: ESO), p. 30.

Portinari, L., Chiosi, C., & Bressan, A. 1998, A&A, 334, 505 (PCB).

Renzini, A., & Voli, M. 1981, A&A, 94, 175 (RV).

Smartt, S.J., & Rolleston, W.R.J. 1997, ApJ, L47.

Woosley, S.E., Langer, N., & Weaver, T.A. 1993, ApJ, 411, 823.

Woosley, S.E., & Weaver, T.A. 1995, ApJS, 101, 181.

THE UPPER END OF THE GALACTIC METALLICITY SCALE

Benjamin J. Taylor
Brigham Young University, USA

Abstract

Research on "super-metal-rich" stars (with [Fe/H] $>$ +0.2 dex) is reviewed. There is good evidence for the existence of SMR dwarfs and subgiants. Despite conclusions drawn in the literature, however, it is not known whether μ Leo or any other class III star is SMR. In addition, the possible SMR status of the galactic cluster NGC 6791 has not been established. In large part, these uncertainties are due to inadequate research strategy, inadequate control of data quality, extensive reliance on unsupported axioms, and rejection of statistical analysis.

Introduction

Spectroscopists who use spectra of nearby stars to interpret the spectra of external galaxies have an obvious interest in high-metallicity stars. Ideally, one would help those spectroscopists by simply listing such "super-metal-rich" stars and discussing their properties. Unfortunately, issues of scholarship–not results–dominate the "SMR" problem. Such issues will be the chief concern of this review.

The scholarship and results considered here are derived from a) analyses of color-magnitude (C-M) diagrams, and b) high-dispersion analyses. When the latter are considered, the abbreviation "NAHDS" will stand for "(mostly) North American high-dispersion spectroscopist(s)." Extramural spectroscopists are included among NAHDS if they have analyzed high-metallicity stars.

1. Statistics and C-M diagrams: assessing a key result

If present-day C-M analysis is compared to its state when Martin Schwarzschild was co-founding the field, a curious contrast emerges. In most aspects (such as stellar opacities), C-M analysis has advanced substantially during the last half-century. This is not true, however, for comparisons of observed C-M diagrams

M. Chávez et al. (eds.),
New Quests in Stellar Astrophysics: The Link Between Stars and Cosmology, 95–102.

with theoretical tracks. It appears that such comparisons are still performed graphically in all cases, just as they were in Martin Schwarzschild's time.

Graphical analysis is non-statistical. For this reason, it does not yield rigorous confidence intervals. The results of graphical analysis are formally meaningless and may be over-interpreted. Among C-M analysts, there seems to be no general awareness of these problems.

A pertinent example of graphical C-M analysis is given by Chaboyer et al. (1999). For the old galactic cluster NGC 6791, those authors find that [Fe/H] $= +0.4 \pm 0.1$ dex. Chaboyer et al. (1999) derive five parameters in all from their analysis. For this reason, they run a special risk posed by the increase in size of confidence intervals as the number of derived parameters increases. Rigorous confidence intervals for the Chaboyer et al. (1999) solutions might be too large to be useful.

One also notes that Chaboyer et al. (1999) do not perform any solutions for metallicities between 0.0 and +0.3 dex. As a result, they do not even establish a prima facie case against metallicities in that range. All told, it is fair to suspect that a 68% confidence interval for their results would be larger than the confidence interval they quote.

2. Statistics and high-dispersion spectroscopy: assessing a second key result

Rigorous statistical analysis is also absent from some parts of the high-dispersion discipline. Examples of such analysis can be found outside–but not inside–the North American branch of high-dispersion work. This contrast is well illustrated by the difference between the Li analyses of Merchant and Bonifacio which appear elsewhere in this volume.

Unlike C-M analysts, NAHDS sometimes apply quasi-statistical analyses. However, these procedures are not rigorous. Peterson & Green (1999) present a good example of this problem in an analysis of a blue horizontal-branch star in NGC 6791.

Peterson & Green (1999) quote their final result as $+0.4 \pm 0.1$ dex. Their confidence interval turns out to have five drawbacks: a) it is not clearly based on rigorous accidental errors in temperature and microturbulent velocity v_t, b) the effects of the errors they quote appear to be combined in one way for [Fe/H] and another for log g, c) no demonstration is given that either combination yields a rigorous confidence interval, d) no statement of the numerical confidence level for their confidence interval is given, and e) no allowance seems to be made for covariance between derived values of [Fe/H] and v_t (see eq. [10.12] of Kendall & Stuart (1977)).

The reader has doubtless noted that the Chaboyer et al. (1999) and Peterson & Green (1999) results are numerically identical. Chaboyer et al. (1999) interpret that identity as meaningful agreement. In fact, this is an over-interpretation made possible by non-rigorous statistical procedures. The two results cannot be compared rigorously, and they cannot be accepted as firm evidence that [Fe/H] $\sim +0.4$ dex for NGC 6791.

3. Interlude: a purported refutation of statistical analysis

It is quite likely that some readers have dismissed the objections given above as an irrelevance based on false premises. It is worthwhile to see what lies behind such a reaction.

The absence of rigorous statistical analysis described above is not simply a matter of habit. In large part, it is instead founded on an active rejection of statistics as such. This attitude is seldom (if ever) displayed openly, so some statisticians refuse to believe in its existence. However, rejection of statistics can be quite common in peer review. In particular, it is a common feature of reviews by NAHDS.

One refutation of statistics urged by NAHDS is as follows:

> "Systematic errors are far more important than accidental errors."

This common belief will be called "misconception S" (for "systematic"). In part, it is based on choosing between accidental and systematic errors instead of considering errors of both kinds. There is also a secondary assumption that systematic errors are always numerically significant. In practice, no numerical support is cited for that assumption. (Its effects may be seen in the review by Kurucz given elsewhere in this volume.)

Misconception S has been cited to the present author as a conclusive reason for barring statistical analysis from C-M and high-dispersion work. It will appear again when [Fe/H] cataloging is discussed (see Section 6).

4. The zero-point issue for high-dispersion analyses

Statistical analysis is one of two "framework issues" for the SMR problem. The other such issue concerns the zero points of high-dispersion values of [Fe/H]. Two zero-point techniques may be contrasted here. In rigorous differential analysis, solar and stellar data produced in closely comparable ways are differenced. However, if "external zeroing" is used, only stellar data are produced. Their zero points are then established by subtraction of a published solar metallicity. In the latter case, the solar and stellar results may be nonuniform in at least four ways. For this reason, external zeroing is fallacious (see Section 3 of Taylor (1999b)). It is convenient to focus here on one of the non-uniformities: improperly matched solar and stellar atmospheres. Stellar atmospheres are almost always drawn from a published grid. By contrast, the "empirical" Holweger & Muller (1974) model is almost always used to determine published solar metallicities.

Trimble & Bell (1981) were among the first to warn against pairing grid and empirical model atmospheres. Afterwards, Castro et al. (1996) (hereafter C96) published a high-dispersion analysis of μ Leo, whose possible SMR status has been debated since Spinrad & Taylor (1969) first suggested it. C96 obtained a high metallicity for μ Leo. However, they used external zeroing, so their result is fallacious. Note that Bell–who is one of the C96 co-authors–had repudiated the C96 zero-point procedure some fifteen years before it was applied.
One also notes that Castro has participated in both fallacious and rigorous zero-point analyses (compare C96 and Castro et al. (1997) with Castro et al. (1999)). These examples reflect a state of confusion about zero-point procedures in high-dispersion spectroscopy. Fallacious but undefended procedures may be used in print without a general acknowledgment that their results are not credible.
For μ Leo specifically, the zero-point problem has led to two contrasting schools of thought. The "Smith-Ruck school" (Smith & Ruck (2000); Feltzing & Gonzalez (2001)) accepts the validity of the arguments given by Taylor (1999b). Their principal datum is a rigorously-zeroed result produced by Smith & Ruck (2000). By contrast, the "C96 school" refuses to consult Taylor (1999b). In their view, the C96 result is an axiom which is not subject to disproof.

5. A contrast between two vectors

At this point, readers may guess that an endorsement of the Smith-Ruck school is forthcoming. However, this will not be done. To see why not, we must return to the issue of statistical analysis. A pivotal part of that issue turns on the character of high-dispersion results.
Consider the following vector:

$$100R = (25, 16, -1, -5, -4, -11, 1, 0, 8, -9, 6, 1, -4, 1, -12) \text{ dex.} \quad (1)$$

This vector contains the first 15 residuals from a difference between two high-dispersion data vectors. Presumably the random character of the residuals is evident at once. When published values of [Fe/H] are compared, vectors of this sort arise repeatedly. A constructive response to them is to use them to derive rms errors. Such errors are quoted in averaged-value [Fe/H] catalogs (Taylor (1995); Taylor (1999a)).
However, NAHDS proceed very differently. First, they tacitly reject by axiom all data vectors like those in eq. (1). Each NAHDS then consults a very different vector (compiled here from peer review): 1) sources of model atmospheres, 2) convection, 3) H$-$ and comparable opacities, 4) line blanketing, 5) non-LTE effects, 6) unnamed subtle model-atmosphere effects, 7) individual temperatures, 8) temperature scales, 9) f-values, 10) damping constants, 11) parameter determinations, 12) equivalent-width scales, 13) the solar [Fe/H] abundance, 14) non-solar abundance ratios, and 15) etc. Each NAHDS adopts a *different*

selection of systematics from this list to explain differences among published values of [Fe/H]. No numerical support is offered, and no admission is made that numerical testing can rule out possible error sources (see, for example, Thoren & Feltzing (2000)). In addition, no notice is taken that some error sources in the list are irrelevant to the catalogs and that others are evaluated (numerically, if possible) as the catalogs are compiled (see Taylor (1994); Taylor (1998)). The entire process follows closely from misconception S.

6. The μ Leo problem

The procedures described above will now be called "the statistical paradigm" and "the default paradigm," respectively. Let us see how those paradigms are applied to μ Leo data.

The statistical paradigm is applied in detail in Sections 5 and 7 of Taylor (1999b). For that reason, only a summary and updated results are given here. To quote from Taylor (1996), "the default option for a questionable datum is to keep it, and the burden of proof must rest with arguments for data rejection." All relevant values of [Fe/H] are therefore considered. Rejected data are then listed with reasons for rejecting them. Accepted data are averaged, using inverse-variance weights derived from rms errors. No "principal" datum is adopted; the risk posed by such a procedure has been known since the Peterson (1976) datum was rejected (see Sections IV–V of Taylor Taylor (1982b)).

Since the appearance of Taylor (1999b), high-dispersion μ Leo metallicities have been published by Smith & Ruck (2000) and by Takeda et al. (1998). However, addition of those data yields little change in the average value of [Fe/H]. The updated average is $\sim +0.23 \pm 0.025$ dex. By applying a t test to this result, one finds that values of [Fe/H] below $+0.2$ dex are not ruled out at 95% confidence.

In contrast, the default paradigm appeals to another common misconception about statistics:

> "Statistical data appraisals are useless and may be dismissed out of hand."

This "misconception U" (for "useless") rejects the statistical data-editing protocol and so opens the door to arbitrary data selection. Data are then adopted for reasons which are explained (if at all) only by using the term "recent," which is ambiguous (see Section 2.1 of Taylor (1999b)). Like selections of alleged causes of data scatter, selections of data themselves differ distinctively between sources. To illustrate: a datum from Branch et al. (1978) is regarded as correct by McWilliam & Rich (1994) and by McWilliam (1997). McWilliam is also a co-author of C96, but C96 omit the Branch et al. (1978) datum without comment. Gratton & Sneden (1990) compare the Branch et al. (1978) datum with a high-dispersion result of their own. Smith & Ruck (2000) then adopt the Gratton-Sneden datum, but they themselves do not mention the Branch et al. (1978) datum.

The values of omitted data may be high or low. For instance, NAHDS uniformly omit data near +0.4 dex and +0.1 dex (from Bonnell & Branch (1979) and Bell (1976), respectively). However, NAHDS apply one more key axiom: only data near +0.3 dex or greater are used to determine μ Leo metallicities. For the Smith-Ruck and C96 schools, the principal data are +0.29 dex and +0.46 dex, respectively.

The default paradigm may be summarized as an arbitrary process which is used to force a high-metallicity deduction from the evidence. The paradigm could fitly be taught in statistics classes to warn students about the consequences of rejecting statistical safeguards.

7. Stars other than μ Leo

Suppose that for SMR stars, one requires [Fe/H] to exceed +0.2 dex with confidence levels ≥ 0.95. As noted above, μ Leo does not satisfy that definition. Are any other stars known to satisfy it?

When Taylor (1996) addressed this issue, he found seven SMR dwarfs and subgiants (see his Table 4). However, he could find no SMR giants in the then-current [Fe/H] data base. An updated data base yields improved evidence for the existence of SMR dwarfs and subgiants, but one still finds no SMR giants. Does this mean that there are no such stars? If this is so, one might blame the deep convective layers in giants. Perhaps a large fraction of SMR stars has ingested planets and now has contaminated surfaces because of convective zones that are not deep enough to prevent the contamination (Gonzalez 1998). This issue can be gauged by comparing high-dispersion values of [Fe/H] for μ Leo and other candidate SMR stars. With 23 such results available, μ Leo is one of the most-frequently analyzed stars in the sky. By contrast, only scanty data exist for other candidate SMR giants. Though Deming & Butler (1979) and Taylor & Johnson (1987) have made note of such stars, they continue to attract little attention.

The existence or nonexistence of SMR giants is a completely open question at present. If μ Leo were used as a strong-lined standard star and compared with strong-lined candidate SMR giants, this problem might be solved. Meanwhile, its current state is largely due to a failure by NAHDS to employ a well-devised strategy for solving it.

8. What is to be done?

Hopefully it is now clear that issues of scholarship do indeed dominate the SMR problem (recall Section 1). What can be done in response?

To answer this question, one can start by rehearsing some of the problem's history. The present author first reviewed the problem in print nearly twenty years ago (see Taylor (1982a); Taylor (1982b)). At that time, its scholarship was seriously deficient, and one aim of the review was to encourage reform of that scholarship. Peer-review responses included two which stand out in retrospect. Two reviewers who failed to grasp the analysis of scholarship used forceful language to reject the premise that scholarship should be discussed in print. A third reviewer argued that the SMR problem had allegedly been solved and that its dubious history should therefore be forgotten.

Unfortunately, reform has not taken place. The SMR problem has deteriorated instead. Its salient characteristics are now

- inadequate research strategy (recall Section 8),
- inadequate data-quality control (recall Section 5),
- rejection of statistical data interpretation, and
- reliance on misconceptions S and U and other unsupported axioms.

The axioms satisfy the definition of "dogma" given by Öpik (1977).

Private judgments about the problem are likewise not encouraging. Some continue to hold a false belief that the problem has been solved (recall Section 5). Others urge chimerical solutions, including use of improved model atmospheres or use of new data *without* improved data interpretation. Among NAHDS, belief is widespread that statistical analysis is unimportant and beneath notice. The most extreme of these reactions are root-and-branch rejections of statistics which can fairly be described as "statistics xenophobia." Not even those judgments, however, are more striking than reactions to scholarship analysis. Such reactions can still extend to outright anger, and they still reflect a widespread belief that analyses of scholarship should not be published even if they are both pertinent and accurate.

All told, it seems very unlikely to the present author that the SMR problem will be reformed quickly–though attempts to reform it should continue. The only improvement that seems within reach at present is a constructive reaction by astronomers other than NAHDS. It would be helpful to refuse to trust conclusions about the SMR problem unless they have been established statistically. In addition, spectroscopists doing extragalactic work who can use candidate SMR stars without well-established metallicities can consult the literature sources named in Section 8. Until such time as C-M analysis is reformed and a major change takes place in the attitudes and practices of NAHDS, these measures seem to be the best available.

References

Bell, R. A. 1976, in IAU Symp. 72, Abundance Effects in Classification, eds. B. Hauck, & P. C. Keenan (Dordrecht: Reidel), p. 49.

Bonnell, J., & Branch, D. 1979, ApJ 229, 175.

Branch, D., Bonnell, J., & Tomkin, J. 1978, ApJ 225, 902.

Castro, S., Rich, R. M., McWilliam, A., Ho, L. C., Spinrad, H., Filippenko, A. V., & Bell, R. A. 1996, AJ, 111, 2439.

Castro, S., Rich, R. M., Grenon, M., Barbuy, B., & McCarthy, J. K. 1997, AJ, 114, 376.

Castro, S., Porto de Mello, G. G., & da Silva, L. 1999, MNRAS, 305, 693.

Chaboyer, B., Green, E. M., & Liebert, J. 1999, AJ, 117, 1360.

Deming, D., & Butler, D. 1979, AJ, 84, 839.

Feltzing, S., & Gonzalez, G. 2001, A&A, 367, 253.

Gonzalez, G. 1998, A&A, 334, 221.

Gratton, R. G., & Sneden, C. 1990, A&A, 234, 366.

Holweger, H., & Müller, E. A. 1974, Solar Phys., 39, 19.

Kendall, M., & Stuart, A. 1977, The Advanced Theory of Statistics, 4th ed. (London: Griffin).

McWilliam, A. 1997, ARA&A, 35, 503.

McWilliam, A., & Rich, R. M. 1994, ApJS, 91, 749.

Öpik, E. J. 1977, ARA&A, 15, 1.

Peterson, R. 1976, ApJS, 30, 61.

Peterson, R. C., & Green, E. M. 1998, ApJ, 502, L39.

Smith, G., & Ruck, M. J. 2000, A&A, 356, 570.

Spinrad, H., & Taylor, B. J. 1969, ApJ, 157, 1279.

Takeda, Y., Kawanamodo, S., & Sadakane, K. 1998, PASJ, 50, 97.

Taylor, B. J. 1982a, Vistas in Astron., 26, 253.

Taylor, B. J. 1982b, Vistas in Astron., 26, 285.

Taylor, B. J. 1994, PASP, 106, 590.

Taylor, B. J. 1995, PASP, 107, 734.

Taylor, B. J. 1996, ApJS, 102, 105.

Taylor, B. J. 1998, PASP, 110, 699.

Taylor, B. J. 1999a, A&AS, 134, 523.

Taylor, B. J. 1999b, A&A, 344, 655.

Taylor, B. J., & Johnson, S. B. 1987, ApJ, 322, 930.

Thoren, P., & Feltzing, S. 2000, A&A, 363, 692.

Trimble, V., & Bell, R. A. 1981, QJRAS, 22, 361.

SPECTRAL INDICES OF STARS AT SUPER-SOLAR REGIME*

Alberto Buzzoni
Telescopio Nazionale Galileo, A.P. 565, 38700 Santa Cruz de La Palma (Tf), Spain, and Osservatorio Astronomico di Brera, Milano Italy

Miguel Chávez
Instituto Nacional de Astrofísica, Optica y Electrónica, A.P. 51 y 216, 72000 Puebla, Mexico

Maria Lucia Malagnini
Dipartimento di Astronomia, Università degli Studi di Trieste, Via G.B. Tiepolo 11, 34131 Trieste, Italy

and

Carlo Morossi
Osservatorio Astronomico di Trieste, Via G.B. Tiepolo 11, 34131 Trieste, Italy

Abstract We derived Lick narrow-band indices for 139 candidate super metal-rich stars of different luminosity class for a sample of which fundamental atmosphere parameters have been obtained from a grid of synthetic reference spectra by Malagnini et al. (2000). Indices include Iron Fe50, Fe52, Fe53, and Magnesium Mgb and Mg_2 features. By comparing models and observations, no evidence is found for non-standard Mg vs. Fe relative abundance (i.e. [Mg/Fe] = 0, on the average, for our sample). Both the Worthey et al. (1994) and Buzzoni et al. (1992, 1994) fitting functions are found to suitably match the data, and can therefore confidently be extended for population synthesis applications also to super-solar metallicity regimes.

*Based on observations collected at the INAOE "G. Haro" Observatory, Cananea (Mexico).

M. Chávez et al. (eds.),
New Quests in Stellar Astrophysics: The Link Between Stars and Cosmology, 103–106.

Table 1. The Databases.

Observations	Synthetic Spectra
139 Pop I Stars	Chavez et al. 1997 collection
300 spectra	693 spectra
4600 – 5500 Å	4850 – 5400 Å
$R = \Delta\lambda/\lambda = 2000$	$R = \Delta\lambda/\lambda = 250\,000$
$[Fe/H] \geq 0.1$	$[M/H] = -1.0 \rightarrow +0.5$
M $\rightarrow$ F spectral type	$4000\ K < T_{eff} < 8000\ K$
I $\rightarrow$ V MK luminosity class	log g = 1.0 $\rightarrow$ 5.0

1. The Databases

As a useful tool to investigate the distinctive properties of stellar aggregates, the fitting-function technique (Buzzoni 1995; Worthey et al. 1995) has been extensively used in population synthesis models to reproduce integrated spectroscopic indices, even in super metal-rich (SMR) systems such as elliptical galaxies or bulges of spirals.

Since these relations mainly derive from a fit of the local stellar population, with $Z \sim Z_{\odot}$, it is important to confidently assess their validity also at super-solar metallicity regimes. For this reason, we derived Lick narrow-band indices from spectroscopic observations of a sample of 139 SMR stars with $[Fe/H] \geq +0.1$, comparing with a suitable theoretical database of synthetic reference spectra. A summary of the main characteristics of the databases is given in Table 1 (see also Chavez et al. 1997, and Malagnini et al. 2000 for further details).

From the observed spectra, Iron Fe50, Fe52, and Fe53, as well as Magnesium Mgb and Mg_2 indices have been computed according to Worthey et al. (1994, hereafter W94). Calibration to the standard system has been done by using the subset of 49 stars in common with the Lick original database, and/or with Buzzoni et al. (1992, 1994; B92 and B94, respectively).

2. Index residuals and atmosphere parameters

The (O–C) index residuals obtained by comparing observations with the corresponding fitting spectra from the theoretical database have been analyzed vs. atmosphere fiducial parameters, according to Mala-gnini et al. (2000). Figure 1 shows the residual distribution for Mg_2 and the combined Iron index $<$ Fe $>= (Fe52 + Fe53)/2$ vs. [Fe/H]. Solid dots mark the "fair" subsample of 73 stars with more confident fundamental parameters (cf. Malagnini et al. 2000).

The four points with the largest (negative) residuals correspond to the coolest stars in our sample ($T_{\rm eff} \leq 4000\ K$), and reflect the partial inadequacy of the theoretical database at this temperature range. As far as the remaining sample of 135 stars is considered, however, the residual distribution for all

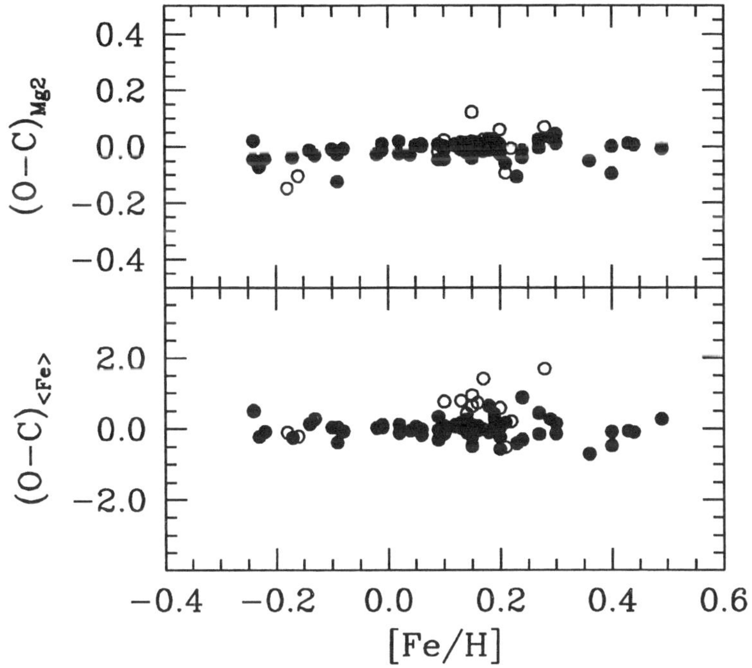

Figure 1. (O–C) index residuals vs. [Fe/H]. The 73 stars in the "fair" subsample, with more confident atmosphere parameters, are singled out ("•" markers). The residuals distribution is consistent with a zero average.

indices is consistent with a zero-average hypothesis. Quite remarkably, we also verified the lack of any negative correlation between Mg and Fe (O–C) residuals. This indicates that a non-solar partition for the [Mg/Fe] relative abundance can confidently be ruled out for our stars.

3. Fitting Functions

Analytical fitting functions, giving the Fe and Mg index strength vs. stellar atmosphere parameters across the H-R diagram, have been provided by B92, B94, and W94. To check self-consistency of these functions also in the SMR regime we studied their index residuals with respect to our observations. Figure 2 summarizes the results for Mg_2 and $< Fe >$.

The (O–C) distribution in both panels is consistent with a zero average residual confirming that the two sets of equations properly account for high-metallicity stars, at least in the range of the atmosphere parameters sampled by our stars. The W94 fitting function is slightly more accurate than the B92 one (points

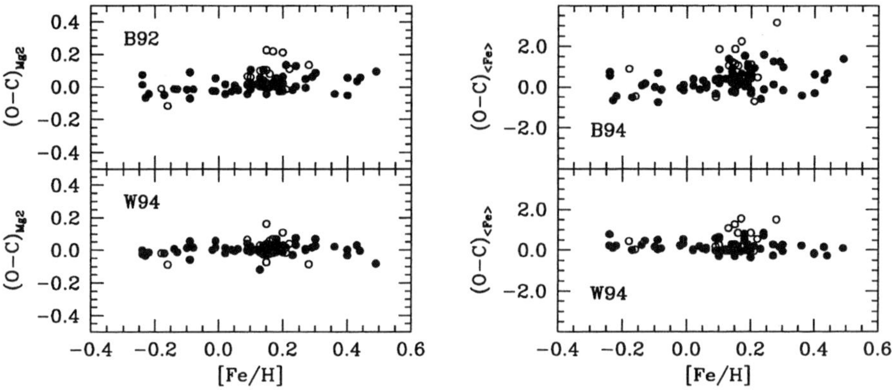

Figure 2. The Mg_2 and < Fe > index residuals (in the sense [Observed – Computed]) with respect to the Buzzoni et al. (1994, 1994) and Worthey et al. (1994) fitting functions are displayed vs. metallicity. Like in Fig. 1, solid dots mark stars in the "fair" subsample. Both (O–C) distributions are consistent with a zero average.

spread for the "fair" sample is $\sigma(\mathrm{Mg}_2) = \pm 0.047$ mag vs. ± 0.052 mag for the latter case), but at cost of a more elaborated multi-branch analytical fit.
A similar behavior is found for the $<$ Fe $>$ index, with $\sigma(<\mathrm{Fe}>) = \pm 0.61$ Å and ± 0.34 Å with respect to B94 and W94, respectively. We verified that the skewed positive residuals with respect to B94, evident from the figure, mostly come from the warmer stars in the sample ($T_{eff} \gtrsim 6700$ K) for which the B94 output predicts a vanishing Fe52 index.

Acknowledgments

This work received partial financial support from the Italian MURST via COFIN '98 02-013, COFIN'00 02-016, and 60% grants, and from the Mexican CONACyT via grant 28506-E.

References

Buzzoni, A. 1995, in Fresh Views of Elliptical Galaxies, eds. A. Buzzoni, A. Renzini & A. Serrano (San Francisco: ASP), p. 189.

Buzzoni, A., Gariboldi, G. & Mantegazza, L. 1992, AJ, 103, 1814 (B92).

Buzzoni, A., Mantegazza, L. & Gariboldi, G. 1994, AJ, 107, 513 (B94).

Chavez, M., Malagnini, M. L. & Morossi, C. 1997, A&AS, 126, 267.

Malagnini, M. L., Morossi, C., Buzzoni, A. & Chavez, M. 2000, PASP, 112, 1455.

Worthey, G., Faber, S.M., Gonzalez, J.J., & Burstein, D. 1994, ApJS, 94, 687 (W94).

Worthey, G., Trager, S. C. & Faber, S. M. 1995, in Fresh Views of Elliptical Galaxies, eds. A. Buzzoni, A. Renzini & A. Serrano (San Francisco: ASP), p. 203.

THE PROPER MOTION OF THE GLOBULAR CLUSTER NGC 6553 AND OF BULGE STARS WITH HST*

M. Zoccali, A. Renzini
ESO, Garching bei München, Germany

S. Ortolani
Università di Padova, Italy

E. Bica
Universidade Federal do Rio Grande do Sul, Brazil

and

B. Barbuy
Universidade de S. Paulo, Brazil

Abstract

WFPC2 images obtained with the Hubble Space telescope 4.16 years apart have allowed us to measure the proper motion of the metal rich globular cluster NGC 6553 with respect to the background bulge stars. With a space velocity of $(\Pi, \Theta, W) = (-3.5, 230, -3)$ km s^{-1}, NGC 6553 follows the mean rotation of both disk and bulge stars at a Galactocentric distance of 2.7 kpc. While the kinematics of the cluster is consistent with either a bulge or a disk membership, the virtual identity of its stellar population with that of the bulge cluster NGC 6528 makes its bulge membership more likely. The astrometric accuracy is high enough for providing a measure of the bulge proper motion dispersion and confirming its rotation. A selection of stars based on the proper motions produced an extremely well defined cluster color-magnitude diagram (CMD), essentially free of bulge stars. The improved turnoff definition in the decontaminated CMD confirms an old age for the cluster ($\sim$ 13 Gyr) indicating that the bulge underwent a rapid chemical enrichment while being built up at in the early Universe.

*This work is based on observations with the NASA/ESA Hubble Space Telescope, obtained at the Space Telescope Science Institute, operated by AURA Inc. under contract to NASA

M. Chávez et al. (eds.),
New Quests in Stellar Astrophysics: The Link Between Stars and Cosmology, 107–110.

Introduction

The study of the bulge globular clusters (GCs) is of particular interest for two main reasons. First, their age and metallicity, compared with that of the halo GCs, can constrain the formation processes and timescales of the inner spheroid. Second, they are metal-rich simple stellar population, ideal laboratory to test the stellar models widely used to investigate the stellar content of elliptical galaxies.

However, the study of such GCs is complicated by at least two problems: they are located in high extinction regions, and also suffer from strong contamination by both disk and bulge stars. In order to minimize these effects it is very useful to restrict the analysis to their innermost regions, now fully resolved by HST+WFPC2. Indeed, in the last $\sim$ 5 years great progress has been made in the study of bulge GCs, in particular for the two less reddened and most populated ones, namely NGC 6528 and NGC 6553.

Using a technique pioneered by King et al. (1998), in the present work we combine HST wide field and planetary camera (WFPC2) data from two epoch observations of the central regions of the bulge cluster NGC 6553 ($l = 5\overset{\circ}{.}25$, $b = -3\overset{\circ}{.}02$). Thanks to the excellent astrometric performances of WFPC2, we measure the relative proper motion of the cluster with respect to the bulge. This allows us to *i*) decontaminate the CMD of NGC 6553 from the background and therefore to obtain a more reliable measure of its age; *ii*) obtain the three components of the cluster space velocity; and *iii*) measure the bulge proper motion dispersions along Galactic longitude and latitude.

1. The Data

We analyze two sets of observations of the central region of the bulge globular cluster NGC 6553, taken with a time interval of 4.16 years. Both sets were collected with HST using the WFPC2 camera, with the filters F555W (V) and F814W (I). The two sets of data were reduced independently, using the DAOPHOT II/ALLFRAME package, while calibration of the HST instrumental magnitudes to the Landolt system was performed following the prescriptions by Dolphin (2000). Since the quality of the first epoch photometry turned out to be better than that of the second epoch, in what follows we will base our analysis on the 1994 photometry, while the spatial information from the 1998 data combined to that of 1994 will be used for the relative proper motion derivation.

Polynomial 20-coefficient coordinate transformations were then calculated in order to take into account both the differences in the pointings and the distortion of the WFPC2 field. Figure 1 shows the residuals differences between the coordinates of the two epochs (epoch2-epoch1). Obviously, since the astrometric transformations were constrained mainly by the much more numerous cluster

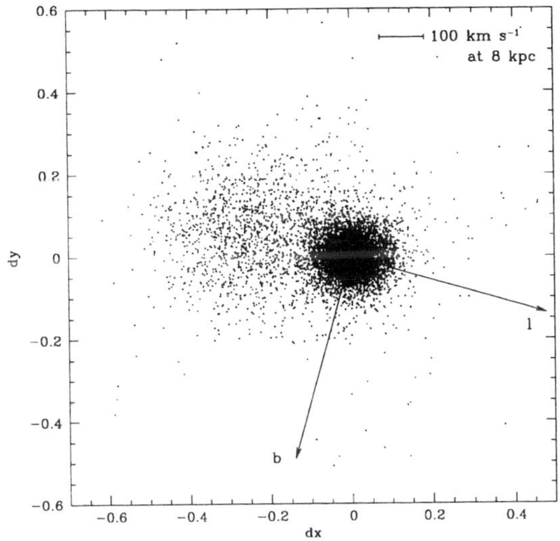

Figure 1. Relative proper motions (dx,dy) in pixels between the two epochs. The arrows show the direction of increasing Galactic longitude (l) and latitude (b).

stars, the residuals are, by definition, distributed around zero. Foreground and background (mostly bulge and some disk) stars, instead, are distributed around $(dx, dy) = (-0.23, 0.05)$ in this plot.
Figure 2 shows the result of the astrometric decontamination of the CMD. The left panel shows the CMD of only those stars which displacement between the two epochs lies inside a radius of 0.1 pixels centered on (0,0) in Fig. 1, which, as expected, are mainly cluster stars. Many cluster stars fall outside this circle, but such a small radius was chosen in order to select the cleanest sample of cluster stars. The central panel of Fig. 2 show the stars having proper motions $dx < -0.15$. Those are mainly bulge stars, with some residual contamination from cluster outliers.

2. Results

The distribution of the residuals shown in Fig. 1 allows us to determine both the relative proper motion of NGC 6553 with respect to the bulge, and the dispersion of the bulge proper motions along the two Galactic coordinates. The former is the difference between the centroids of the bulge and cluster proper motion distributions, and corresponds to a relative proper motion of NGC 6553 with respect to the bulge of $\mu_l = 5.89$ and $\mu_b = 0.42$ mas yr^{-1}. Combining this result with the solar motion, the cluster distance (5.3 kpc), and its radial velocity we obtain a space velocity (Π,Θ,W)=(–3.5, 230, –3) km s^{-1}.
The σ of the distribution of bulge stars shown in Fig. 1 also allow the determination of the bulge proper motion dispersion: $\sigma_l = 0.119 \pm 0.012$ and

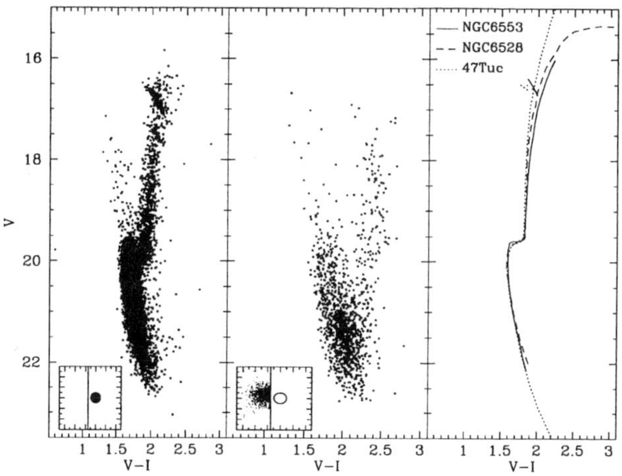

Figure 2. Left panel: NGC 6553 stars, selected as those lying inside a radius of 0.1 pixels centered on (0,0) in Fig. 1 (reproduced in the panel inset). Central panel: CMD of (mainly) bulge stars, selected as those having $dx < -0.15$ in Fig. 1. Cluster stars are still present in this CMD because, being much more numerous, a still appreciable number of them fall well outside the selected cluster circle. Right panel: comparison among the fiducial lines of NGC 6553, NGC 6528, and 47 Tuc

$\sigma_b = 0.098 \pm 0.009$. More details on these calculations together with an extended discussion on the results can be found in Zoccali et al. (2001a).

The relative proper motion decontamination has provided a cleaner cluster CMD, allowing a measure of age and distance with higher accuracy with respect to the previously available data. The estimated apparent magnitude of the cluster HB is $V^{\rm HB} = 16.6 \pm 0.05$. We adopt $M_{\rm V}^{\rm HB} = (0.16 \pm 0.10) \times [{\rm Fe/H}] + (0.98 \pm 0.10)$ and [Fe/H]=-0.5 ± 0.3, giving $M_{\rm V}^{\rm HB} = 0.90 \pm 0.12$. The cluster apparent distance modulus is therefore $(m - M)_V = 16.60 - 0.90 = 15.70 \pm 0.13$. A comparison among the fiducial lines of NGC 6553, NGC 6528, and 47 Tuc Fig. 2 shows differences in the TO and HB locations well inside the uncertainty in the construction of the fiducial lines, from which we conclude that the three clusters are coeval to within $\sim$ 1 Gyr, and therefore the absolute age of NGC 6553 (and NGC 6528) is $\sim$ 13 Gyr as recently determined for 47 Tuc (Zoccali et al. 2001b). This demonstrates that the Galactic bulge was enriched to solar abundance to within a rather short time scale (less than a few Gyr).

References

Dolphin, A. E. 2000, PASP, 112, 1397.
King I. R., Anderson J., Cool A. M.,& Piotto G., 1998, ApJ, 492, L37.
Zoccali M., et al. 2001a, ApJ, 553, 733.
Zoccali M., Renzini A., Ortolani S., Bica E., & Barbuy B., 2001b, AJ, 121, 2638.

UPDATE ON GLOBULAR CLUSTER AGES

Pierre Demarque

Yale University, New Haven, CT 06520, USA

Young-Cheo Kim

Yonsei University, Seoul 120-749, Korea

and

Sukyoung Yi

California Institute of Technology, Pasadena, CA 91125, USA

Abstract

We update globular cluster age estimates and discuss their implications in the context of present knowledge in cosmology. On the basis of the best information now available, the ages of the oldest globular clusters are estimated at 13.0±1.5 Gyr. If this result, whose main uncertainty is primarily in the distance scale, continues to hold, the long standing discrepancy between the stellar evolution timescale and the universal expansion timescale will have been resolved.

Introduction

Globular clusters hold a special place in studies of stellar chronology, because they contain the oldest known stars in the Galaxy. As such, the ages of the oldest globular clusters set a lower limit to the age of the Universe. Equally important, the distribution of the ages of globular clusters throws light on the formation process, evolution and chronology of the Galactic bulge and halo.

Already fifty years ago, we knew, and gained comfort from, the fact that the ages derived for the oldest star clusters were of the same order of magnitude as the Hubble time derived from the expansion of the Universe. However the uncertainties in observations of globular clusters and in stellar evolution calculations remained very large, and estimates of the Hubble constant H_0 oscillated between 50 and 100 km/s/Mpc. The problem was seen as a competition between the stellar evolution timescale and the expansion timescale derived from

M. Chávez et al. (eds.),
New Quests in Stellar Astrophysics: The Link Between Stars and Cosmology, 111–118.

the value of H_0 in a universe with $\Lambda = 0$. And until recently, the large ages derived for globular clusters from color-magnitude diagrams (CMD) and stellar evolution calculations have been difficult to reconcile with the usually shorter expansion age derived for the Universe. At the Symposium on Observational Tests of Cosmological Inflation held ten years ago in Durham (Shanks et al. 1991), the age for the oldest globular clusters was quoted in the range 13-15$\pm$3 Gyr by Renzini (1991), while Demarque et al. (1991) estimated their minimum age at 14 Gyr (1991). The situation had not much changed only five years ago, when on the basis of a Monte Carlo study of the ages of the oldest globular clusters (Chaboyer et al. 1996), the conclusion was reached that to the best of our knowledge, the mean age of the oldest globular clusters is 14.6 Gyr, with a 5% probability that it is below 12.1 Gyr or above 17.1 Gyr (Demarque 1997).
In this paper, we survey briefly the progress made in the last few years. Better globular cluster observations are available. The physics of stellar evolution has improved. The Hipparcos satellite has forced a revision of the accepted distance scale. The combination of all these factors has resulted in a reduction in the age estimate for the oldest globular clusters to about 13.0$\pm$1.5 Gyr. At the same time, the HST key project has led to a concensus on the value of H_0, with H_0 = 72 $\pm$8 km/s/Mpc (Freedman et al. 2001).
The most surprising result has been the evidence for a non-zero cosmological constant (in the vicinity of 0.7), on the basis of distances based on SNIa observations. And from these observations, two independent groups have derived a dynamical age for the Universe of 14.2$\pm$1.7 Gyr (Riess et al. 1998), and 14.9+1.4 $- 1.1(0.63/h)$ Gyr (Perlmutter et al. 1999), respectively. The particular role of the cosmological constant Λ, given a value of H_0, is to increase the age estimate of the Universe. Consistency with a globular age of 14 Gyr would require that H_0 must be less than 45 km/s/Mpc in a Universe with $(\Omega = 1.0, \Lambda = 0.0)$, and $H_0 \leq 58$ km/s/Mpc in a Universe with $(\Omega = 0.2, \Lambda = 0.0)$. On the other hand, with a cosmological constant $\Lambda \simeq 0.7$, a maximum globular cluster age as high as 14 Gyr is marginally compatible with $H_0 \simeq 70$ km/s/Mpc.
One of the consequences of these developments is that although many uncertainties remain, for the first time, *there is no apparent inconsistency between the ages derived for the oldest globular clusters and the age of the Universe.*

1. Advances in globular cluster research

In the last ten years, the quality and quantity of globular cluster CMD's has made rapid strides, thanks to CCD detectors and to the HST. Chemical abundances are also better known within globular clusters and among field halo stars. In particular, more evidence has been gathered concerning variations in $[\alpha/Fe]$ for stars of different $[Fe/H]$ in the Galactic halo and elsewhere (Carney 1996).

On the theoretical side, we have seen great advances in the microscopic physics of stellar interiors and stellar evolution (for references, see the review by Demarque 1997). Most significant has been the remarkable agreement between two independent sets of interior opacities using different physical approaches, *i.e.* the opacities from the Opacity Project (Seaton 1996), and those from the Livermore OPAL opacities (Rogers & Iglesias 1996). Low temperature opacities, which include the effects of molecules, have also made impressive advances. This agreement, and the success of the new opacities in resolving several long standing problems in pulsation theory lends much greater credibility in the details of stellar models. The spectacular success of helioseismology (Guenther & Demarque 1997) further strengthen our confidence in the available opacity tables, and in recent advances in the equation of state of stellar matter relevant to globular cluster stars (Rogers, Swenson & Iglesias 1996).

2. The YY isochrones

The YY (Yale-Yonsei) isochrones (Yi et al. 2001) use the OPAL (Rogers & Iglesias 1995) interior opacities, the low temperature opacities of Alexander & Ferguson (1994). Particularly important are the improvements in the equation of state of stellar matter (Rogers, Swenson & Iglesias 1996). Helium diffusion (Thoul, Bahcall & Loeb 1994) and convective core overshoot have also been taken into account. The energy generation rates are primarily taken from Bahcall & Pinsonneault (1992), with some revisions. The chemical compositions used in the YY isochrones are: Z = 0.00001, 0.0001, 0.0004, 0.001, 0.004, 0.007, 0.01, 0.02, 0.04, 0.08 and 0.1. ($Z_\odot$ = 0.0181), Y = 0.23 + $2Z$ and the ages range from 0.001 to 20 Gyr.

Two significant features of the YY isochrones are (1) that the models start their evolution from the pre-main sequence birthline (Palla & Stahler 1990) instead of the age-zero main sequence, and (2) the color transformation has been performed using both the tables of Lejeune, Cuisinier & Buser (1997, 1998), and an improved version of the Green, Demarque & King (1987) table. Although there are differences, satisfactory agreement has generally been found with the recent work of Girardi et al. (2000). Several comparisons of the YY isochrones with observational data can be found in Yi et al. (2001). Figures 1 shows the metal-poor halo cluster M68 as an example.

It is noteworthy that the change in the derived ages of star clusters caused by recent updates in stellar models alone is a sensitive function of metallicity. Relative to RYI-based studies (Green, Demarque & King 1987), the globular cluster ages are decreased by amounts varying from 15% (for Z=0.0004) to 11% (for Z=004). On the other hand the ages of solar metallicity clusters (Z=0.02) are increased by about 10%. When a simple scaling for α-element enhancement similar to that used by Chaboyer et al. (1992) is taken into account as well, the

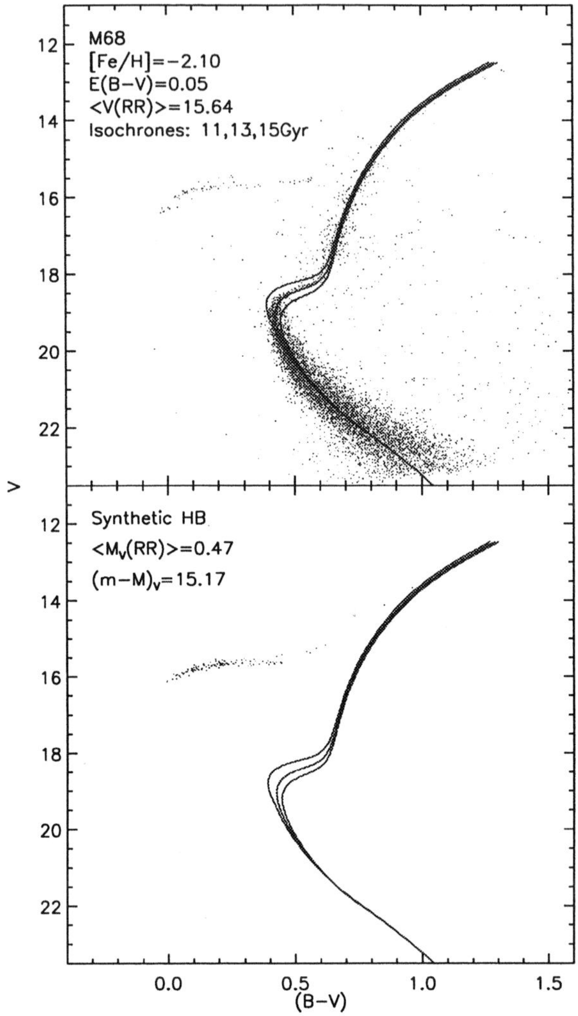

Figure 1. Fit to halo cluster M68. Photometry from Walker (1994). $[Fe/H]$ and reddening from Harris (1996), and $[\alpha/Fe] = 0.3$ adopted.

reduction in the ages of the most metal-poor globular clusters is approximately 20%.

3. Quantifying the main uncertainties in dating globular clusters; a Monte Carlo approach

In addition to the uncertainties in the theoretical isochrones themselves, the fitting of globular cluster CMD's involves many steps which are sources of error in age determinations. Chaboyer et al. (1996, 1998) have tried to quantify the various source of uncertainties in the age determinations due to the fitting procedure, as well as in the physics input in the stellar models and the composition input parameters. Naturally, the fitting procedure introduces uncertainties of its own. One example is in using the ΔV technique, and assuming that the distance indicator $M_v(RR)$ depends uniquely on metallicity. Similarly, the errors in the treatment of convection were calculated using a fixed mixing length parameter along each evolutionary track. This last assumption, which is standard in stellar evolution calculations, could introduce a bias in the ΔV technique.

With these provisos, the Monte Carlo study by Chaboyer et al. (1998) provides useful quantitative estimates of the relative importance of the uncertainties in globular cluster dating. The paper contains quantitative corrections that can be applied as input parameters improve with time. As suspected, the dominant source of error lies in the distance estimate of the cluster (in this case the appropriate value of $M_v(RR)$), which in this study is responsible for a 16% uncertainty in the derived ages. The quantity $[\alpha/Fe]$ is next by introducing a 7% age uncertainty, followed by the treatment of convection (5%).

3.1 The distance scale and Hipparcos

As emphasized above, the distance scale remains the main question mark in the dating of the globular clusters. The Hipparcos mission has led some authors to surprisingly large increases in distance estimates. For illustration, the post-Hipparcos age estimate for the oldest globular clusters derived by Chaboyer et al. (1998) is 3 Gyrs larger than the pre-Hipparcos age derived on the basis of the same stellar models (Chaboyer et al. 1998). This revision, which may have been too extreme, illustrates how sensitive globular cluster ages depend on distances, all else being equal.

The YY isochrones agree with the position of subdwarfs in the HR-diagram observed by Hipparcos (Fig. 2). Data from Reid (1998) were used, following the criterion that $[Fe/H] < -0.55$, $M_v > 4.5$, and $\sigma_\pi/\pi < 0.07$. The adopted chemical composition was $[\alpha/Fe] = +0.4$ and $+0.2$ for $[Fe/H] \leq -1.0$ and $[Fe/H] = -0.5$, respectively. The YY isochrones are also compatible with the theoretical synthetic HB models of Demarque et al. (2000). This distance scale agrees with the Walker (1992) and Freedman et al. LMC distance modulus of 18.5 ± 0.15. It is therefore internally consistent with the derived cosmological dynamical ages of the Universe quoted earlier. Future observations will test whether these conclusions are correct.

4. Summary and future prospects

To summarize, recent advances lead to an age estimate for the oldest Galactic globular clusters of $13.0{\pm}1.5$ Gyr. Important issues regarding globular cluster

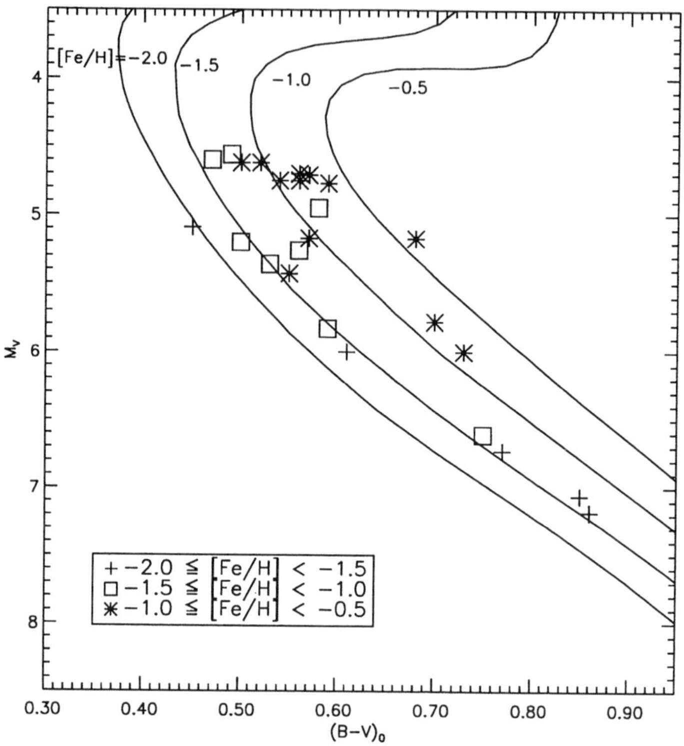

Figure 2. A test of the subdwarf main sequence stars whose distance have been determined by Hipparcos observations.

ages will be resolved during the next few years. A more precise absolute age for the oldest clusters, will be derived with improvements in the distance scale (first from the space mission FAME, then from GAIA and SIM). Better abundances for main sequence stars in globular clusters will soon be available from the 10-meter class ground-based telescopes. More realistic models of convection will improve atmospheric models and surface boundary conditions of cool star models (Kim & Chan 1998; Demarque, Guenther & Kim 1999). Much work remains to be done on stellar atmospheres and on the conversion of observable parameters to effective temperatures and surface gravities.

Beyond determining the ages of the oldest star clusters, refining age determinations in globular clusters is also needed for improving the chronology of the Galaxy and understanding its formation and evolution. Answers to fundamental and still controversial questions such as the chronology of halo, disk and bulge formation will guide us in exploring how other galaxies form (Stetson, Van-

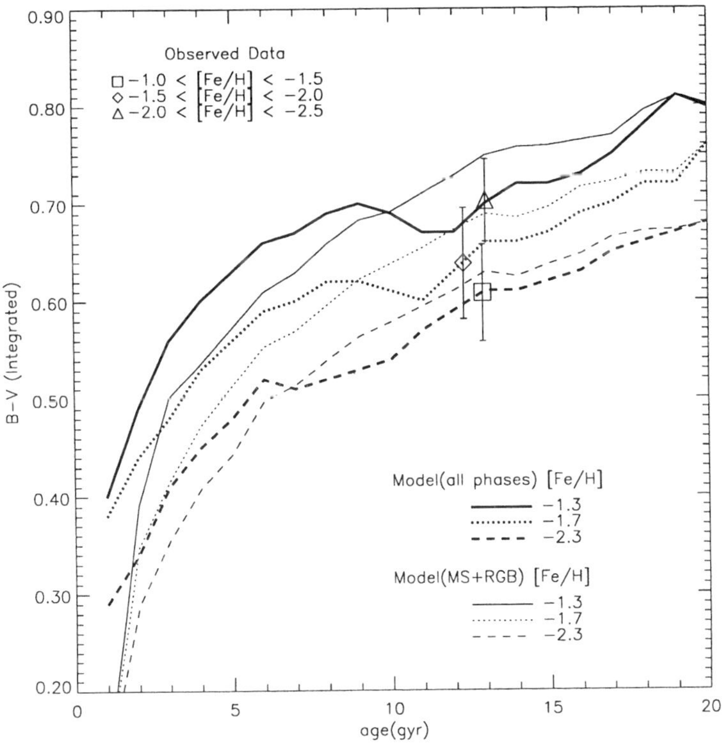

Figure 3. Evolution in integrated color of metal-poor populations. Compared are the observed integrated colors of Galactic globular clusters with $E(B-V) < 1.0$. The cluster data are from Harris (1996).

denBerg & Bolte 1996; Sarajedini, Chaboyer & Demarque 1997; McNamara 2001).

Finally, one should note the importance of extragalactic globular clusters in the Local Group. Globular clusters in the LMC and other Local Group galaxies now serve as stepping stones to the more distant stellar populations in which individual stars cannot be resolved. It is encouraging to note that when the post-RGB stages of evolution are included, the ages of the globular clusters derived from integrated colors are consistent with the isochrone fitting ages. This point is illustrated in Figure 3.

Acknowledgments

This research was supported in part by NASA grant NAG5-8406 (PD), and the Creative Research Initiative Program of the Korean Ministry of Science & Technology (YCK, SY).

References

Bahcall, J. N. and M. H. Pinsonneault, M.H. 1992, Reviews of Modern Physics, 64, 885 and private communications.

Chaboyer, B., C. P. Deliyannis, P. Demarque, M. H. Pinsonneault & A. Sarajedini 1992, ApJ, 388, 372.

Chaboyer, B., P. Demarque, P. J. Kernan & L. M. Krauss 1996, Science, 271, 957.

Chaboyer, B., P. Demarque, P. J. Kernan & L. M. Krauss 1998, ApJ, 494, 96.

Demarque, P. 1997, in The Extragalactic Distance Scale, eds. M. Livio, M. Donahue & N. Panagia (Cambridge: University Press), p. 30.

Demarque, P., C. P. Deliyannis & A. Sarajedini 1991, in Observational Tests of Cosmological Inflation, eds. T. Shanks, A. J. Banday, & R. S. Ellis, NATO ASI Series, p. 111.

Demarque, P., D. B. Guenther & Y.-C. Kim 1999, ApJ, 517, 510.

Demarque, P., R. Zinn, Y.-W. Lee & S. Yi 2000, ApJ, 119, 1398.

Freedman, W. et al. 2001, ApJ, 553, 47.

Girardi, L., A. Bressan, G. Bertelli, G. & C. Chiosi 2000, A&AS, 141, 371.

Green, E. M., P. Demarque & C. R. King 1987, The Revised Yale Isochrones and Luminosity Functions, (New Haven: Yale Univ. Obs.).

Harris, W. E. 1996, ApJ, 112, 1487.

Iglesias, C. A. & F. J. Rogers 1996, ApJ, 464, 943.

Kim, Y.-C. & K. L. Chan 1998, ApJ, 496, L121.

Lejeune, Th., F. Cuisinier & R. Buser 1997, A&AS, 125, 229.

Lejeune, Th., F. Cuisinier & R. Buser 1998, A&AS, 130, 65.

McNamara, D.H. 2001, PASP, 113, 335.

Palla, F. & S. W. Stahler 1990, ApJ, 360, L47

Perlmutter, S. et al. 1999, ApJ, 517, 565

Reid, N. 1998, AJ, 115, 204

Renzini, A. 1991, in Observational Tests of Cosmological Inflation, eds. T. Shanks, A. J. Banday, R. S. Ellis, NATO ASI Series, p. 131.

Riess, A. G. et al. 1998, AJ, 116, 1009

Rogers, F. J., F. J. Swenson & C. A. Iglesias 1996, ApJ, 456, 902

Sarajedini, A., B. Chaboyer & P. Demarque 1997, PASP, 109, 1321.

Seaton, M. 1996, MNRAS, 279, 95.

Stetson, P. B., D. A. VandenBerg & M. Bolte 1996, PASP, 108, 560

Thoul, A. A., J. N. Bahcall & A. Loeb 1994, ApJ, 421, 828

Walker, A.R. 1992, ApJ, 390, L81

Walker, A.R. 1994, AJ, 108, 555

Yi, S., P. Demarque, Y.-C. Kim, Y.-W. Lee, Y.-W., C.-H.Ree, Th. Lejeune & S. Barnes 2001, ApJS (in press).

A SURVEY OF OPEN CLUSTERS AT SAN PEDRO MÁRTIR

A. Moitinho, W. Schuster, and C. Chavarría
Observatorio Astronómico Nacional, UNAM, Apdo. Postal 877, 22800, Ensenada B.C., México

Abstract In this contribution we introduce a UBVRI survey of open clusters that is being undertaken at the Observatório Astronómico Nacional at Sierra de San Pedro Mártir (SPM), Mexico. Most of the clusters visible from SPM and will be surveyed using the same instrumentation (telescope, CCD, filters) and reduction procedures. The survey will, therefore, provide a highly homogeneous database of open cluster photometry.

Introduction

We present a CCD UBVRI survey of open clusters that is being undertaken at Pedro Mártir (SPM), Mexico. The survey is being performed using the same instrumental setup (telescope, CCD, filters), reduction procedures, and system of standard stars. The UBVRI system was chosen for its capability of providing cluster reddenings, distances and ages, even in situations where the contamination from field stars is severe, without large observing overhead. With the photometric database we intend to: (1) Provide a common UBVRI photometric scale for open cluster studies. (2) Build an atlas of color-color and color-magnitude diagrams for the cluster sample. (3) Determine a homogeneous set of cluster reddenings, distances and ages. (4) Increase the number of distant and old open cluster studies. (5) Provide selection criteria for further studies (reddening, distance and age ranges; clusters containing certain types of stars etc.).

1. The Sample

All clusters visible from SPM ($\delta > -15°$) were selected from the WEBDA open cluster database (*http://obswww.unige.ch/webda*) of Mermilliod (1995). Duplicate entries were eliminated and clusters with very large angular diameters ($d > 30'$) were also left out. We have checked the remaining clusters against the Palomar plates (using the digital sky survey) and were not able to identify

M. Chávez et al. (eds.),
New Quests in Stellar Astrophysics: The Link Between Stars and Cosmology, 119–120.

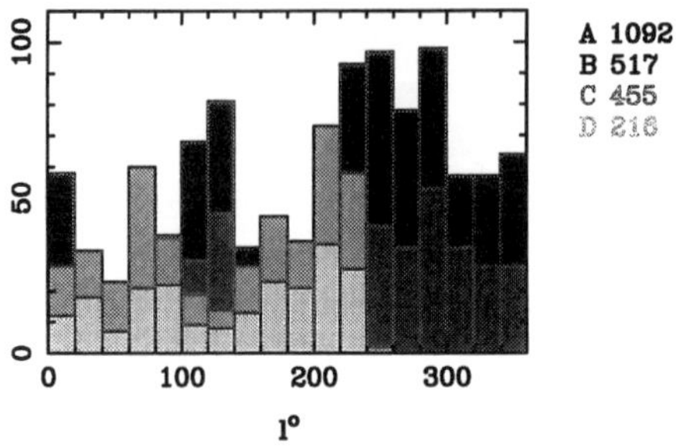

Figure 1. Sample statistics as a function of Galactic longitude. The grayscale has the following codes: A - clusters in WEBDA; B - clusters in WEBDA that have not been studied; C - selected sample; D - clusters in the selected sample that have not been studied. The total numbers in each category are presented.

approximately 5% of them. The final sample consists of 455 clusters, of which 215 have not been previously studied (see Fig. 1).

2. Instruments and reductions

Data are being gathered at the SPM 84cm telescope, that provides images of fine quality (the typical seeing is around 1.2″ reaching 0.9″ in many of the acquired images).[1] A set of UBVRI filters following the recipe of Bessel (1990) has been bought for this survey. All filters have excellent peak transmissions, ranging from 68% in U to 96% in I. A grade 1 Site 1024x1024 CCD with a 300,000 e^- full well capacity is being used. The gain is set at 1.2 e^-/ADU, and the readout noise is 7.8 e^-. The plate scale is 0.394″/pix which results in a 6.7′× 6.7′ field of view.

Image processing is done within IRAF. All images are subject to the usual bias, shutter timing, flat field corrections. Data from bad pixels are not used, and cosmic rays are removed. Photometry is performed using IRAF/DAOPHOT tasks. Calibration to the standard system is assured by the observation of Landolt (1992) fields.

Acknowledgments

This work is financially supported by DGAPA (Mexico) through the project IN111500, and by CONACyT (Mexico) through the project I33940-E.

References

Bessel, M. S. 1990, PASP, 102, 1181.

Landolt, A. U. 1992, AJ, 104, 340.

Mermilliod, J. -C. 1995, in Information and On-Line Data in Astronomy, eds. D. Egret & M. A. Albrecht (Dordrecht: Kluwer), p. 127.

[1]For details see *http://www.astrosen.unam.mx/Telescopes/telescopes.html#84cm.*

ABOUT THE IMF IN YOUNG STELLAR CLUSTERS*

Alicia Porras

Instituto Nacional de Astrofísica, Optica y Electrónica, Tonantzintla, Puebla 72840, Mexico

Irene Cruz-González, and Luis Salas

Instituto de Astronomía, UNAM Apdo. Postal 70-264, C.U., México D.F. 04510, Mexico

Abstract The initial mass function obtained from an infrared study of 38 young stellar clusters in the Perseus Arm is presented. The main results are that star formation processes are able to create clusters with the same number of stars but either massive or low-mass, with $M_{cluster}$ in the massive case about three times larger. An "average" IMF for young stellar clusters is constructed, which results flatter ($\Gamma = -1.00 \pm 0.10$) than both Salpeter's and Scalo's IMF.

Introduction

Studies of age and mass distribution of single star-forming events, associated with the individual molecular cloud cores at the earliest stages of stellar evolution, are crucial to our understanding of young stellar clusters (e.g. Meyer et al. 2000). Several studies of the stellar content in nearby YSCs have characterized their physical parameters such as: size (~0.2 - 0.8 pc), spatial distribution (centrally concentrated), stellar content (~100 stars), age ($\leq$ 10 Myr) and mass (20 - 160 $M_{\odot}$). The age and mass are the most important parameters and require special techniques to derive them. Rough mass and age distributions can be obtained from near-IR photometry alone. While more reliable values of stellar masses require combining photometry and spectral types from near-IR spectroscopy to place stars directly in the H-R diagram.
One global characteristic of YSCs is how the parent cloud mass gets distributed into stars. This initial distribution of stellar masses is known as the initial mass function (IMF). The IMF was first studied in the solar neighborhood and in this case evolutionary corrections to the observed mass distribution were required

*Based on observations obtained at the Observatorio Astronómico Nacional at San Pedro Mártir, B.C., México.

M. Chávez et al. (eds.),
New Quests in Stellar Astrophysics: The Link Between Stars and Cosmology, 121–124.

(e.g. Salpeter 1955, Miller & Scalo 1979). In young stellar clusters the observed mass distributions reveal the history of star formation from the cloud fragmentation up to now. Young stellar clusters provide the best laboratories to study the IMF in a range of 0.2 to 20 $M_{\odot}$ through infrared observations. The observational advantages are (Meyer et al. 2000): 1) there are minimum evolutionary corrections to the present day mass distribution, and so, the present mass distribution can be considered as the IMF; 2) youth precludes wandering of cluster members far from their birth-sites due to peculiar motions; 3) observations are more sensitive to low-mass objects because the M-L relationship for stars in the PMS phase is not as steep function of mass as for stars on the main sequence; 4) they are not contaminated by foreground stars due to their compactness; 5) they are not contaminated by background stars due to the presence of molecular gas which provides a natural veil to avoid them.

1. Young Stellar Clusters in the Perseus Arm

Our IMF study of embedded clusters is based on an observational study in the near-infrared of 38 young stellar clusters along the Perseus Arm (Porras 2001). The observations were obtained with the infrared camera CAMILA at the 2.1 m telescope in San Pedro Mártir, operated by the Observatorio Astronómico Nacional (OAN/SPM) in Baja California, México. Photometric images were obtained in the J-, H- and K-bands on five observing runs held from 1996 to 1998.

We propose a multi-color photometrical method to estimate cluster sizes, stellar densities and ages, as well as individual extinction and masses of a total of 5781 cluster stars. We use luminosity functions to estimate cluster age and diagnostic diagrams to distinguish stellar populations in clusters. Stellar masses are obtained to construct the mass distributions that yield the cluster IMF, relying on a uniform treatment of the data. A comparison of clusters is discussed both in the IMF index and in the spatial distribution of masses.

With distances obtained from the literature (mean value 3.4±1.6 kpc) and the spatial distribution of stars we obtain cluster parameters such as surface density, cluster radius and the average concentration parameter: $3< \Sigma <220$ stars/pc^2; $0.26< r_c <2.89$ pc; c_{YSC}=0.45±0.11, which means that their stellar concentration is ~3 times smaller than in globular clusters.

2. IMF

We develop a method to estimate ages based on a comparison between observed and theoretical luminosity function (LF) histograms. The estimated ages for clusters yield a mean value of 2.55 ± 1.37 in an interval from 1.1 to 5.6 Myr, which is consistent with the expected lifetime of a GMC (10^{6-7} Myr). The construction of isochrones plotted in the observational J/J-H diagram are an es-

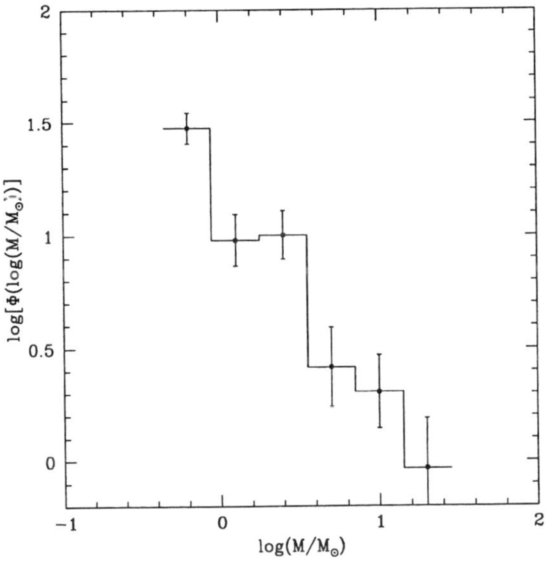

Figure 1. IMF histogram of the "average" cluster constructed with the averaged values bin by bin of 0.3 in log M of all the 38 clusters.

sential tool for the determination of stellar masses. The isochrones are obtained from pre-main sequence evolutionary tracks by D'Antona & Mazzitelli 1994, assuming Alexander opacities and the Canuto & Mazzitelli convection model. Using the estimated ages, the main sequence stellar masses can be obtained from the J/J-H diagram by de-reddening them to the previously estimated isochrone. For the masses of reddened CTT and Ae/Be stars, we use the relations (Carpenter et al. 1993) $\log M = -0.25\ M_H + 0.44$ and $\log M = -0.24\ M_K + 0.24$, where M_H and M_K are the uncorrected for extinction H- and K-band magnitudes.

The large number of stars with derived masses in the YSCs studied here gave us a statistically significant sample to construct a single IMF, called "average" cluster IMF, which contains the average value, bin by bin, of the overall cluster sample. The number of stars per bin in this average cluster is defined as: $N_i = \frac{1}{N_c}\sum_{j=1}^{41} N_c N_j$ where i is the mass bin, j the cluster number and N_c the number of complete clusters at bin i. The resulting "average" cluster is presented in the histogram of Fig. 1. The IMF index obtained is $\Gamma_{all} = -1.00 \pm 0.10$.

The spatial distribution of stellar masses inside the cluster area was obtained and show that 46% of the sample contain massive stars. Furthermore, three different mass compositions are possible: regions with an even balance of low and intermediate mass stars, and regions where intermediate mass stars or low mass stars are dominant. The star formation processes involved are able to create two types of clusters with different cluster mass: massive and low-mass. For the same number of formed stars the massive one has a total cluster mass about three times that in the low-mass case.

A rather useful diagnostic is the ratio of intermediate-to-low-mass stars. We define $C = N(1.2 - 20\ M_{\odot})/N(0.6 - 1.2\ M_{\odot})$, based in the minimum and maximum mass values in our sample. We can calculate C only for 13 YSCs, complete in all the mass intervals. The mean and mode values are 1.4 ± 0.7 and 1.0 ± 0.5, which result larger that $C_{-1.35l}$= 0.63, appropriate for a Salpeter IMF. If we consider that the mode and mean difference may be due to a change of slope in the low-mass regime, we found $\Gamma = -0.3$ for $M < 0.9\ M_{\odot}$ and $\Gamma = -1.1$ for $M > 0.9\ M_{\odot}$. It is significant to note that the Γ = -0.3 part is in close agreement with the low mass regime proposed by Kroupa et al. 1993.

Acknowledgments

AP wishes to thank CONACYT for providing Graduate student scholarship. This project is partly supported by DGAPA (UNAM) IN-120198.

References

Carpenter, J.M., Snell, R.L., Schloerb, F.P., & Skrutskie, M.F. 1993, ApJ, 407, 657.

D'Antona, F., & Mazzitelli, I. 1994, ApJS, 90, 467.

Meyer, M. R., Adams, F. C., Hillenbrand, L. A., Carpenter, J. M., & Larson, R. B. 2000, in Protostars and Planets IV, eds. V. Mannings, A. P. Boss, & S. S. Russell (Tucson: Univ. of Arizona Press), p. 121.

Kroupa, P., Tout, C. A., & Gilmore, G. 1993, MNRAS, 262, 545.

Miller, G.E., & Scalo, J.M. 1979, ApJS, 41, 513.

Porras, A. 2001, PhD thesis, Instituto Nacional de Astrofísica, Optica y Electrónica, Mexico.

Salpeter, E. E. 1955, ApJ, 121, 161

IV

STARS IN LOCAL GALAXIES

RECONSTRUCTING STAR FORMATION EPISODES IN THE LOCAL GROUP

D. Valls–Gabaud

UMR CNRS 5572, Observatoire Midi-Pyrénées, 31400 Toulouse, France

X. Hernandez

Instituto de Astronomía, U.N.A.M., A.P. 70-264, 04510 México

and

G. Gilmore

Institute of Astronomy, Madingley Road, Cambridge, UK

Abstract The color-magnitude diagrams (CMDs) of resolved stellar populations within the Local Group provide unique constraints on the cosmological models of galaxy formation. The unprecedented quality of recent CMDs, probing environments as different as the Hipparcos solar neighborhood, the dense cores of globular clusters and the diffuse, galaxy-wide samples of nearby dwarf spheroidal galaxies, requires proper techniques to infer their star formation histories. Our state-of-the-art method, based on a coupling of maximum likelihood with variational calculus, allows us to infer -for the first time- a robust, non-parametric reconstruction of the evolution of the star formation rate which gave rise to the observed CMDs. We apply our method to the best CMDs available, where crowding effects and blending are minimal. In the solar neighborhood we find evidence for a regularity of about 50 Myr in the star formation activity over the past 3 Gyr, possibly linked to interactions with spiral arms. The case of the five Local Group dwarfs studied so far shows an unexpected wide variety in their star formation histories.

Introduction

The integrated properties of galaxies have been so far interpreted in terms of smoothly varying star formation histories (e.g. Sandage 1986, for a classic summary). For instance, stellar populations in ellipticals have been thought to be formed nearly on a free-fall time scale, a massive burst of star formation with little else happening afterwards. The properties of spiral galaxies, on the contrary, seem to be consistent with exponentially decreasing functions of

M. Chávez et al. (eds.),
New Quests in Stellar Astrophysics: The Link Between Stars and Cosmology, 127–130.

time, with perhaps stronger or nearly constant recent activity for the later types. In dwarf galaxies –which dominate by far the number density of galaxies– there are several competing theories which account for their properties, from early star formation within collapsing dark matter haloes, to tidal tails formed by the interaction with massive galaxies or even late formation or delayed star formation due to various feedback effects. To go beyond these simple schemes requires first of all excellent data sets, and secondly state-of-the-art methods to interpret them. The issue is important: today's dwarf galaxies may be the best counterparts to the high redshift objects with early star formation, so understanding this fossil record may hold the clue to understanding galaxy formation, especially if the current paradigm of galaxy formation within CDM haloes is correct.

1. Reconstructing star formation histories in dwarf galaxies

The excellent quality of the CMDs of nearby dwarfs provided by HST (see Aparicio 1998, Mateo 1998 for recent reviews) requires powerful tools to invert them and obtain the star formation histories. The methods used so far are based on comparisons between the observed CMD and synthetic CMDs computed assuming a given SFH, and then looking for the best matching SFH. For instance Tolstoy & Saha (1996) use a Bayesian likelihood technique, Mighell (1997) a classical χ^2 optimization, Ng (1998) a Poisson merit function, while Gallart et al. (1999) and Hurley-Keller et al. (1999) minimize a counts in cells statistic for well-defined boxes in the CMD. Alternatively, Dolphin (1997) makes a linear decomposition in terms of fiducial CMDs and solves for the best matching final CMD, in terms of a χ^2 statistic. There are problems both in terms of proper statistics for Poisson variables and in terms of uniqueness. In contrast, the method we have developed makes use of variational calculus to derive, non-parametrically (that is, without making any a priori assumption on the shape of the SFR) the functions that maximize the likelihood that the observed CMD is produced by that SFR history (Hernandez, Valls-Gabaud & Gilmore, 1999).
An example of reconstruction is shown on Fig. 1 (see also Hernandez, Gilmore & Valls-Gabaud 2000, for further examples), where the central line gives one possible function that makes the likelihood maximal. Any SFR function within the boundaries of the envelope does it as well (hence answering the issue of uniqueness) given the data. Of course, functions with higher frequency content are also allowed, given the errors in the photometry. It is surprising that this dwarf galaxy, despite having formed stars for a large fraction of a Hubble time, still presents a low metallicity, as measured spectroscopically in stars of the upper giant branch. Could the metallicities of the turn off and subgiant stars be different? How is it possible that in such shallow potential wells the effects of

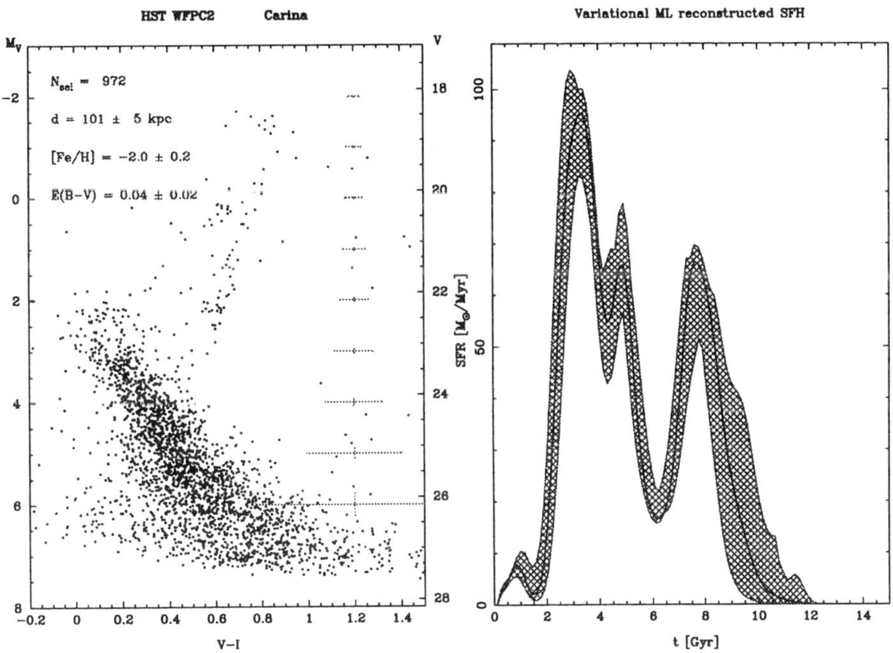

Figure 1. WFPC2 CMD diagram of Carina (left panel) with 2σ error bars. Note the three thin subgiant branches which hint to three main episodes of star formation. The variational maximum-likelihood reconstructed star formation rate (right panel) is the envelope which gives the range where possible functions can exist maximizing the likelihood, given uncertainties in distance, metallicity and color excess.

supernovae explosions appear to be negligible, since several important bursts do happen after the first one at about 8 Gyr ?

2. The recent SFR in the solar neighborhood

Previous attempts at measuring the SFH in the galactic disc have relied on indirect methods (see Rocha-Pinto et al. 2000ab for a recent one) whilst the quality of the Hipparcos CMD of the solar neighborhood provides a first glimpse at the recent evolution of the SF rate (see Hernandez, Valls-Gabaud & Gilmore 2000 for details). Figure 2 shows the main result of the inversion, a surprising regularity in the star formation episodes, perhaps superimposed on a somewhat constant level of activity. A possible interpretation, besides having a stochastic star formation activity, is that because of the differential speed between the solar neighborhood and the pattern speed of the spiral arms, star formation is triggered roughly periodically every time an interaction happens. Could these interactions leave a kinematical signature in the stars formed during these

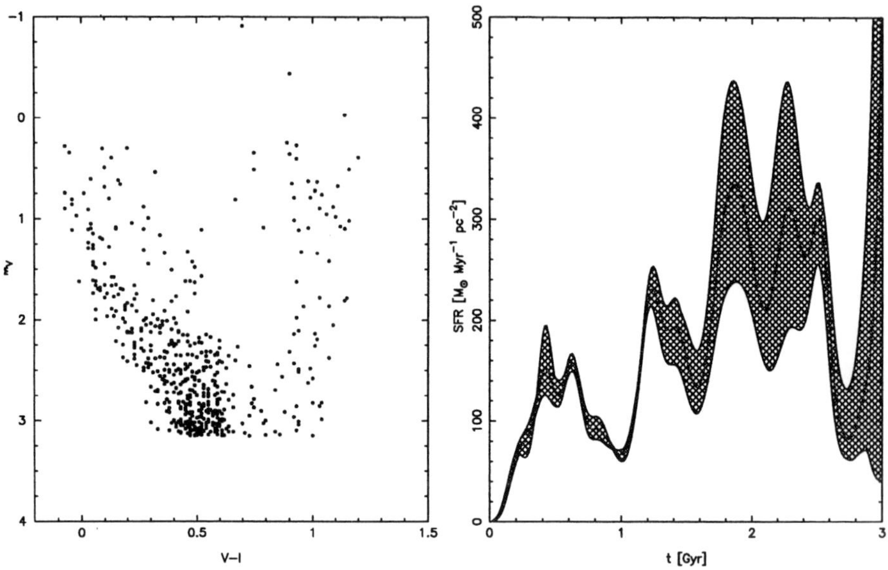

Figure 2. A sub-sample of the solar neighborhood, as observed by HIPPARCOS (left panel). The reconstructed SFR (right panel) shows several well-defined bursts, features which are not obvious from the morphology of the CMD.

episodes? Are star clusters the usual products of the interaction, which then slowly dissolve into the field?

References

Aparicio A. 1998, in IAU Symp 192, The stellar content of Local Group galaxies, ed. P. Whitelock & R. Cannon (San Francisco: ASP), p. 20.
Dolphin A. 1997, New Ast., 2, 397.
Gallart C. et al. 1999, AJ, 118, 2245.
Hernandez X., Valls-Gabaud D., & Gilmore G. 1999, MNRAS, 304, 705.
Hernandez X., Gilmore G., & Valls-Gabaud D. 2000a, MNRAS, 317, 83.
Hernandez X., Valls-Gabaud D., & Gilmore G. 2000b, MNRAS, 316, 605.
Hurley-Keller D., Mateo M., & Nemec J. 1998, AJ, 115, 1840.
Mateo M. 1998, ARA&A, 36, 435.
Mighell K.J. 1997, AJ, 114, 1458.
Ng Y.K., 1998, A&AS, 132, 133.
Rocha-Pinto H.J., Scalo J., Maciel W.J., & Flynn C. 2000a, ApJ, 531, L115.
Rocha-Pinto H.J., Maciel, W.J., Scalo J., & Flynn C. 2000b, A&A, 358, 869.
Sandage A. 1986, A&A, 161, 89.
Tolstoy E., & Saha A. 1996, ApJ, 462, 672.

DEEP STUDIES OF THE RESOLVED STELLAR POPULATIONS IN THE OUTSKIRTS OF M31

Annette M. N. Ferguson
Kapteyn Institute, PO Box 800, 9700 AV Groningen, The Netherlands

Abstract

We discuss the first results from ongoing studies of the resolved stellar populations in the outskirts of our nearest large neighbor, M31. Deep HST/WFPC2 archival observations are used to construct color-magnitude-diagrams which reach well below the horizontal branch at selected locations in the outer disk and halo, while a panoramic ground-based imaging survey maps spatial density variations through resolved star counts to a projected radius of $\sim$ 50kpc.

Introduction

The fossil record of star formation and galaxy evolution is imprinted on the spatial distribution, ages and metallicities of galactic stellar populations. Surprisingly little is known about the old and intermediate-age stellar populations in massive galaxies outside our own Milky Way. Located at $\sim$780 kpc, M31 provides our best opportunity to explore the stellar populations across the face of a large disk galaxy. Furthermore, its favorable inclination means that the disk and halo components at large radii can be easily distinguished and independently studied.

Most previous studies of M31 have either been based on observations of single fields (eg. Holland et al 1996, Durrell et al 2001) or on large-area surveys limited to the bright disk (eg. Walterbos & Kennicutt 1988). It is generally thought that M31 resembles the Milky Way in many ways, although it is somewhat larger, more luminous and has a denser stellar halo (see Freeman 1999 for a review). The disparity between the mean metallicity of M31's field halo and that of the Milky Way ($\approx$ factor of 10 lower) was first recognized by Mould & Kristian (1986) but remains to this day poorly understood.

1. Probing M31 with Deep HST/WFPC2 Pointings

We searched the HST archive for all deep (T>4000s) WFPC2 pointings towards M31 in the F555W and F814W filters. We focus here on those fields which sample the disk and halo at large radius ($R_{deproj} \gtrsim$ 20kpc).

M. Chávez et al. (eds.),
New Quests in Stellar Astrophysics: The Link Between Stars and Cosmology, 131–134.

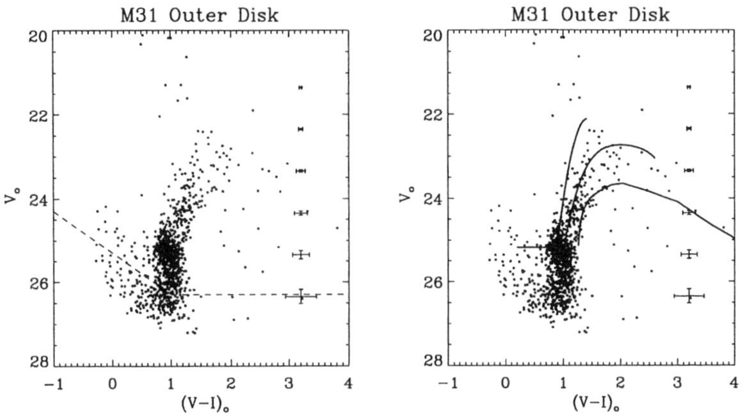

Figure 1. CMDs of the M31 outer disk field (30 kpc or 5 disk scalelengths). (Left) the dashed line indicates the 75% completeness level. (Right) GC fiducials are overlaid corresponding to NGC 6397 ([Fe/H]= −1.9), 47 Tuc ([Fe/H]= −0.7) and NGC 6553 ([Fe/H]= −0.3). Also indicated is the mean V magnitude of the blue horizontal branch detected in M31 halo by Holland et al (1996).

Figure 1 shows the WFPC2 CMD of a field at $\sim$ 30 kpc (or 5 disk scale lengths) along the major axis. Based on extrapolation of the structural parameters of Walterbos & Kennicutt (1988), we expect 95% of the stars in this field to belong to the disk. The prominence and morphology of the red giant branch (RGB) and red clump (RC) indicate an old-to-intermediate age, fairly metal-rich population. Comparison of the outer disk RGB with globular cluster fiducials indicates a mean metallicity comparable to 47 Tuc ([Fe/H]= -0.7). The significant width of the RGB exceeds that of photometric errors, and is most easily explained by an intrinsic dispersion ($\simeq$ 2 dex) in the metallicity of the stellar population. While these properties have been noted before for the M31 halo (eg. Holland et al 1996), we find here that they also characterize the outer disk. The RC properties, when coupled with the mean metallicity from the RGB color, suggest a mean age for the population of $\gtrsim 7-8$ Gyr (Cole 1999, Girardi & Salaris 2001). This is also consistent with the apparent lack of asymptotic giant branch stars above the tip of the RGB (I$\sim$20.5), which would represent young-to-intermediate age (2–6 Gyr) shell He-burning stars. Finally, we note the marginal detection of horizontal feature in the CMD connecting the blue plume at V$\sim$ 25 to the RC. While this could represent the subgiant branch of a $\sim$ 1 Gyr population, it is more likely to be an old ($\geq$ 10 Gyr), metal-poor ([Fe/H]$\sim$ -1.7) horizontal branch (see Figure 1).

Figure 2 shows CMDs for a representative halo-dominated field and a field in which disk and halo are expected to contribute in roughly equal amounts. Apart from stellar density (which reflects both intrinsic variations as well how much of the WF area was usable in our analysis), the dominant stellar population in

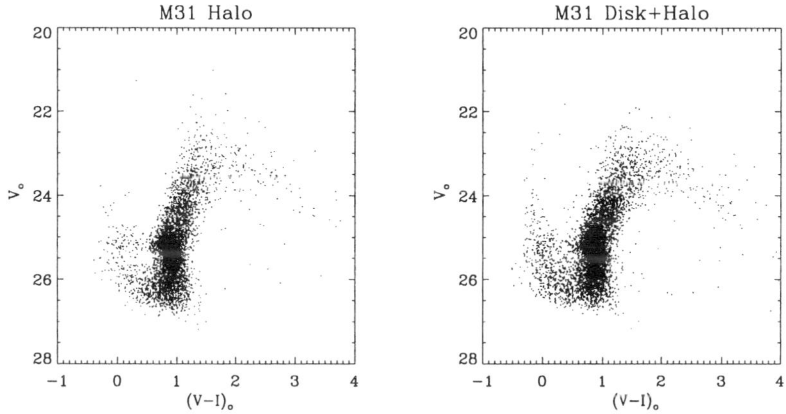

Figure 2. CMDs for a halo-dominated field (left) and a field which is expected to contain equal numbers of disk and halo stars (right). Both fields sample roughly the same halo radius (10.5 kpc) but the disk contribution is greater in the latter.

these fields appears strikingly similar (and indeed, both resemble the outer disk CMD in Figure 1). The homogeneity of halo and outer disk populations was first remarked upon by Morris et al (1994) from shallow ground-based data, but we show here the similarity holds to well below the horizontal branch.

2. Probing M31 with Panoramic Ground-based Imagery

While providing detailed information on the stellar populations, the HST pointings tell us nothing about the large-scale structure of M31 (WFPC2 FOV $\sim$ 0.2 kpc^2). We have therefore embarked upon a ground-based, moderate-depth panoramic imaging survey of the outskirts of M31 with the INT Wide Field Camera. To date, 58 contiguous fields (0.3 deg^2 per pointing) have been observed in the SE half of the galaxy, mapping out to a projected radius of $\sim$ 50 kpc.

An exciting first result has been the discovery of a giant stream of stars (overdensity a factor of two) in the halo of M31 near the southern minor axis (Ibata et al 2001). The stream extends beyond the limit of our imagery, or $\sim$ 40 kpc at this position angle, and has an average V-band surface brightness of $\sim$ 30 mag arcsec^{-2}. If an old coeval population is assumed, the excess stream population is found to be of similar or slightly higher metallicity than the M31 field halo and outer disk.

The stream appears to lie along a line which connects M32 and NGC 205, the two dwarf elliptical companions of M31. Both these satellites display rather odd properties for their morphological class – including young and/or intermediate-age stars and in the case of NGC 205, cold gas – and both exhibit distorted isophotes in the outer regions, suggestive of tidal distortion and possibly disruption. The similarity between the metallicities of these dwarfs and that of the stream suggests that one or both of them may be the origin of the feature.

3. Discussion

We are exploring the stellar populations in the outskirts of M31 using deep HST archival pointings in combination with a ground-based panoramic imaging survey. We summarize our first results as follows:

♣ the extended disk of M31 appears to have a significant mean age ($\gtrsim 7-8$ Gyr). For current cosmologies, this lookback time corresponds to a redshift of $\simeq 1$ at which about half the stellar disk was in place at 30 kpc. This may be problematic for theories which invoke delayed disk formation as a way to circumvent the angular momentum problem seen in numerical simulations of disk formation (eg. Weil et al 1998). Equally puzzling is our finding that the dominant populations in the outer disk and halo are remarkably similar, suggesting these components have experienced similar formation epochs and evolutionary histories.

♣ the discovery of a giant stellar stream in the outer halo of M31 attests to the fact that the hierarchical process of galaxy formation continues to the present-day and that at least some fraction of the M31 field halo has not formed *in-situ*, but has been accreted from lower mass objects. Could this interaction have polluted the outer regions of M31 to such an extent that it explains that apparent uniformity of the stellar populations in these parts? Future observations will help us assess this possibility more throughly.

Acknowledgments

I gratefully acknowledge the input of my collaborators on the work discussed here—specifically, Rachel Johnson and Nial Tanvir on the HST studies and Mike Irwin, Rodrigo Ibata, Geraint Lewis and Nial Tanvir on the INT/WFC survey. Thanks also to the Leids Kerkhoven-Bosscha Fonds for financial assistance to attend this conference.

References

Cole, A. A. 1999, PhD Thesis, University of Wisconsin.

Durrell, P. R., Harris, W. E., & Pritchet, C. J. 2001, AJ, 121, 2557.

Freeman, K. C. 1999, in ASP Conf. Ser. 165, The Galactic Halo, eds. B. K. Gibson, T. S. Axelrod, & M. E. Putman (San Francisco: ASP), p. 167.

Girardi, L., & Salaris, M. 2001, MNRAS, 323, 109.

Holland, S., Fahlman, G. G., & Richer, H. B. 1996, AJ, 112, 1035.

Ibata, R., Irwin, M., Lewis, G., Ferguson, A. & Tanvir, N. 2001, Nature, 412, 49.

Morris, P. W., Reid, I. N., Griffiths, W. K., & Penny, A. J. 1994, MNRAS, 271, 852.

Walterbos, R. A. M., & Kennicutt, R. C. 1988, A&A, 198, 61.

Weil, M. L., Eke, V. R., & Efstathiou, G. 1998, MNRAS, 300, 773.

COMMENTS ON BINARY STAR EVOLUTION AND THE SMALL MAGELLANIC CLOUD SYSTEM HD 5980

Gloria Koenigsberger, and Edmundo Moreno

Instituto de Astronomía, Universidad Nacional Autónoma de México,

Apdo. Postal 70-264, 04510 México, D.F., México

Abstract

Studies of the effects that close binary stars may produce in the evolution of stellar populations are starting to suggest that the role of binary stars cannot be neglected when deducing the properties of starburst regions from population synthesis models. It is thus important to study massive binary systems such as HD 5980, located in the Small Magellanic Cloud young stellar cluster NGC 346, in this context. The Luminous Blue Variable type eruptions that the system has suffered suggest that it may be presenting the results of strong interaction effects due to tidal forces. Because of the eccentricity of the orbit and due to the oscillations that are excited in such a non-equilibrium configuration, its evolution is likely to be proceeding differently than predicted by models based on the assumption of an equilibrium configuration. In particular, standard models that assume Roche Lobe Overflow are not applicable.

1. Massive Binaries and Stellar Evolution

Over the past few years considerable progress has been made in the development of theoretical models of massive stars extending from the Main Sequence (MS) to the supernova (SN) explosion. The theoretical evolutionary tracks provided by these models are used in studies of a wide range of topics, including chemical enrichment of the ISM, galaxy evolution, and the age and evolution of starburst regions in extragalactic objects. Massive stars play a significant role in the processes implied by these topics, due to their short life-times and the impact on their environments produced by stellar winds and the final supernova (SN) explosion. According to the standard evolutionary scenarios (Langer et al. 1994; Schaller et al. 1992; Maeder & Meynet 1994), after spending $< 10^7$ years on the MS and reaching central H-exhaustion, very massive ($\sim$ 50- 120 $M_{\odot}$) O-stars undergo a fast increase in radius, evolving towards lower temperatures. At this stage, the H-rich layers of their atmospheres which have not already

M. Chávez et al. (eds.),
New Quests in Stellar Astrophysics: The Link Between Stars and Cosmology, 135–138.

been removed by the stellar winds are believed to be violently ejected during what is known as the luminous blue variable (LBV) phase of evolution. A number of LBV's have been observed through various phases of their activity (comprehensive reviews are given by Davidson & Humphreys 1994), and the list of LBVs includes such illustrious members as η Carinae, P Cygni, and R127. The "major" eruptions which have occurred in some of these objects begin to approach supernovae in terms of luminosities and ejected masses (Gallagher 1987). Once the H-rich outer layers have been removed, the star becomes a Wolf-Rayet (WR) star. WR stars are characterized by extraordinarily massive (dM/dt$\sim 10^{-5}$ $M_\odot$ yr^{-1}) and rapid (2000-3000 km s^{-1}) stellar winds. In these winds, the relative abundances of the chemical elements such as He, N, C and O are enhanced with respect to the abundances of stars on the MS, due to the presence of nuclear processed material at the stellar surface. The WR phase is very short-lived ($\sim 10^5$ yrs), and is soon followed by the supernova event. It is important to note that the post-MS evolution depends strongly on the stellar mass, rotation, and on the degree to which the star is able to mix its interior layers containing nuclear processed material with outer, H-rich layers (Langer et al. 1994; Langer et al. 1999; Heger & Langer 2000; Maeder & Zahn 1998). All the evolutionary studies mentioned above refer to single stars only. The evolution of stars in binary systems (see for example, Wellstein et al. 2001; Vanbeveren 1999 and references therein) has been assumed to proceed as in single stars until the post-MS expanding stellar radius attains the dimensions of the equilibrium equipotential surface within the system (i.e., the "Roche Lobe"), at which time a phase of massive and very rapid mass loss takes place. Depending on the orbital separation, the "Roche Lobe overflow" (RLO) may occur, for example, either after core H-exhaustion or core He-exhaustion. Once the star returns to a core nuclear-burning phase, the evolutionary models once again assume single star evolution. Thus, the current models of evolution in binary systems assume that the only effect that a binary companion has on a star is through the RLO process and this occurs at individual and very specific points along the evolutionary history of the star. However, the concept of the Roche Lobe carries implicitly the assumption that the binary system is in an equilibrium state; i.e., the orbit is circular, the rotation axes of the two stars are parallel, and the stellar rotation periods are equal to the orbital period. In general, all three of these conditions will not be satisfied in a massive system. Since the timescales for circularization of the orbit and for achieving co-rotation are larger than the evolutionary times (although the post-MS timescales for circularization of the orbit can become much shorter Savonije et al. 1995). Hence, in general, an evolving star in a close binary system is likely to leave the MS before having achieved an equilibrium configuration, thus making the use of the RLO assumption questionable. We argue that the star enters a phase of enhanced mass-loss long before it reaches Roche Lobe dimensions due to the effects of tidal forces (Koenigsberger et al. 2001).

2. The effects of tidal forces in binary systems

If the binary system is not in an equilibrium configuration as, for example, when the orbit is eccentric or the orbital and rotation periods are not synchronized, the tidal forces act to move the system closer to equilibrium. As part of this process, a large number of oscillation modes are driven (Zahn, 1975,1977). These oscillations are not expected to affect the convective core of the star and thus, the evolution of the central region during the MS should not differ from that of a single star. However, virtually the entire net tidal torque is exerted in the immediate vicinity of the boundary between the convective core and the radiative envelope in massive stars (Zahn 1977), and the outer layers of the star may be synchronized with the orbital motion much sooner than the core, leading to a differentially-rotating outer structure (Goldreich & Nicholson 1989). Hence, a massive star in a binary system is likely to have an envelope structure that differs significantly from that of a single star of the same mass, chemical composition, and convective core rotation rate. In addition, because the amplitudes and dominant frequencies of the tidally-driven oscillations depend strongly on the stellar radius and on the ratio of rotation to orbital angular velocities, as the star starts to leave the MS, the growth of the oscillation amplitudes may drive the star into an enhanced mass-loss phase. This phase is non-existent in single-star evolutionary models.

3. The WR/LBV system HD 5980 in the Small Magellanic Cloud

The binary system HD5980 is located on the outskirts of the young cluster NGC 346 and consists of an eclipsing binary pair (Star A+Star B; $P = 19.266$ days) in an eccentric ($e = 0.27$) orbit, and a third component (Star C; Koenigsberger et al. 2000) which may itself be a binary system. HD 5980 has been known to be a WR since the 1950's, but in 1993 and 1994 it underwent two relatively fast LBV-type eruptive events (Bateson and Jones 1994; Barbá et al. 1995). This is the first, and up to now only, case of an observed transition between a WR and an LBV and back again to a WR (another such object could be η Carinae). Star A has an enriched helium surface composition (Koenigsberger et al. 1998), implying that it has suffered considerable previous mass loss, exposing nuclear processed layers and/or has undergone considerable mixing. We have proposed that the peculiar behavior of HD 5980 may be linked to the perturbations produced by the binary companion through tidal forces (Koenigsberger et al. 1999). According to this scenario, star A is a massive O-star that is in the process of leaving the MS. Because of the orbital eccentricity, it is subjected to variable tidal forces throughout the orbital cycle. As its radius increases, the amplitude of the oscillation increases until part of the outer layers are blown away in what we observe to be an eruptive event. This is followed by a decrease in the radius, and return to temporary stability. If this scenario is correct, then HD 5980 must have undergone eruptions in the past, and should do so again in the near future.

References

Barbá, R., Niemela, V., Baume, G., & Vazquez, R.A 1995, ApJ, 446, L23.

Bateson, F.M., & Jones, A.F. 1994, Pub. RAS New Zealand, 19, 50.

Davidson, K., & Humphreys, R.M. 1997, ARA&A, 35, 1.

Gallagher, J. 1989, in IAU Coll. 113, Physics of Luminous Blue Variables, (Dordrecht: Kluwer), p. 185.

Goldreich, P., & Nicholson, P.D. 1989, ApJ, 342, 1079.

Heger, A., & Langer, N. 2000, ApJ, 544, 1016.

Koenigsberger, G., Peña, M., Schmutz, W. & Ayala, S. 1998, ApJ, 499, 889.

Koenigsberger, G., Moreno, E., Cantó, J., & Raga, A. 1999, in IAU Symp. 193, Wolf-Rayet Phenomena in Massive Stars and Starburst Galaxies, eds. K.A. van der Hucht, G. Koenigsberger, & P. R. J. Eenens, (San Francisco: ASP), p. 38.

Koenigsberger, G., Georgiev, L., Barbá, R., Tzvetanov, Z., Walborn, N.R., Niemela, V.S., Morrell, N., & Schulte-Ladbeck, R. 2000, ApJ, 542, 428.

Koenigsberger, G., Cervantes, F., Moreno, E., & Georgiev, L. 2001, Rev. Mex. Astron. Astrof. Conf. Series, 10, 199.

Langer, N. et al. 1994, A&A 290, 819.

Langer, N., Garcia-Segura, G., & MacLow, M. M. 1999, ApJ, 520, L49.

Maeder, A., & Meynet, G. 1994, A&A, 287, 803.

Maeder, A., & Zahn, J.-P. 1998, A&A, 334, 1000.

Savonije, G.J., & Papaloizou, J.C.B., & Alberts, F. 1995, MNRAS, 277, 471.

Schaller, G., Schaerer, D., Meynet, G., & Maeder, A. 1992, A&ASS, 96, 269.

Vanbeveren, D. 1999, in IAU Symp. 193, Wolf-Rayet Phenomena in Massive Stars and Starburst Galaxies, eds. K.A. van der Hucht, G. Koenigsberger, & P. R. J. Eenens, (San Francisco: ASP), p. 196.

Wellstein, S., Langer, N., & Braun, H. 2001, A&A, 369, 939.

Zahn, J.P. 1975, A&A, 41, 329.

Zahn, J.P. 1977, A&A, 57, 383.

STELLAR METALLICITIES BEYOND THE LOCAL GROUP WITH VLT/FORS

Norbert Przybilla, Fabio Bresolin, Roberto H. Méndez
Universitätssternwarte München, Scheinerstraÿe 1, D-81679 München, Germany

and

Rolf-Peter Kudritzki
Institute for Astronomy, University of Hawaii, 2680 Woodlawn Drive, Honolulu, HI 96822, USA

Abstract

Luminous BA-type supergiants are the visually brightest stars in spiral and irregular galaxies. Substantial progress in their theoretical modeling has been achieved by the application of sophisticated non-LTE spectrum synthesis. After testing of the method on Local Group objects at high resolution, the next step has been taken. First results from medium-resolution spectroscopy of supergiants in NGC 3621 at a distance of 6.7 Mpc, obtained with the FORS instrument on the VLT, are presented. Fundamental parameters of individual stars are determined at such distances for the first time. This allows the reduction of systematic errors of the Cepheid distance scale by providing information on the metallicity, color excess and extinction of objects adjacent to Cepheids. Moreover, the supergiants themselves will act as an independent distance indicator through the application of the wind momentum–luminosity relationship after a proper empirical calibration.

Introduction

High-resolution spectroscopy of individual blue supergiants in Local Group galaxies has been performed with ground-based telescopes in recent years. Non-LTE analyses of these luminous stars (up to $M_V = -10$) provide a wealth of data on stellar parameters, elemental abundances and the properties of the intervening interstellar matter. In addition, the objects show signatures of strong mass outflow, which are successfully interpreted in the framework of radiation-driven wind theory (Kudritzki & Puls (2000); the corresponding wind momentum-luminosity relationship (WLR) promises to develop into an independent extragalactic distance indicator (Kudritzki et al. 1999). It is our goal

M. Chávez et al. (eds.),
New Quests in Stellar Astrophysics: The Link Between Stars and Cosmology, 139–142.

to obtain spectra of blue supergiants out to the Virgo and Fornax clusters of galaxies with existing large telescopes and instrumentation.
Here, we report on a test of our spectrum synthesis technique for photospheric analyses and to its application on two supergiants in NGC 3621, for which medium-resolution spectra have been obtained recently (Bresolin et al. 2001), using FORS on the VLT. A reliable distance to this galaxy is known from a study obtained within the HST Key Project on the Cepheid distance scale (Rawson et al. 1997).

1. Test of the spectrum synthesis technique

In order to perform reliable studies of distant supergiants, the spectrum synthesis – which is mandatory for the analysis of medium-resolution data – has to be tested on nearby objects at high resolution. Photospheric analyses of supergiants can be successfully performed on the basis of hydrostatic and plane-parallel LTE line-blanketed model atmospheres, as obtained with the ATLAS9 code (Kurucz 1991) after proper modifications, and subsequent non-LTE spectrum synthesis. Stellar parameters are determined spectroscopically by standard techniques, using ionization equilibria (e.g. Mg I/II, Przybilla et al. 2001) and Balmer wing fitting. The entire observed spectra of early A- and late B-type supergiants are reproduced, using DETAIL (statistical equilibrium/radiative transfer) and SURFACE (line formation, Butler & Giddings 1985), accounting for lines from 28 chemical species. Non-LTE effects turn out to be essential for the interpretation by removing systematic discrepancies and reducing the statistical scatter of individual abundances, as a test on galactic supergiants has shown (Przybilla et al., in preparation). Further testing by modifying parameters is shown in Fig. 1, leading to the conclusion, that stellar metallicities can be well determined from medium-resolution spectra.

2. First results on blue supergiants in NGC 3621

Two early A-type supergiants from the multi-object spectroscopy in NGC 3621 (Bresolin et al. 2001) are analyzed. The main results are summarized in Table 1, the spectrum synthesis for solar and LMC abundances is displayed in Fig. 2. $T_{\rm eff}$ is estimated from the spectral classification and $\log g$ determined from the Balmer line strengths and values for the microturbulence ξ and the helium content y (by number) are assumed to be the same as in galactic comparisons. The color excess is determined from comparing the photometry with the model atmosphere predictions; most of the effect is consistent with being due to Galactic foreground reddening. This value is smaller than that obtained by the HST Key Project ($E(B-V) = 0.23$), but the WFPC2 field is located much closer to the center of NGC 3621. Finally, the luminosities L, stellar radii R and masses M are derived from these parameters and from comparison with recent stellar

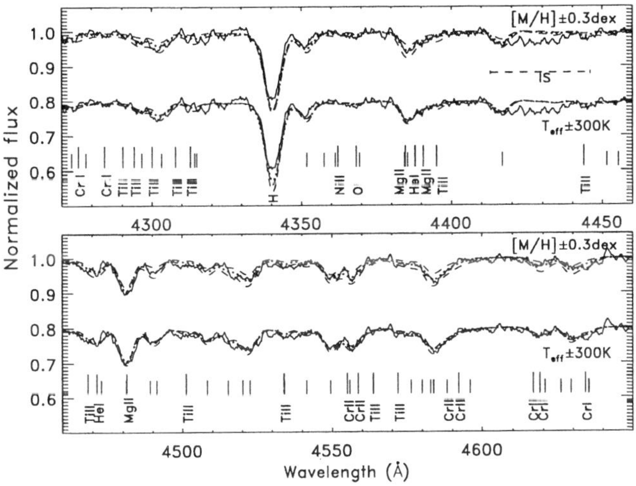

Figure 1. Spectrum synthesis for the bright galactic A0 Iae supergiant HD 92207 ($T_{\rm eff}=9500$ K, $\log g=1.20$ (cgs), $\xi=8\,{\rm km\,s^{-1}}$, $y=0.12$, $[{\rm M/H}]=-0.09$ dex). Full line: observations, downgraded to 5 Å resolution and $S/N \simeq 50$; dotted line: synthetic spectrum as obtained from the high-resolution analysis. Excellent agreement is found, except for the region around H_γ, which is filled by wind emission and the interstellar band around 4430 Å. Tests for scaled abundances and different temperatures (dashed lines) show that the only important parameter for the modeling is metallicity; $\log g$ is determined from fitting the higher Balmer lines. Unlabeled short marks are used for the identification of Fe II lines, all others are marked explicitly.

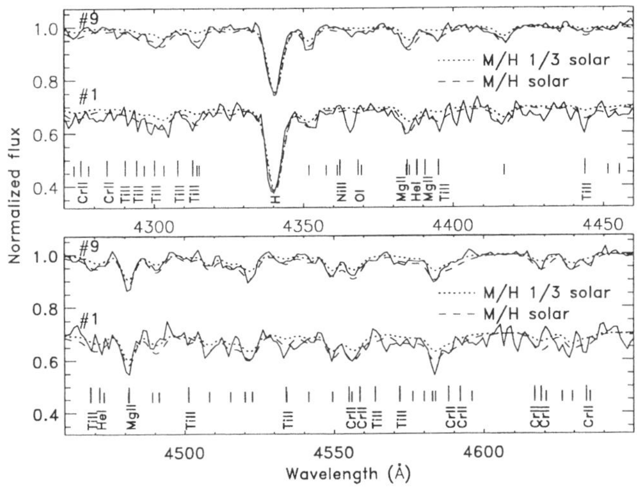

Figure 2. Spectrum synthesis for two NGC 3621 A1 Ia supergiants (slits #1 and #9, full lines) at two different metallicities for the stellar parameters listed in Table 1.

Table 1. NGC 3621 supergiants

Star	#1	#9
$T_{\rm eff}$ (K)	9000±300	9000±300
log g (cgs)	1.20±0.10	1.05±0.10
ξ (km s^{-1})	8±1	8±1
y	0.12	0.13
[M/H] (dex)	−0.1±0.3	−0.1±0.2
V	21.43	20.47
$B-V$	0.14	0.17
$E(B-V)$	0.12	0.12
log L/L$_\odot$	5.2±0.1	5.6±0.1
R/R$_\odot$	160±20	250±30
M/M$_\odot^{ZAMS}$	21±3	30±5
M/M$_\odot^{spec}$	15±4	26±7

evolution tracks accounting for rotation (Meynet & Maeder 2000). Metallicities for both objects are estimated from the comparison with the spectrum synthesis. For #9, this metallicity is consistent with the result of the analysis of a nearby H II region by Ryder (1995), who found [O/H] = −0.28±0.19 at a slightly larger galactocentric distance. This study demonstrates the feasibility of quantitative analyses of individual stars in galaxies well beyond the Local Group. Work on the other targets from the same observing run is in progress and will be extended to supergiants in other distant galaxies, accompanied by further calibration work on galactic, LMC and SMC supergiants.

Acknowledgments

NP would like to acknowledge financial support by the BMBF through grant 05AV9WM12.

References

Bresolin F., Kudritzki R.P., Méndez R.H., & Przybilla N. 2001, ApJ, 548, L159.

Butler, K., & Giddings, J. 1985, Newsletter on Analysis of Astronomical Spectra, No. 9, University of London.

Kudritzki, R.P., & Puls, J., 2000, ARA&A, 38, 613.

Kudritzki, R.P., et al. 1999, A&A, 350, 970.

Kurucz, R.L. 1991, in Stellar Atmospheres: Beyond Classical Models, NATO ASI Ser. C-152, ed. Crivellari L., et al., p. 441.

Meynet, G., & Maeder, A. 2000, A&A, 361, 101.

Przybilla, N., Butler, K., Becker, S.R., & Kudritzki, R.P., 2001, A&A, 369, 1009.

Rawson, D.M., et al. 1997, ApJ, 490, 517.

Ryder, S.D. 1995, ApJ, 444, 610.

AN ESTIMATE OF H_0 USING CEPHEID PERIOD LUMINOSITY RELATIONS AT MAXIMUM LIGHT

S. M. Kanbur, and S. Nikolaev
University of Massachusetts, USA

Abstract

We develop the technique of using Cepheid period luminosity relations at maximum light to estimate distance. This is different to the standard technique which uses the correlation between period and mean light to estimate distance. We present the Cepheid distance to 5 galaxies observed by the Key Project using this method. Generally the distances we calculate agree with the KP results though in some cases our results differ by as much as 0.15 mags. In a crude application of the method, we compute the distance to 1365 whose recession velocity is known. This leads to an estimate of H_0 of 73 ± 17 km/sec/Mpc. The large error on this estimate is due to the large error on the recession velocity.

Introduction

Simon, Kanbur and Mihalas (1993, hereafter SKM), using radiative hydrodynamic models of classical Cepheids showed that such stars occupy a narrower range of photospheric temperatures at maximum light than at mean light. Since the range of photospheric temperatures at given period is one cause for scatter in the Cepheid period luminosity (PL) relation, Kanbur and Hendry (1996) investigated the existence of PL relations at maximum light (PL(max)) in the LMC. They found that such relations do exist and have comparable scatter to their counterparts at mean light (PL(mean)). Since then the OGLE team has produced high quality light curves in the V and I bands for a large number of Cepheids in the LMC/SMC. Kanbur and Nikolaev (2001) has found that with this OGLE data, the PL(max) relation in the LMC has smaller scatter than the PL(mean) relation.

The Key Project (Freedman et al. 2001, hereafter KP) has produced distances to 17 galaxies using the Hubble Space Telescope (HST) to observe Cepheids in these galaxies. The KP observations consisted of 12 or so points per V band and 3 or 4 points per I band per Cepheid. They used this data to estimate V and I band means using template light curves and together with fiducial PL(mean)

M. Chávez et al. (eds.),
New Quests in Stellar Astrophysics: The Link Between Stars and Cosmology, 143–146.

relations in the LMC to estimate the distance to these 17 galaxies. These distances were then used to calibrate a range of secondary distance indicators so that the distances to galaxies far out in the Hubble flow could be calculated. This led to an estimation of Hubble's constant: 73 ± 7 km/sec/Mpc (Freedman et al. 2001).

In this study we used Fourier decomposition and a variation of the template light curves technique used by the KP (Stetson 1996) to reconstruct the V and I band light curves of Cepheids in 5 galaxies observed by the KP. From these light curves we estimated V and I band maxima. This together with PL(max) relations for the LMC constructed using OGLE fundamental mode Cepheid data enabled us to calculate the distances to 5 galaxies studied by the KP.

1. Interrelations and Intrarelations

Using high quality Galactic Cepheid data as templates, we perform a Fourier expansion of the V and I band light curves:

$$X = A_0 + \sum_{k=1}^{k=N} (A_k cos(k\omega t) + \phi_k), \tag{1}$$

where X refers to the V or I band, N is the order of the fit, typically 4, and $\omega = 2\pi/P$, P the published period in days. We then look for interrelations between the Fourier coefficients of the V and I band:

$$A_k(I) = \alpha A_k(V) + \beta, \phi_k(I) = \gamma\phi_k(V) + \delta\phi_k(I), \tag{2}$$

and

$$A_k(X) = xA_{k-1}(X) + y\phi_k(X) = a\phi_{k-1}(X) + y. \tag{3}$$

Nikolaev, Kanbur, Hendry and Tanvir (2001) discuss the properties of such relations in detail. For the case of the KP Cepheid data which consist of poorly sampled V and I band light curves, such relations allow us to reconstruct the I band light curve given the V band curve. This is accomplished in the following manner: a fourth or second order fit is constructed for the 12 points in the V band. If its a second order fit, this is expanded to a fourth order fit using the intrarelations in equation (3). The I band light curve is then obtained using the interrelations in equation (2) and existing I band points. Kanbur, Nikolaev, Hendry and Tanvir (2001) show that such a procedure allows the reconstruction of the V and I band curves and the estimation of mean and maximum magnitudes with an accuracy of about 0.05 magnitudes.

2. Maximum Light distance estimation

We used maximum light PL relations for the LMC (Kanbur and Nikolaev 2001) obtained from OGLE data (Udalski et al. 1999) and PL(max) relations for target

Table 1. PL(mean) and PL(max) distances to four galaxies

Galaxy	$\mu_{V_{mean}}$	$\mu_{I_{mean}}$	μ_{mean}	$\mu_{V_{max}}$	$\mu_{I_{mean}}$	μ_{max}
NGC 2090	30.634	30.511	30.323 $\pm$ 0.063	30.530	30.470	30.379 $\pm$ 0.064
NGC 2541	30.802	30.648	30.412 $\pm$ 0.069	30.659	30.585	30.472 $\pm$ 0.070
NGC 3319	30.818	30.714	30.555 $\pm$ 0.078	30.730	30.684	30.613 $\pm$ 0.078
NGC 925	30.163	29.991	29.728 $\pm$ 0.036	30.094	29.972	29.785 $\pm$ 0.037

galaxies, constructed as described above, to solve jointly for the true distance modulus and reddening of the target galaxy. Like the KP, we assume a LMC distance modulus of 18.5. Our procedure differs from the KP in a number of ways:

1 We use PL relations at maximum light

2 We use template light curves and Fourier inter and intrarelations as described above to estimate mean and maximum magnitudes

3 Rather than assuming the slope of the PL relation to be fixed to that in the LMC, we let both the LMC and *all* the target galaxies determine the slope of the PL relation at both mean and maximum light.

3. Results

Our results are presented in Table 1. Note that once the distance moduli in the V and I band are obtained, a reddening free distance modulus is calculated by

$$\mu = \mu_V - R(\mu_V - \mu_I), \tag{3}$$

where $R = 2.53$ (Willick and Batra 2001). The reddening free estimates using PL(mean) and PL(max) are listed in columns 4 and 7 of Table 1. In Table 2, we compare our results with those from the KP and Willick and Batra (2001). In a crude application of the method, we estimate the Hubble constant using recession velocity data from a single galaxy, NGC 1365. This galaxy is a member of the Fornax cluster, which has a recession velocity of $v = 1306 \pm 300$km/sec (Madore et al. 1998). The derived distance modulus for this galaxy is $\mu = 31.253 \pm 0.040$, which produces a distance estimate of $r = 17.807 \pm 0.328$Mpc. Substituting these values into the Hubble law gives a crude estimate of $H_0 = 73.3 \pm 16.9$ km/sec/Mpc.
Kanbur et al. (2001) give full details of the error budget in this procedure. In future work, we plan to carry out this procedure for all Cepheids observed by the HST and KP and recompute the distances using PL(max) relations. We will expand the calibrating set to include longer period stars and OGLE data.

Table 2. Comparison of recent estimates from other studies

Galaxy	Willick and Batra	KP
NGC 2090	30.298 ± 0.076	30.27 ± 0.05
NGC 2541	30.319 ± 0.049	30.26 ± 0.05
NGC 3319	30.583 ± 0.055	30.64 ± 0.09
NGC 925	29.782 ± 0.050	29.80 ± 0.04

Acknowledgments

SMK and SN acknowledge support from the STSCI HST Cycle 9 Archive grant 08752.01-A.

References

Hendry, M. A., Tanvir, N. R., & Kanbur, S. M. 1999, in ASP Conf. Series 167, Harmonizing Cosmic Distance Scales in a Post-HIPPARCOS Era, eds. D. Egret & A. Heck, (San Francisco: ASP), p. 192.

Freedman, W et al., 2001, ApJ, 553, 47

Kanbur, S. M., & Hendry, M. A. 1996, A&A, 305, 1.

Kanbur, S. M., & Nikolaev, N. 2001, BAAS, no. 46.13.

Kanbur, S. M., Nikolaev, N., Hendry, M., & Tanvir, N., 2001 (in preparation).

Madore, B, et al. 1998, Nature, 395, 47.

Nikolaev, S. & Kanbur, S. M. 2001, BAAS, no. 46.12.

Nikolaev, S., Kanbur, S. M., Hendry, M., & Tanvir, N., 2001 (in preparation).

Simon, N. R., Kanbur, S. M. & Mihalas, D. 1993, ApJ, 414, 310.

Stetson, P. 1996, PASP, 108, 851.

Udalski, A., Soszynski, I., Szymanski, M., Kubiak, M., Pietrzynski, G., Wozniak, P., & Zebrun, K. 1999, AcA, 49, 223.

Willick, J. A. & Batra, P. 2001, 548, 564.

Hα LUMINOSITY FUNCTION OF H II REGIONS: PRELIMINARY CALIBRATION OF A POWERFUL STANDARD CANDLE

Maitc Rozas

Observatorio de San Pedro Mártir, México and Instituto de Astrofísica de Canarias, Spain

Andrew Cardwell

Instituto de Astrofísica de Canarias, Spain and The Open University, United Kingdom

Almudena Zurita

Isaac Newton Group of Telescopes, La Palma, Spain and Instituto de Astrofísica de Canarias, Spain

and

John E. Beckman

Instituto de Astrofísica de Canarias, Spain and CSIC, Spain

Abstract

We propose the use of the H II region luminosity function (LF) as a refined powerful standard candle for extragalactic distances. We have found evidence of a change in the properties of H II regions in spirals at a fixed luminosity in Hα which we term the Strömgren luminosity ($L_{H\alpha}$ = 3.39 10^{38} erg s^{-1}). The key piece of evidence is the presence in the Hα H II region LF of a change in gradient and a local sharp peak. The invariance and high luminosity of the change measured in the LF was proposed in previous studies as a possible secondary standard candle, but it is calibrated here for the first time using the LFs of NGC 925, NGC 4535 and M100, whose Cepheid distances have been determined in the HST Extragalactic Distance Scale Key Project. Using this method we derive a value of H_0 of 73±7 km s^{-1} Mpc^{-1}.

M. Chávez et al. (eds.),
New Quests in Stellar Astrophysics: The Link Between Stars and Cosmology, 147–150.

Introduction

Our studies of the Hα H II region luminosity function (LF) of galaxies have revealed the presence of an invariant high-luminosity feature in the LFs of a number of galaxies NGC 6814 (Knapen et al. 1993), NGC 157, NGC 3631, NGC 6764, NGC 6951 (Rozas et al 1996), NGC 7479 (Rozas et al. 1999) & NGC 3359 (Rozas et al. 2000). To obtain a preliminary calibration of the luminosity at which this feature occurs we have constructed LFs for NGC 925, NGC 4535 and M100 for which Cepheid distances are available from the HST Extragalactic Distance Scale Key Project. Once fully calibrated this method will provide a useful extension to the extragalactic distance scale for spiral galaxies. While Cepheid variables and Planetary Nebulae Luminosity Functions can only give distances out to ~20 Mpc due to limitations of resolution and signal to noise, this method should be capable of giving accurate distances out to ~100 Mpc from ground based observations taken under good (sub-arcsecond) seeing conditions, and considerably further using space based observations.

1. The Strömgren luminosity and the LF peak

The calibration galaxies were observed with the TAURUS camera on the 4.2m William Herschel Telescope (NGC 925 & M100) and with the 2.56m Nordic Optical Telescope (NGC 4535). Good seeing (~ 1 arcsec) and photometric conditions ensured images of excellent quality. The luminosity functions all show a change in gradient accompanied by a peak at a luminosity close to $L_{H\alpha}$= 3.39 10^{38} erg s^{-1} (see Fig. 1). In Beckman et al. (2000) we term this the Strömgren luminosity, L_{Str}, and suggest that this feature is due to the transition between ionization bounding and density bounding of the H II regions.

2. First calibration

Using the mean L_{Str} from the Cepheid calibrated galaxies we recomputed the distances to all the galaxies we have studied, using the changes in slope or glitches found in their LFs. To estimate H_0 from heliocentric redshifts one can either correct for Galactic rotation and Local Group centered motion, or go directly to the cosmic microwave background (CMB) rest-frame. We used the latter approach, computing cz/d for each object based on the CMB kinematic data from Fixsen et al. (1996) (see Zurita 2001). We derive a mean value of H_0=73±7 km s^{-1} Mpc^{-1}.

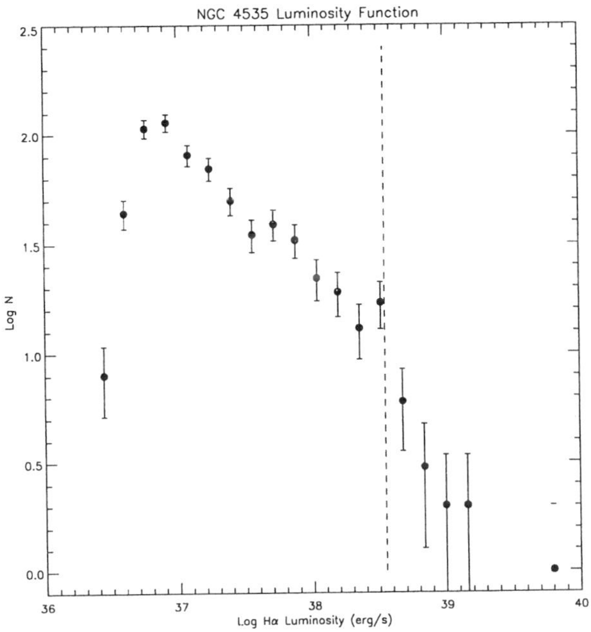

Figure 1. The $H\alpha$ H II region luminosity function of NGC 4535.

References

Ayres, T.R. & Testerman, L. 1981, ApJ, 245, 1124.

Beckman J. E., Rozas M., Zurita A., & Knapen J.H. 2000, AJ, 119, 6.

Fixsen D. J., Cheng E. S., Gales J. M., Mather J. C., Shafer R. A., & Wright, E. L. 1993, ApJ, 473, 576.

Knapen J.H., Arnth-Jensen N., Cepa J., & Beckman J. E. 1993, AJ, 106, 56.

Rozas M., Beckman J.E., & Knapen J.H. 1996, A&A, 307, 735.

Rozas M., Zurita A., Heller C.H., & Beckman J.E. 1999, A&ApS, 135, 145.

Rozas M., Zurita A., & Beckman J.E. 2000, A&Ap, 354, 823.

Zurita A. 2001, PhD thesis, Univ. La Laguna.

ON THE BEHAVIOR OF THE IONIZED GAS IN NGC 4449

Margarita Valdez–Gutiérrez
INAOE, México

Margarita Rosado
IA–UNAM, México

and

Ivânio Puerari
INAOE, México

Abstract In this contribution we report some preliminary results derived from the global content of ionized gas in the nearby irregular galaxy NGC 4449.

We have performed an Hα kinematical analysis of the ionized gas in NGC 4449 (Valdez–Gutiérrez et al., in preparation) as part of a long term project focused on the study of the interrelationship between gas and stars in irregular galaxies (Valdez–Gutiérrez et al. 2001; Borissova et al. 2000; Rosado et al. 2001). The kinematics of NGC 4449, a nearby irregular Magellanic type galaxy, is highly peculiar and so, worth to study: besides displaying numerous HII region complexes, superbubbles, a faint Hα filament population and a diffuse emission that embraces almost all the optical body, it has a HI velocity gradient opposite to the velocity gradient of the ionized gas in the inner disk (Hunter et al. 1998). The observations were carried out with the UNAM Scanning Fabry-Perot Interferometer PUMA attached to the f/7.9 Ritchey-Chrètien focus of the 2.1 m telescope at the Observatorio Astronómico Nacional at San Pedro Mártir, B.C., México. Details of this kind of observations and data reductions are described in Valdez–Gutiérrez et al. (2001). The field of view covered by our observations was $10'$ in diameter, embracing the totality of the optical extension of NGC 4449.

M. Chávez et al. (eds.),
New Quests in Stellar Astrophysics: The Link Between Stars and Cosmology, 151–152.

Our preliminary results show that, on global scales (i.e. galactic ones), the radial velocity field looks well behaved: we detected Hα emission ranging from 130 to 350 km s^{-1} (heliocentric velocities). At first glance a decreasing velocity gradient going from NE to SW along the optical bar can be seen, suggesting solid body rotation (Valdez-Gutiérrez et al. 2000). Such gradient agrees with Sabbadin et al. (1984) report. On smaller scales, the HII region population was kinematically analyzed in order to characterize the local behavior of the ionized gas. In this form we could discern a complex behavior compared to the global scales. The HII region radial velocities show a range going from 181 to 254 km s^{-1}. However, there were some regions where we found line splitting in the profiles and so, some extreme velocities appeared (188 to 352 km s^{-1}). The identification of those locations coincides with filaments in projection. To emphasize even more the great activity of the ionized gas in NGC 4449, the velocity dispersions measured on the whole sample (83 regions) have a mean value of 22.8 km s^{-1}, being supersonic in all but one region. On the other hand, to characterize the size of the HII regions we got the diameter distribution of this population. The characteristic diameter, D_0= 1.97$''$ or 35.4$\pm$1.7 pc at the adopted distance, is in agreement with previous values (Fuentes-Masip et al. 2000). With all this information concerning the HII regions, the rotation curve was calculated. Our preliminary results point towards a low degree of rotation in this galaxy. The implications of our preliminary findings in NGC 4449 will be published elsewhere (Valdez–Gutiérrez et al., in preparation).

Acknowledgments

MVG thanks support from a CONACyT scholarship under register number 114735. This work was partially supported by the grants IN122298 of DGAPA-UNAM and 27984-E of CONACYT.

References

Borissova, J., et al. 2000, Rev. Mex. Astron. Astrof., 36, 151.

Fuentes-Masip, O., Castañeda, H. O., & Muñoz-Tuñón, C. 2000, AJ, 119, 2166

Hunter, D. A., Wilcots, E. M., van Woerden, H., Gallagher, J. S., & Kohle, S. 1998, ApJ, 495, L47.

Rosado, M., Valdez-Gutiérrez, M., Georgiev, L., Arias, L., Borissova, J., & Kurtev, R. 2001, AJ, 122, 194.

Sabbadin, F., Ortolani, S., Bianchini, A. 1984, A&A, 131, 1.

Valdez–Gutiérrez, M., Rosado, M., Georgiev, L., Lacey, C., Borissova, J., & Kurtev, R. 2000, in ASP Conf. Ser. 195, Imaging the Universe in Three Dimensions, eds. W. van Breugel & J. Bland-Hawthorn (San Francisco: ASP), p. 346.

Valdez–Gutiérrez, M., Rosado, M., Georgiev, L., Borissova, J., & Kurtev, R. 2001, A&A, 366, 35.

NGC 5044-N50: A LINK BETWEEN BLUE COMPACT GALAXIES AND DWARF ELLIPTICALS*

Sergio A. Cellone

Facultad de Ciencias Astronómicas y Geofísicas, Universidad Nacional de La Plata, Argentina

and

Alberto Buzzoni

Telescopio Nazionale Galileo, La Palma, Spain, and Osservatorio di Brera, Italy

Abstract

We present new optical observations of the dwarf galaxy N50 in the NGC 5044 Group, showing that this object is probably at an intermediate $BCD \to dE$ evolutionary stage, after a relatively recent burst of star formation.

The possible connection between dwarf elliptical (*dE*), dwarf irregular (*dI*), and blue compact dwarf (*BCD*) galaxies is matter of sustained interest because of its implications in studies of galaxy evolution at low redshift (Doublier et al. 1999).

In this regard, we present here new optical observations of the dwarf galaxy N50, classified as *dEpec,N/BCD ring* in the NGC 5044 Group catalogue of Ferguson & Sandage (1990). While its integrated colors are consistent with a fully normal *dE*, its surface brightness profile shows, on the contrary, some peculiarity compared with ellipticals of similar absolute luminosity (Cellone 1999).

Direct imaging was obtained with the ESO 3.6 m telescope at La Silla (Chile), under sub-arcsec seeing conditions, in the Gunn g, r, i, z system. A ring of bright knots appears surrounding the nucleus, out to a $3''.5$ ($\sim$ 400 h_0^{-1} pc) radius (Fig. 1, *left*).

Fig. 1 (*right*) compares N50 integrated colors (big open dot) with the locus of main sequence stars ("$\star$" markers), and with Buzzoni (1998) 15 Gyr galaxy

*Based on observations taken at ESO Observatory, La Silla (Chile)

M. Chávez et al. (eds.),
New Quests in Stellar Astrophysics: The Link Between Stars and Cosmology, 153–154.

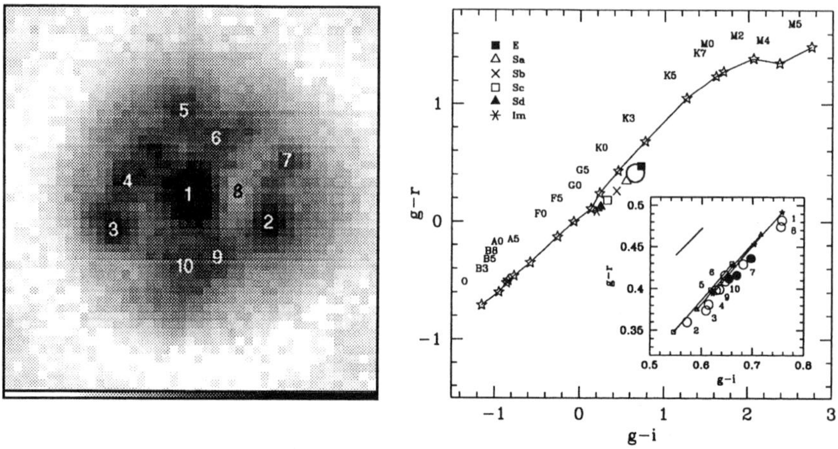

Figure 1. *Left*: Central 15″ × 15″ (1″ $\simeq 110\,h_0^{-1}$ pc) g band image of N50 (after subtraction of main component), with knotty structures numbered. *Right*: Results of Gunn system aperture and selected knots photometry.

models for different Hubble types. N50 integrated colors [i.e. $(g - r) = 0.41$ and $(g - i) = 0.65$] appear to be intermediate between those of E and Sa galaxies, with the main bulk of stars consisting of a 10-15 Gyr old population with slightly sub-solar metallicity ($[Fe/H] \simeq -0.2$). Aperture photometry (solid dots, *insert panel*) and bright knots colors (open dots) match the Buzzoni (1989) simple stellar population models. Ages $\sim$ 5 Gyr are inferred for the bright knots, while a mild radial metallicity gradient is also evident.
This composite stellar population is indeed affecting the morphology of N50 with no significant perturbation of its spectral energy distribution. While this galaxy did likely share much of the *nE BCD* features, like e.g. in Mrk 996 (Thuan et al. 1985), its current photometric properties rather point to a later evolutionary scenario with the galaxy fading to become a (nucleated?) *dE*.

References

Buzzoni, A. 1989, ApJS, 71, 817.

Buzzoni, A. 1998, in IAU Symp. 183, Cosmological Parameters and the Evolution of the Universe, ed. K. Sato (Dordrecht: Kluwer), p. 134.

Cellone, S. A. 1999, A&A, 345, 403.

Doublier, V., Caulet, A. & Comte, G. 1999, A&AS, 138, 213.

Ferguson, H., & Sandage, A. 1990, AJ, 100, 1.

Thuan, T. X., Izotov, Y. I., & Lipovetsky, V. A. 1996, ApJ, 463, 120.

STAR FORMATION IN RING GALAXIES

Y. D. Mayya, and R. Romano
Instituto Nacional de Astrofísica, Optica y Electrónica, Puebla, Mexico

Abstract We present an analysis of the star formation rate (SFR) in ring galaxies using Hα imaging. Global SFRs calculated from Hα and IRAS data are moderate, but the median Hα+[NII] equivalent width is higher compared to that of normal galaxies, suggesting enhancement in SFRs in ring galaxies.

Introduction

Ring galaxies are believed to have formed as a result of on-axis collisions between an intruder galaxy and a gas-rich disk-galaxy. Such a collision sets off a stellar density wave in the disk of the gas-rich galaxy (Lynds & Toomre 1976). Compression of the gas by the density wave is expected to enhance the level of star formation in ring galaxies. Appleton & Struck-Marcell (1987) addressed this question using the IRAS data for a sample of 26 ring galaxies. They found evidence for enhanced dust temperature, which is an indicator of an enhanced SFR.

Hα photometry is one of the best techniques to address the question of massive star formation. Marston & Appleton (1995) carried out Hα imaging observations of eight ring galaxies and indeed confirmed that the majority of the star forming regions are located at the position of the ring. The same was found to be true in the prototypical ring galaxy, the Cartwheel (Higdon 1995). However the question of enhancement of star formation in ring galaxies remains unresolved.

1. Observations and Results

Ring galaxies in the northern sky from a sample of Appleton & Struck-Marcell (1987) were selected for an Hα photometric study using the 2.1-m telescope at the Guillermo Haro Astrophysical Observatory, Cananea, México. Hα filters corresponding to the red-shifted wavelength of the galaxies were used. Images were also taken in the R-band to facilitate subtraction of the in-band continuum from the Hα-filter images. Throughout this analysis we assume $H_0 = 75\,\mathrm{km\,s^{-1}\,Mpc^{-1}}$.

M. Chávez et al. (eds.),
New Quests in Stellar Astrophysics: The Link Between Stars and Cosmology, 155–156.

Table 1. Derived Properties of the Sample Galaxies.

Galaxy	log f(Hα+[NII]) erg cm^{-2} s^{-1}	EW(Hα+[NII]) Å	SFR(Hα) M$\odot$ yr^{-1}	SFR(FIR) M$\odot$ yr^{-1}
Arp 291	−12.47	76	0.11	0.05
NGC 2793	−12.38	79	0.27	0.16
Arp 141	−13.01	28	0.22	0.30
Arp 143	−12.21	57	2.66	3.74
Arp 142	−12.47	51	4.20	7.82
I Zw 45	−12.43	142	6.88	8.38
II Zw 28	−12.81	156	3.10	2.65
II Hz 4	−13.10	65	3.58	6.07
VII Zw 466	−13.06	86	4.93	7.47

We carried out synthetic aperture photometry on the Hα+[NII] and R-band images to obtain total fluxes of each of the nine ring galaxies. The resulting values are given in the second column of Table 1. Hα+[NII] equivalent widths (EWs) are given in column 3. We used the relations from Kennicutt (1998) to derive SFRs from Far Infrared (FIR) and Hα luminosities, the values are given in the last two columns. Current SFRs of our sample galaxies are moderate and fall well below the rates experienced by typical starburst galaxies. However moderate SFRs can still imply enhancements of star formation if the past SFRs were very low. EW(Hα+[NII]) measures the current SFR with respect to the past SFR and should be used to infer enhancements of SFR. EW(Hα+[NII]) had been successfully used by Kennicutt et al. (1987) to infer enhanced SFRs in interacting pairs of galaxies. They found a median value of 46 Å for the interacting galaxies drawn from Arp (1966) catalog, while the corresponding value was 21 Å in a sample of normal galaxies. The median value for our sample of galaxies is 76 Å with only one galaxy having EW<46 Å. Hence our sample of ring galaxies show enhancements of SFRs, the enhancements being even higher than in the Arp pairs.

Acknowledgments

This work was partly supported by CONACyT research grant 211290-5-25869E.

References

Appleton P. N., & Struck-Marcell, C. 1987, ApJ, 312, 566.

Arp, H., 1966, ApJS, 14, 1.

Higdon, J. L., 1995, ApJ, 455, 524.

Kennicutt, R. C., 1998, ARA&A, 36, 189.

Kennicutt, R. C., Roettiger, K. A., Keel, W. C., van der Hulst, J. M., & Hummel, E. 1987, AJ, 93, 1011.

Lynds, R., & Toomre, A., 1976, ApJ, 209, 382.

Marston, A. P., & Appleton, P. N., 1995, AJ, 109, 1002.

V

POPULATION SYNTHESIS AND GALAXY MODELS

RECENT STAR FORMATION IN GALAXIES

Alessandro Bressan
Osservatorio Astronomico di Padova, Padova, Italy

Abstract In recent years the study of starburst galaxies has become a very popular subject because of their intimate connection with the global star formation history of the Universe. Current estimates of the star formation rate of the Universe have been interpreted on the basis of our understanding of local analogous galaxies, in particular through UV continuum and optical line emission. On the mean time, recent observations in the IR/mm have revealed that dusty star burst galaxies constitute a major cosmological component during the past epochs of the universe. In spite of the short duration, the transient IR- active phases of these luminous and ultra luminous IR galaxies (LIRGs and ULIRGs) are responsible for a large fraction of the energy emission and metal production at high redshifts. On one side, because of their transient nature, it is of paramount importance to delineate their recent star formation history with the highest possible accuracy. On the other, these IR phases are characterized by a very complex distribution of stars and the dusty interstellar medium (ISM), and it has become clear that the evaluation of dust extinction and star formation is impossible based on optical-UV data only. We review here some of the methods adopted for reconstructing the recent star formation in such galaxies, focusing on two main aspects, namely the degeneracy induced by age-selective absorption in the UV-optical-near infrared wavelength range and the possibility of using the far-infrared to radio correlation to investigate on the early stages of the star-burst phenomenon.

1. Age selective extinction and the degeneracy of optical spectra

Star burst galaxies are characterized by a substantially enhanced ongoing star formation with respect to the past average. A measure of this phenomenon is provided by the time-scale needed to exhaust the observed gas mass under the current star formation rate. While in normal spirals this time-scale turns out to be of the order of the Hubble time, in star burst galaxies it is much shorter and possibly as short as a few tens of million years. Star burst galaxies were initially selected for the prominence of their optical emission lines, but now it is clear that this criterion excludes other very peculiar objects and possibly

M. Chávez et al. (eds.),
New Quests in Stellar Astrophysics: The Link Between Stars and Cosmology, 159–166.

limits our understanding to a small phase of their evolution. In fact, after the IRAS satellite, it become clear that star formation is also highly enhanced in very and ultra luminous infrared galaxies -VLIRGs- that are otherwise highly attenuated in the optical. Recent optical spectroscopic surveys (Wu et al. 1998, Poggianti and Wu 2000) reveal that the spectra of a large fraction of VLIRGs display a peculiar combination of spectral features in the optical: Hα in emission (EW $\simeq$ 60 Å), Hβ practically filled (EW $\simeq$ 0 Å), a strong Hδ line in absorption (EW $>$ 4 Å) and a moderate [OII] emission (EW $<$ 40 Å). The equivalent width of their Hδ line exceeds that of typical, quiescent spirals at low-z and their low [Hβ/Hα ratios are consistent with the emission line fluxes being highly extinguished by dust. Galaxies with this type of spectra were named "e(a)" galaxies and were found to be quite numerous in the cluster and field environments at $z = 0.4 - 0.5$ (Poggianti et al. 1999). The combination of moderate emission lines and strong Balmer absorption has been also detected in the spectra of high-z IR luminous ISO galaxies (e.g. Flores et al. 1999). A possible explanation for this unusual combination of spectral features is selective extinction. In recent years it has become more and more popular the concept that dust is age-selective, i.e. HII regions, where the emission lines and UV originate, are affected by more extinction compared to the older stellar population which are responsible for the optical continuum and Balmer absorption (Calzetti et al 1994). Silva et al. 1998, first modeled the effects of age selective extinction by introducing the *escape time*, namely the time by which young star clusters get rid of their parental molecular clouds. Granato et al. (2000) showed that this type of models were able to account for the peculiar attenuation law observed in UV selected star-burst galaxies (Calzetti et al 1994) without any particular assumption on the dust properties. Along this way, Poggianti, Bressan and Franceschini (2001) have recently investigated in some detail the optical spectra of VLIRGs, to constrain the recent history of star-formation and the dust extinction characteristic of various stellar populations. By examining different star formation patterns Poggianti et al. concluded that only a starburst, selective extinction scenario could explain the e(a) spectra. In this scenario the extinction is larger in the younger populations, a fact that mimics the progressive escape of young stars from their parental molecular clouds, as they age. Further evidences of the peculiarity of VLIRGs, as their offset from the relation between the FIR to UV light vs. the continuum UV slope or between the ratio of extinction-corrected Hα emission to the bolometric emission -L(Hα_0)/L_{Bol}- vs. the color excess derived from the Balmer lines and their anomalous, both designed by UV selected starbursts (Meurer et al. 2001) are discussed by Panuzzo et al. in this volume. It is clear that in such powerful local starbursts a significant fraction of the ongoing star formation may be hidden to optical-near infrared star formation estimators, as already noticed by Poggianti and Wu (2000). The observed FIR/V luminosity ratio can be

explained only by models where a significant fraction of the FIR luminosity originates in regions that are practically obscured at optical wavelengths.
In conclusion, the optical-near UV spectrum by itself, even complemented with the information on the far-IR flux, is not enough to identify univocal solutions. In particular, the degeneracy of the optical spectrum hinders any quantitative estimate of the duration of the burst and/or of the wavelength dependence of the extinction.

2. Infrared Emission

A significant step forward in the modeling of star forming galaxies has been made in recent years by Silva and collaborators. Their population synthesis code (GRASIL, Silva et al. 1998) was explicitly designed to perform population synthesis in presence of dust. In brief, the star formation history, the metal enrichment and the current gas fraction are provided by a chemical evolution code. Stars and dust are distributed either in a disk or in a bulge or both, with suitable scalelengths. The volume emissivity is computed by considering the light of young stars, absorbed by the molecular clouds, and the light of older populations which are already outside, both propagated through the diffuse dust component. For a given past star formation history, gas fraction and metallicity, one of the parameter that largely affect the FIR emission is the escape time t_{esc}. Silva et al (1998) and Granato et al (2000) have shown that the UV, optical and FIR properties of local spirals are well reproduced with a typical escape time, $t_{esc} \simeq 3$ Myr while in the case of starburst this characteristic time is much longer, $t_{esc} \geq 10$ Myr. Similar obscuration times seem to be required by the interpretation of the optical spectra of very luminous infrared galaxies (Poggianti et al. 2001). The current situation is summarized by Silva et al and Panuzzo et al in these proceedings, but we may outline here some important results of model computations. The FIR luminosity is a sensitive function of the ongoing SFR ($t < 10^7$yr). As the star- burst evolves, the FIR emission reaches the maximum intensity when about the 30% of the gas mass is transformed into stars, thereafter it declines more or less rapidly, following the decline of the SFR. Consequently, any estimate of the mass formed during the burst based on the FIR emission is affected by the unknown recent SFR history. As an example, two suitable different star-burst models, with age of 50 and 25 Myr respectively, may provide indistinguishable fits to M82, from the far UV to the sub millimetric range (Silva et al, these proceedings). Only at radio wavelengths the two models differ significantly because, as we will see below FIR and radio emission have different fading times.

3. Radio emission

A clear picture of radio emission from normal galaxies is still missing. On one side thermal emission is related to the number of H ionizing photons, Q(H) by $\frac{LT_\nu}{erg/s/Hz} \simeq \frac{Q(H)}{5.495\times10^{25}}(\frac{T_e}{10^4K})^{0.45}(\frac{\nu}{1.49GHz})^{-0.1}$. On the other little is known about the source of non-thermal emission which, in normal star forming galaxies may account for the 90% of the radio emission (Condon 1992). Observations indicate that FIR and radio emission are strongly correlated over a wide range of infrared luminosities. At 1.49 GHz (Sanders & Mirabel, 1996)

$$q = \log \frac{FIR/(3.75 \times 10^{12} W cm^{-2})}{L\nu(1.4GHz)/(W cm^{-2} Hz^{-1})} \simeq 2.35 \pm 0.2 \tag{1}$$

where $FIR = 1.26\ 10^{-14}[2.58\ S(60\mu m) + S(100\mu m)]$. This correlation suggests that non-thermal emission is related to the recent star formation, but, among the possible candidates, radio supernova remnants (RSN) may only account for about the 5% of the non- thermal radio emission and other sources such as radio SNae, pulsars and plerions provide a negligible contribution. The most likely mechanism is constituted by synchrotron emission from relativistic electrons accelerated into the shocked interstellar medium following core-collapse supernova explosions (CCSN). But the poor knowledge of the accelerating mechanism hinders any quantitative prediction of this phenomenon which can only be formulated in an empirical way (Condon 1992). The observed average luminosity per supernova event, can be obtained by considering the ratio between NT radio emission and CCSN rate in our Galaxy (eg. Condon and Yin 1990). By adopting a CCSN rate $\nu_{SN} \simeq 0.015$ (Turatto, private communication) and after converting $L_{0.4GHz} \simeq 6.1\times 10^{21} W Hz^{-1}$ (Berkhuijsen, 1984) to 1.49 GHz by assuming a radio slope of $\alpha = -0.8$, Bressan, Silva & Granato (in preparation) obtain

$$E^{NT}_{1.49} = \frac{L_{1.49GHz}/(10^{30} erg\ s^{-1}\ Hz^{-1})}{\nu_{SN}/yr^{-1}} \simeq 1.44 \tag{2}$$

It is still a matter of debate why the above correlations is unaffected by the dependence of the electrons lifetime and luminosity on the magnetic and radiation density fields, which may change by significant factors in the different environments. Lisenfeld, Volk and Xu (1996) showed that, in normal star forming galaxies, the existence of the FIR/Radio correlation requires the existence of a correlation between the radiation and magnetic energy density fields. However the radiation field changes dramatically in starburst galaxies so that there must be a significant fine tuning between the parameters regulating the intensity of the radiation field and the magnetic field, because they must scale in such a way that they give rise to the FIR/Radio correlation and, at the same time, they must prevent Inverse Compton to dominate over synchrotron losses. As a possible way out Bressan, Silva & Granato suggested that the FIR/Radio correlation naturally originates whenever synchrotron losses happen on timescales that are shorter than the fading time of the SN rate. Assuming that cosmic ray electrons are injected during the adiabatic phases of SN explosions ($t \leq 10^4$–10^5 yrs), ie in a characteristic timescale which is short compared to the star formation time scale, the synchrotron luminosity at an epoch T is given by

$$L_{NT} = \int_0^{min(T,\tau_s^{el})} \nu_{SN}(T-t) l_{NT}(t) dt \tag{3}$$

where $l_{NT}(t)$ is the NT luminosity of the injected electrons after a time t has elapsed, T is the age of the galaxy and τ_{el}^s is the lifetime of electrons against synchrotron losses and depends on the magnetic field intensity. In the case of normal galaxies τ_s^{el} may be as large as several 10^7yr but, because the star formation rate is almost constant over the last Gyr (by definition of normal spiral), the above integral becomes

$$L_{NT} = \nu_{SN} \int_0^{\tau_s^{el}} l_{NT}(t) dt = \nu_{SN} \int_0^{\tau_s^{el}} \frac{dE}{dt} dt = \nu_{SN} E^{el} \tag{4}$$

where we have also assumed that the spectral shape is not affected by environmental conditions. It thus seems justified to assume that the non thermal luminosity scales linearly with the supernova rate and the constant is nothing but the energy of the electrons injected per supernova (E^{el}). In the case of star burst galaxies the supernova rate cannot be considered constant over such a large timescale. However τ_{el}^s must be much shorter to avoid significant losses from Inverse Compton on the intense stellar radiation field, possibly τ_s^{el} <<1-0.1 Myr (Condon 1992). Thus we may still use the approximation 4 because we may consider ν_{Sn} almost constant over such a small timescale. In brief the non thermal radio luminosity of the galaxy is proportional to the integral of the synchrotron power over the electron lifetime and an increase of the former in a larger magnetic field is compensated by a shortening of the latter. This could explain why the FIR/Radio correlation is so robust. No fine tuning is necessary, apart from the requirement that the magnetic field be large in star burst galaxies, which is an independent observational fact. Bressan, Silva & Granato have thus adopted for the NT radio emission

$$\frac{L_{NT}(\nu)}{10^{30} erg\ s^{-1}\ Hz^{-1}} = [E_{1.49}^{SNR}(\frac{\nu}{1.49})^{-0.5} + E_{1.49}^{el}(\frac{\nu}{1.49})^{-\alpha}] \times \frac{\nu_{SN}}{yr^{-1}} \tag{5}$$

Because the NT contribution of SNRs is found to be $E_{1.49}^{SNR} \simeq 0.06$, then eq. 2 requires that $E_{1.49}^{el} \simeq 1.38$. Finally, in order to reproduce the observed average slope of the NT radio emission in normal spiral galaxies ($\simeq$0.8), $\alpha \simeq$0.9. An independent check comes from reproducing the observed ratio between FIR and radio emission, in normal spirals (eq. 1). We have evolved several models with GRASIL for different values of the critical parameters such as, the chemical evolution (different age, gas infall time scale and star formation efficiency), the escape time from molecular clouds (2 Myr and 3 Myr, Granato et al. 2000) and the scale length of the dust distribution (3 Kpc and 4 Kpc). The values of q turn out to be quite independent from the adopted parameters and cluster around $q = 2.35$, in excellent agreement with the observations. We emphasize that, in our models, the SFR depends linearly on the gas fraction (Schmidt law), while the FIR emission depends on the SFR, gas fraction and metallicity and the SN rate on the recent SFR. The consistency between FIR emission, radio emission and supernova rate, is thus remarkable and should be considered as a successful test of the model. We have also obtained the following calibrations between radio emission and star formation rate: SFR $= 2.810^{-28}\ F(8.4GHz)$, SFR $= 0.7510^{-28}\ F(1.49GHz)$. These relations hold only for the case of a constant SFR for more than 100 Myr and are in good agreement with the one obtained by Carilli (2001). It is worth noticing that by assuming a supernova rate of 0.023 -the value adopted by Condon (1992)- we would have obtained a SFR calibration higher by about the 30%.

4. The FIR/Radio in compact ULIRGs

Bressan, Silva & Granato have analyzed the FIR/Radio properties of a sample of compact ULIRGs (Condon et al. 1991), showing that a careful combination of FIR and Radio data may be sufficient to determine the recent star formation history. Figure 1 contrasts the ratio q against the slope $\alpha = \delta \log f_\nu / \delta \log \nu$ of the radio emission between 1.49 GHz and 8.44 Ghz. The evolution of the slope is caused by the variation of the dominant source of radio emission, from thermal during the first 3 to 4 Myr to non thermal when supernova explosions feed relativistic electrons into the galaxy magnetic field. At the same time the total radio power increases and the q ratio shows a significant decline. At the end of the evolution q rises again but this phase can be discarded because the model is no more IR luminous. The effects of free-free absorption are evident in the left panel, where we notice a trend with increasing q at decreasing slope (see also Condon et al 1991). At 8.4 GHz the free-free optical depth is only a few percent and radio emission is not more affected by free-free absorption. The above trend disappears and the quantity $q_{8.4GHz}$, which still shows a range of about 0.6 dex, is a genuine measure of the age of the compact starbursts. Since the slope is still affected by free-free absorption, adding further observations at 23 GHz would possibly provide a clean picture of the critical evolutionary parameters of the obscured starburst. At present we may conclude that compact ULIRGs are compatible with starbursts where about 10^{10} $M_\odot$ of gas are being consumed on short timescales with peak SFR ranging from 100 to 600 $M_\odot$/yr.

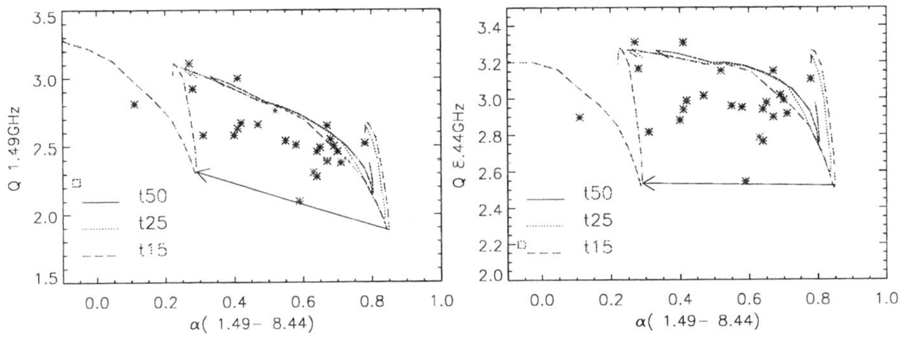

Figure 1. Left panel: evolution of the ratio q=log(FIR/3.75E12/F_ν(1.4GHz)) vs. radio slope ($d\ \log F_\nu / d\ log\nu$ between 8.4 and 1.4 GHz) in ULIRG models where about $7{\times}10^9 M\odot$ of gas are consumed in different e-folding SF timescales (50,25,15 Myr), compared with observed values in compact IRAS sources (Condon et al 1991). The model t15 has been recomputed accounting for free-free absorption with $\tau_{\nu=1.49GHz} = 1$. Characteristic starburst ages range from log t_b=6.5 on the left to log t_b=7.8 on the right side. Right panel: ν=8.4 GHz. Note that the observed q is not more affected by free-free absorption.

Some other implications on the use of the q ratio for the determination of the redshift of SCUBA sources are discussed by Silva et al. in these proceedings. Finally let us notice that the only object in our sample with clear signatures of a central AGN (see eg Soifer et al. 2000), -MKN 231, the open square- occupies a region which is distinctly different from that of the other compact ULIRGs. The value of $q_{1.4GHz}$ is not by itself outstanding but, when corrected for free-free absorption or equivalently by looking at the higher frequency data $q_{8.4GHz}$, it is at least three times more radio powerful (relative to the FIR) than all the other sources and than what allowed by our compact starburst models. Perhaps this is an indication that the parameters plotted in figure 1 (possibly making use of a higher frequency domain to avoid free-free contamination of the radio slope) could be used as a diagnostic diagram to evaluate the relevance of the central AGN.

Acknowledgments

This work is the result of a fruitful collaboration with A. Franceschini, G.L Granato, P. Panuzzo, B. Poggianti and L. Silva. We thank the *Osservatorio di Collurania* at Teramo (Italy), for its hospitality and financial support by the Italian Ministry for University and Research (MURST) under grant Cofin 9902763542_004.

References

Berkhuijsen, E.M. 1984, A&A, 140, 431.

Calzetti, D., Kinney, A. L., & Storchi-Bergmann, T. 1994, ApJ, 429, 582.

Condon, J.J. 1992, ARA&A, 30, 575.

Condon, J.J. & Yin, Q.F. 1990, ApJ, 357, 97.

Condon, J.J., Huang, Z.P., Yin, Q.F., & Thuan, T.X. 1991, ApJ, 378, 65.

Flores, H., et al. 1999, ApJ, 517, 148.

Granato, G.L., Lacey C. G., Silva L., Bressan A., Baugh C. M., Cole S., & Frenk C. S. 2000, ApJ, 542, 710.

Lisenfeld, U., Voelk, H. J., & Xu, C. 1996, A&A, 314, 745.

Meurer, G. K. Heckman, T. M., Seibert, M., Goldader, J., Calzetti, D., Sanders, D., & Steidel, C. C., 2000, in Cold Gas and Dust at High Redshift, ed. D. J. Wilner, in press (astro-ph/0011201).

Poggianti, B. M., & Wu, H. 2000, ApJ, 529, 157.

Poggianti, B. M., Bressan, A., & Franceschini, A. 2001, ApJ, 550, 195.

Silva, L., Granato, G.L., Bressan, A., & Danese, L. 1998, ApJ, 509, 103.

Sanders, D.B. & Mirabel, I. F. 1996, ARA&A, 34, 749.

Soifer, B.T., et al. 2000, AJ, 119, 509.

Thean, A., Pedlar, A., Kukula, M.J., Baum, S.A., & O'Dea, C.P., 2000, MNRAS, 314, 573.

Wu, H., Zou, Z. L., Xia, X. Y., & Deng, Z. G., 1998b, A&AS, 132, 181.

EXTINCTION AND STAR FORMATION HISTORIES OF NEARBY STARBURST GALAXIES

Y. D. Mayya
Instituto Nacional de Astrofísica, Optica y Electrónica, Tonantzintla, Puebla 72840, México

A. Bressan
Osservatorio Astronomico di Padova, vicolo dell'Oseervatorio 5, 35122, Padova, Italy

M. Rodríguez, and J. R. Valdes
Instituto Nacional de Astrofísica, Optica y Electrónica, Tonantzintla, Puebla 72840, México

Abstract Optical spectroscopic observations of a sample of M 82 type starburst galaxies were carried out and analyzed to infer extinction and star formation histories in them. The majority of the starburst regions in the sample are found to have more than three episodes of star formation over the past 1 Gyr. Younger populations show a trend to have higher reddening than those formed just before them.

Introduction

Starburst galaxies are known to form stars at copious amounts at present, but little is known about their past star formation events, in spite of growing observational data. Even for the proto-type starburst M 82, we still neither know the detailed star formation history (SFH), nor the duration of the present starburst activity. This can be attributed to the following two facts: (1) the most recent burst often dominates the observed quantities, making them insensitive to the presence of relatively older stellar populations, and (2) there is a lack of starburst models that can predict a set of observables for a combination of stellar populations of widely differing ages. For example even in the most comprehensive study of a starburst by Rieke at al. (1993), multi-band data of M 82 were fitted with a model that included only two bursts, with the older burst of 30 Myr age; but there is already evidence for the existence of populations older than 60 Myr in the same region (Gallagher et al. 1999). This calls for models

M. Chávez et al. (eds.),
New Quests in Stellar Astrophysics: The Link Between Stars and Cosmology, 167–170.

that can predict a variety of observational quantities with a combination of more than two stellar populations.

Poggianti et al. (2001) have recently devised a more sophisticated method to analyze the SFH in a starburst region. The method relies on a simultaneous fit of optical continuum, hydrogen emission/absorption line strengths and far infrared (FIR) fluxes, and allows for age dependent extinction. It has been successfully applied to derive the SFH in a sample of very luminous IR galaxies showing Hδ in absorption, and Hα in emission. In this work, we investigate the SFH in a sample of less obscured nearby starburst galaxies using the models Poggianti et al. (2001).

1. Sample, Observations and Analysis

Galaxies for this study were chosen from a sample of M82 type starburst galaxies compiled by Devereux (1989). The sample galaxies are at distances between 15–40 Mpc. The FIR luminosities as estimated from IRAS fluxes lie between 10^{10}–10^{11} $L_\odot$. Spectra ($\approx$3500–6800 Å) of 9 galaxies in the sample were taken at the Guillermo Haro Astrophysical Observatory, Cananea, Mexico, during two observing runs in 1999 February and December. Grating with 150 lines/mm was used, resulting in a resolution of 10 Å. Slit widths were typically of 1.6″, centered on the starburst nucleus. Spectra were extracted over a region of 9″ along the spatial direction, which correspond to physical scales of 0.77 to 1.76 kpc in the sample galaxies. Multi-aperture photometric data from Prugniel & Heraudeau (1998) in the V-band were used to calculate FIR/V ratios within 9″ aperture of program galaxies.

2. The Starburst Model and Results

Following Poggianti et al. (2001), we adopted up to nine relevant stellar populations over the last 1 Gyr, superimposed to an old disk, to model our starburst galaxies. We used 10 continuum bands and four spectral lines (Hα, Hβ, Hδ and [O II]λ3727) for fitting purposes. In addition, the models were constrained to produce the observed ratio of FIR to V-band luminosity. The best-fit solution is a combination of ten populations with ages 1, 3, 8, 10, 50, 100, 300, 500, 1000 Myr and the old disk (constant SFR from 12 to 2 Gyr). Interstellar extinction is allowed to vary freely for each population. The stars of all populations are assumed to have a Salpeter-like Initial Mass Function (IMF), with masses between the limits 0.15 and 100 $M_\odot$.

Results of the fit for NGC 4194, the most intense starburst in the sample, are illustrated in Fig. 1. SFRs of all recent episodes of star formation are normalized to the SFR of the old disk (SFR_{disk}), which in turn is calculated as the average SFR of all stars older than 2 Gyr. The left axis and the histograms represent these relative SFRs. The color excess E(B-V) of the different episodes of star

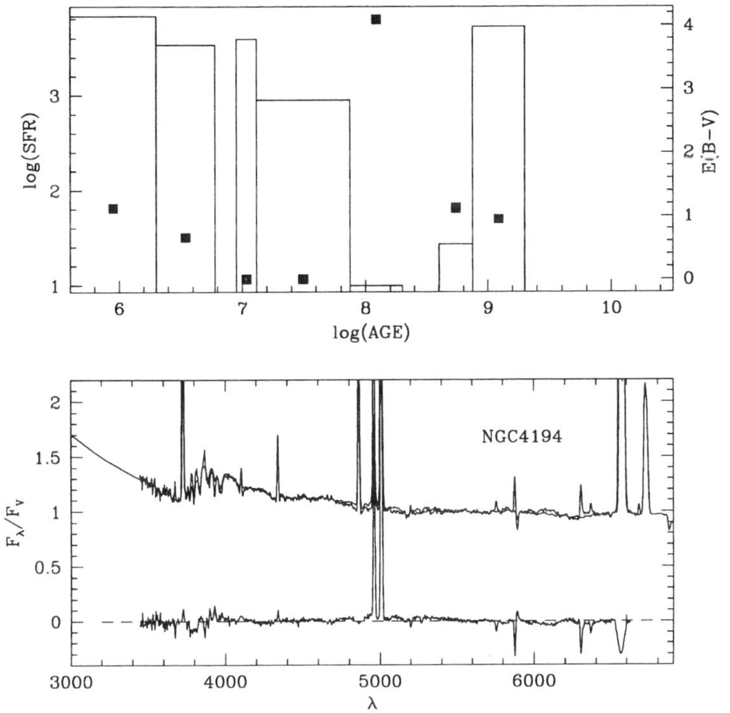

Figure 1. Comparison of the observed nuclear spectra of NGC 4194 with the model spectra. The residuals (obs−model) are shown below the spectrum. The inferred SFH is represented by the histogram in the panel above the spectral panel. The filled squares correspond to the color excess E(B-V) (numbers on the axis to the right) for each burst.

formation are represented on the right axis and by filled squares. The spectrum of this galaxy is blue and is dominated by emission lines. Hδ and higher lines of the Balmer series are seen in absorption, which is the case in majority of the galaxies in our sample. The best-fit solution to the spectrum suggests star formation going on almost continuously for the past 50 Myr. Note the systematic increase of extinction for younger bursts during this period. The nuclei also had an intense burst around 1 Gyr ago, that are responsible for the observed Balmer absorption lines.

Results for all the sample galaxies are presented in Tab. 1. Star formation rates derived using the FIR luminosities are given in Col. 2. Observed and model computed values of FIR/V are in Col. 3. Col. 4 contains the number of star forming episodes over the past 1 Gyr with SFRs above SFR_{disk}. Parameters (age, relative SFRs defined as $\psi = \log_{10}(SFR(t)/SFR_{disk})$ and color excess) for the three most important events are tabulated in the remaining columns. A

Table 1. Properties of three most important populations of sample galaxies

NGC	SFR $M_{\odot}$/yr	FIR/V Obs, Mod	N	Pop 1 Age, ψ, E_{B-V}	Pop 2 Age, ψ, E_{B-V}	Pop 3 Age, ψ, E_{B-V}
0470	1.9	12.4, 12.8	3	8E6, 0.85, 0.43	1E6, 0.84, 0.97	3E8, 0.36, 0.44
2273	1.5	5.9, 7.1	1	1E6, 0.13, 1.10	None	None
2782	3.4	9.1, 10.7	4	1E6, 0.76, 1.03	1E9, 0.71, 1.08	1E7, 0.21, 0.00
3504	4.1	11.2, 12.4	3	1E6, 0.52, 1.02	1E7, 0.36, 0.00	3E6, 0.15, 0.89
4102	3.9	18.4, 25.8	4	1E9, 2.07, 1.08	1E6, 1.97, 1.39	5E8, 0.21, 0.00
4194	9.2	17.4, 19.8	9	1E6, 3.83, 1.12	1E9, 3.72, 0.95	1E7, 3.6, 0.00
4273	4.2	14.5, 14.6	7	1E9, 0.60, 1.12	5E8, 0.56, 1.43	1E7, 0.45, 0.30
4818	2.5	> 3.2, 7.4	1	1E6, 0.10, 0.79	None	None
4984	1.4	4.3, 4.4	0	1E6, -0.15, 0.50	None	None

'None' entry in the table implies that the SFRs of the second and third most important populations are lower than the SFR_{disk}.

3. Summary

We have analyzed the SFH in a sample of starburst galaxies in the framework of a recent theoretical approach, particularly suited to the study of dusty starbursts. Six of the nine starburst nuclei that we studied, show evidence for at least three star formation episodes over the past 1 Gyr, with SFR of each episode well above the underlying disk SFR. In the remaining three galaxies, the current burst is the only significant event. Galaxies with multiple bursts, in general, are found to have higher current SFRs and FIR/V ratios, as compared to the single burst systems. A 1 Gyr old population is mainly responsible for noticeable absorption Balmer jumps that are observed in some of our starburst systems. Reddening seems to systematically decrease towards older bursts, over the past 100 Myr, as has been earlier reported by Poggianti et al. (2001) for dusty starburst galaxies. This suggests escape of older stars from the star-forming molecular clouds.

Acknowledgments

This work was partly supported by CONACyT research grant 211290-5-25869E.

References

Devereux, N. 1989, ApJ, 346, 126.

Gallagher, J. S., & Smith, L. S. 1999, MNRAS, 304, 540.

Poggianti, B.M, Bressan, A. & Franceschini, A. 2001, ApJ, 550, 195.

Prugniel, P., & Heraudeau, P. 1998, A&ApS, 128, 299.

Rieke, G. H., Loken, K., Rieke, M. J., & Tamblyn, P. 1993, ApJ, 412, 99.

DUST AND NEBULAR EMISSION IN STAR FORMING GALAXIES

Pasquale Panuzzo
S.I.S.S.A., Via Beirut 4, I-34014 Trieste, Italy

Alessandro Bressan, Gian Luigi Granato
Osservatorio Astronomico, Vicolo dell'Osservatorio 5, I-35122 Padova, Italy

Laura Silva
Osservatorio Astronomico, Via G. B. Tiepolo 11, I-34131 Trieste, Italy

and

Luigi Danese
S.I.S.S.A., Via Beirut 4, I-34014 Trieste, Italy

Abstract Star forming galaxies exhibit a variety of physical conditions, from quiescent normal spirals to the most powerful dusty starbursts. In order to study these complex systems, we need a suitable tool able to analyze the information available from observations at all wavelengths. We present a new spectro-photometric model which considers in a consistent way starlight as reprocessed by gas and dust. We discuss preliminary results of its application to a sample of local VLIRGs.

Introduction

Ongoing star formation in galaxies is traced by the presence of emission lines, produced by the ionized gas surrounding young star clusters. In the case of hydrogen (H) recombination lines the intensity is a direct measure of the number of massive stars still alive, hence their importance as tracers of the current star formation rate (SFR). These regions however, are often embedded into dusty environments, and the effects of dust may be significant, particularly in the most obscured galaxies such as those discovered by IRAS. It is thus essential to model the nebular emission within a context that consistently takes into account star-

M. Chávez et al. (eds.),
New Quests in Stellar Astrophysics: The Link Between Stars and Cosmology, 171–174.

light reprocessing by dust. To this purpose we present a new implementation of nebular emission modeling into our spectrophotometric code GRASIL Silva et al. (1998), which has been built to simulate the spectral energy distribution (SED) of galaxies in presence of dust, from the far ultraviolet (UV) to the far infrared (FIR). Besides an estimate of the SFR, evaluated simultaneously over a number of different indicators such as UV, line emission, FIR and Radio emission, the study of emission lines with GRASIL will provide us with many further constraints on the dust obscuration, because we could finally compare the optical thickness of star- forming regions derived from line ratios at different wavelengths from the optical to the near and far IR.

1. The model

Our galaxy model consists of two components with different geometrical parameters and star formation history: the bulge and the disk. Within the disk, dust is distributed in molecular clouds (MCs), with active star formation, and as a diffuse medium (the cirrus component). To account for the gradual escape of young stars from their parental MCs, we consider that all SSP younger than a given age (escape time) are attenuated by the MC component -A)-, while stars older than that, being already located outside MCs, do not suffer of such attenuation -B)-. Radiation transfer within the MC is solved by considering a thick spherical shell of dense gas (and dust) illuminated by a central point source. Finally the light emitted by MCs (from far UV to the FIR), as well as that of old stars outside the MCs, is propagated within the diffuse medium. More details on the model be found in Silva et al. (1998) or Silva et al., this conference. Emission lines are calculated in the hypothesis that HII regions are affected by the same kind of attenuation of young stars. Making use of the photoionization code CLOUDY 94 (Ferland 1996), we have first generated an extended library of HII regions (Table 1), spanning a wide range of ionizing SEDs, density, metallicity and radius. The SEDs of ionizing sources are supposed to encompass the variety of star formation scenarios possibly encountered. We have found that they may be suitably parameterized by the number of ionizing photons for HI, HeI and OII (Q(H), Q(He) and Q(O), respectively). Q(H) gives the mass of ionized gas, while the ratios Q(He)/Q(H) and Q(O)/Q(H) are related to the hardness of ionizing flux and to the degree of ionization. This allows a convenient generalization of our HII lines library to whatever kind of star formation history. At any time, the Qs are computed for each population (inside and outside the MCs) and then, after interpolation on the corresponding library of HII regions, nebular emission is attenuated in the same way as the ionizing sources.

Table 1. Lines computed in our HII region models.

Lyα 1216	Lyβ 1025	Lyγ 972	Lyδ 949	Ly 937	Ly 930
Ly 926	Ly 922	Hα6563	Hβ 4861	Hγ 4340	Hδ 4102
H 3970	H 3889	H 3835	H 3798	Paα 18752	Paβ 12819
Paγ 10939	Paδ 10050	Pa 9546	Pa 9229	Pa 9015	Pa 8863
Brα 40515	Brβ 26254	Brγ 21657	Brδ 19447	Br 18175	Br 17363
Br 16808	Br 16408	Pfα 74585	Pfβ 46529	Pfγ 37398	Pfδ 32964
Pf 30386	Pf 28724	Pf 27577	Pf 26746	Huα 123690	Huβ 75011
Huγ 59071	Huδ 51277	Hu 46716	Hu 43756	Hu 41700	Hu 40201
HeII 1640	HeII 1217	HeII 1085	HeII 4686	HeII 3203	HeII 2733
HeII 2511	HeI 4471	HeI 5876	HeI 6678	HeI 10830	HeI 3889
HeI 7065	[CI]9850	[CI]8727	[CI]4621	[CI]609μm	[CI]369μm
[CII]157.7μm	CII]2326	CIII]1908	[NI]5199	[NI]3466	[NI]10400
[NII]6584	[NII]6548	[NII]5755	[NII]122μm	[NII]205μm	NII]2141
[NIII]57μm	[OI]6300	[OI]6363	[OI]5577	[OI]63μm	[OI]145μm
[OII]3727	[OII]7325	[OII]2471	OIII]1663	[OIII]5007	[OIII]4959
[OIII]4363	[OIII]2321	[OIII]88μm	[OIII]52μm	[OIV]26μm	[NeII]13μm
[NeIII]15.5μm	[NeIII]36μm	[NeIII]3869	[NeIII]3967	[NeIII]3343	[NeIII]1815
[NeIV]2424	[NeIV]4720	MgII2800	[SiII]35μm	[SII]10330	[SII]6731
[SII]6717	[SII]4070	[SII]4078	[SIII]19	[SIII]33.5	[SIII]9532
[SIII]9069	[SIII]6312	[SIII]3722	[SIV]10.4μm	[ArII]69850	[ArIII]7135
[ArIII]7751	[ArIII]5192	[ArIII]3109	[ArIII]3005	[ArIII]22μm	[ArIII]9μm

2. Spectrophotometric properties of VLIRGs

Very Luminous InfraRed Galaxies (VLIRGs), with luminosities $L_{\rm IR} \geq 10^{11.5}$ $L_\odot$, were revealed in the IR as the most powerful star forming galaxies at low redshift, and are nowadays considered local examples of the high redshift population of active star forming galaxies. Analysing their properties is thus fundamental for our understanding of the galaxy evolution at high redshifts. In particular, recent observations show that VLIRGs do not follow the Meurer's relation $L_{\rm FIR}/L_{\rm FUV}$ versus UV spectral index β found for UV selected starbursts (Meurer et al. 1999, 2000). Furthermore their SFR computed from Hα, even corrected for the extinction derived from the Balmer decrement Hα/Hβ, is always less than the SFR derived from the FIR emission (Poggianti & Wu 2000). Our simulation of a starburst galaxy consists of a quiescent star formation lasting for 12 Gyr (baryonic mass $5 \cdot 10^{10} M_\odot$) and a final burst. At the beginning of the burst half of total ($1.2 \cdot 10^{10} M_\odot$) gas mass is supposed to be under molecular form, and a fraction (variable from 70% to 99%) of it is then converted into stars. As the molecular gas is consumed during the burst, the MCs become more and more transparent. This is to mimic the consumption of the gas and the feedback of SNe. The initial optical thickness of the MCs is chosen to reproduce the observed equivalent width of Hα+[NII] and Hβ (-62.5 Å and 0.7 Å) and the ratio $L_{\rm FIR}/L_{\rm V} = 88$ (Poggianti et al. 2001). The time

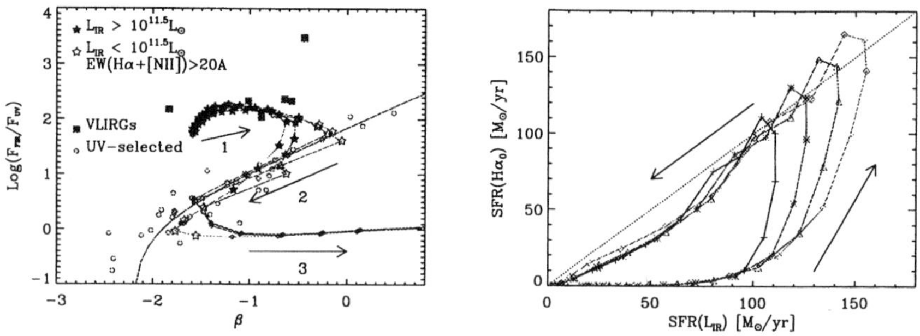

Figure 1. *Left:* Time evolution of the model in the β-FIR/FUV plane. ULIRG (solid squares) and UV-selected starburst observations (open circles) are from Meurer et al. 2000). Different paths correspond to different fraction of gas converted in stars. *Right:* SFR derived from Hα corrected for extinction with Balmer decrement versus SFR from FIR.

evolution of the model is shown in Fig. 1. In the β-FIR/FUV plane we can disentangle three different phases: 1) VLIRG phase: the burst is very obscured; IR emission comes from the burst while UV comes from the disk. There is no correlation between UV and IR because they come from different component of the galaxy. In this phase the Hα intensity, even corrected for extinction, is not a good estimate of the SFR, (right panel of Fig.1). In fact in this phase the emission can be dominated by the population outside MCs which is affected by a lower extinction. This effect can explain the low $L_{\mathrm{H}\alpha_0}/L_{Bol}$ observed in VLIRGs (Meurer & Seibert 2001) 2) UV-bright phase: the burst has almost consumed the gas and the MCs become transparent; the model moves along Meurer's relation. 3) Final phase: the burst is aging and the model departs from the relation, but the galaxy can no longer be selected as a starburst, because of its low Hα emission. In conclusion, we have presented a new spectrophotometric model able to interpret different UV, optical and IR properties of obscured galaxies within a unique star formation scenario, dominated by the effects of age-selective extinction. This constitute the first step toward a consistent picture of galaxy evolution at high redshift.

References

Ferland, G. J. 1996, Hazy, a brief introduction to CLOUDY, Univ. of Kentucky, Dept. of Physics & Astronomy Internal Report.

Meurer, G. R., Heckman, T. M., & Calzetti, D. 1999, ApJ, 521, 64.

Meurer, G. R. Heckman, T. M., Seibert, M., Goldader, J., Calzetti, D., Sanders, D., & Steidel, C. C. 2000, in Cold Gas and Dust at High Redshift, ed. D. J. Wilner, in press (astro-ph/0011201)

Meurer, G. R., & Seibert, M. 2001, in Starburst Galaxies Near and Far, eds. D. Lutz & L. Tacconi, (Heidelberg: Springer Verlag), in press (astro-ph/0101479).

Poggianti, B. M., Bressan, A., & Franceschini, A. 2001, ApJ, 550, 195.

Poggianti, B. M., & Wu, H. 2000, ApJ, 529, 157.

Silva, L., Granato, G. L., Bressan, A., & Danese, L., 1998, ApJ, 509, 103.

MODELING THE RADIO TO X-RAY SED OF GALAXIES

Laura Silva
Osservatorio Astronomico di Trieste and SISSA, Trieste, Italy

Alessandro Bressan, Gian Luigi Granato
Osservatorio Astronomico di Padova, Padova, Italy

and

Pasquale Panuzzo
SISSA, Trieste, Italy

Abstract We present our model to interpret the SED of galaxies. The model for the UV to sub-mm SED is already well established ?1998. We remind here its main features and show some applications. Recently we have extended the model to the radio range (Bressan et al. 2001), and we have started to include the X-ray emission from the stellar component.

1. Modeling the UV to sub-mm SED of galaxies

We have developed a model (GRASIL, *http://grana.pd.astro.it*) for the SED of galaxies including dust effects (Silva et al. 1998; Silva 1999). We consider a geometry more complex and realistic than previous treatments, and take into account 3 distinct environments where stellar radiation interacts with dust (Fig. 1): (1) **molecular clouds (MCs) + young stars** (clumped component), (2) **diffuse ISM (cirrus) + older stars** (diffuse component), (3) AGB envelopes (included in the SSP library, Bressan et al., 1998). Therefore both stars and dust have a *realistic 2-phase distribution*. The association between young stars and MCs is modeled to mimic the birth of stars in MCs and their gradual escape as they grow old: the fraction of each SSP energy radiated inside MCs is decreased with its age (Fig. 1). It is the first time that such an *age-dependent or selective extinction* is modeled (see Granato et al., 2000 for interesting applications).

M. Chávez et al. (eds.),
New Quests in Stellar Astrophysics: The Link Between Stars and Cosmology, 175–180.

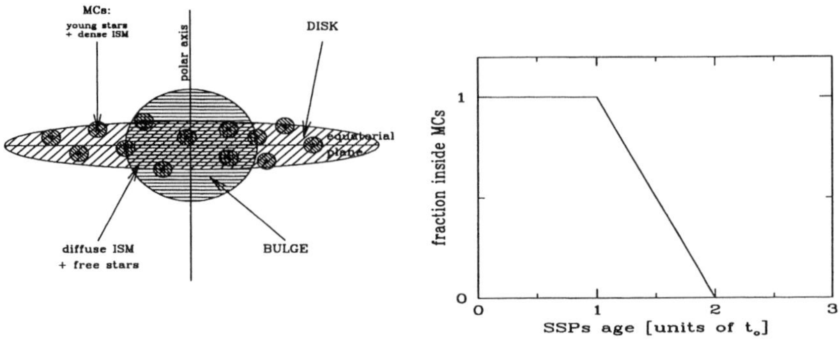

Figure 1. *Left*: Scheme and components of our model galaxy. *Right*: Fraction of SSP energy inside MCs vs SSP age.

The dust model consists of normal grains in thermal equilibrium, small grains fluctuating in temperature and PAH molecules.

2. Modeling radio emission in star-forming galaxies

We have extended our model to the radio. The thermal radio emission is proportional to the production rate of ionizing photons:

$$\frac{L_{Th}(\nu)}{erg/s/Hz} \simeq \frac{Q(H)}{5.495 \times 10^{25}} \left(\frac{T_e}{10^4\,K}\right)^{0.44} \left(\frac{\nu}{1.4GHz}\right)^{-0.1} \tag{1}$$

The non thermal radio emission is the sum of a dominating term, proportional to the supernova rate (calibrated on the Galactic value) and one due to supernova remnants (based on the $\Sigma-$D relation):

$$\frac{L_{NTh}(\nu)}{erg/s/Hz} = K \times \nu_{SN} \left(\frac{\nu}{1.4}\right)^{-\alpha} + L_{1.4}^{SNR} \left(\frac{\nu}{1.4}\right)^{-0.5} \tag{2}$$

See Bressan (these proceedings) and Bressan et al. (2001) for details.

3. Applications

3.1 Fitting the SEDs

As an example, the fit to the starburst galaxy M82 is shown in Fig. 2. Note that the inclusion of the radio emission potentially (with the adopted calibration) helps in constraining the starburst age, due to the different timescales involved in the IR (ongoing SFR, last few Myrs) and radio (recent SFR, a few 10^7 yrs ago) emission.

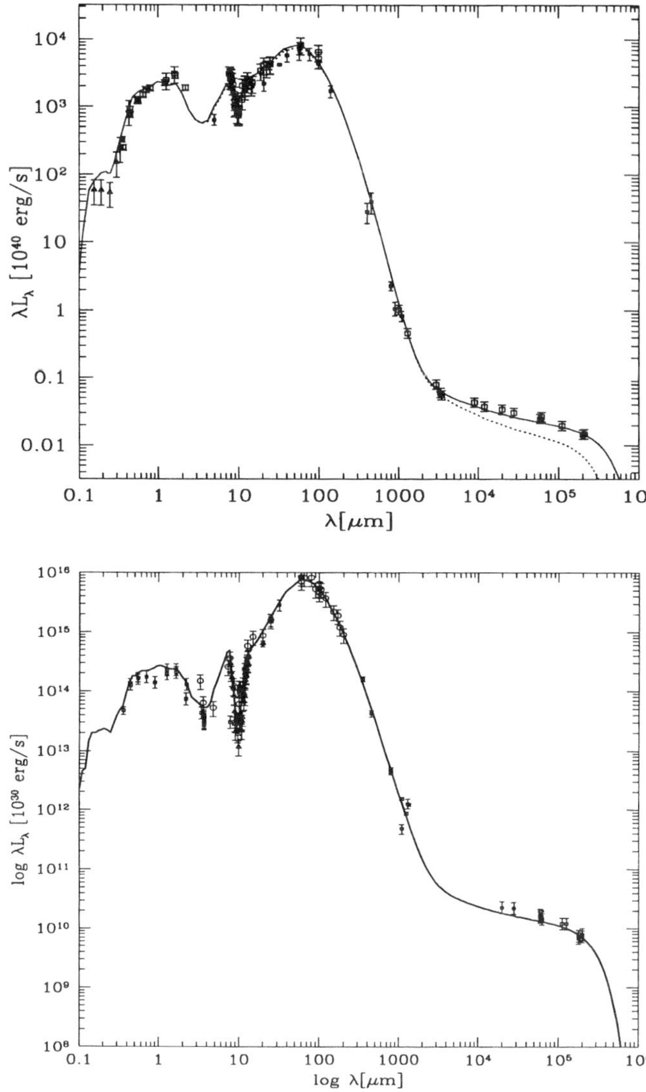

Figure 2. *Upper panel*: Fit to the SED of M82. The starburst age and e-folding time are respectively 25 Myr and 10 Myr (continuous line) and both 50 Myr (dotted line). An older starburst is compatible with the UV to sub-mm SED but not with the radio emission. A free-free absorption $\tau_{1.4\mathrm{GHz}} = 0.2$ has been applied. *Lower panel*: Fit to the SED of Arp220.

3.2 The redshift of sub-mm selected galaxies

The sub-mm to radio spectral index has been recently used to estimate the redshift of sub-mm selected galaxies (e.g. Carilli & Yun 2000). Exploiting

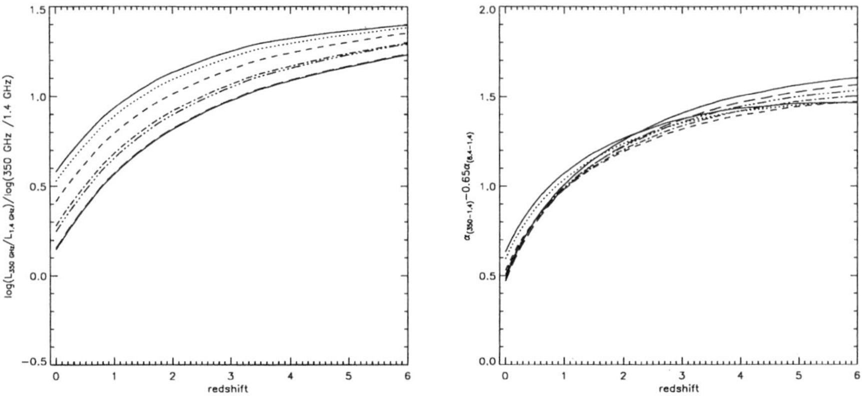

Figure 3. *Upper panel*: 850μm-1.4 GHz spectral index vs. z for the ARP220 model observed at different ages during the starburst. *Lower panel*: linear combination of the (850μm-1.4 GHz) and (8.4-1.4 GHz) spectral indexes vs. z for the same models.

our model it is possible to investigate the dependence of this index on the evolutionary status of galaxies. In Fig. 3 the 850μm-1.4 GHz spectral index is plotted vs redshift using the model fitting the SED of ARP220 (Fig. 2, this seems to be representative of the SCUBA galaxies, e.g. Barger et al., 1999), but for different ages of the starburst. The redshift dispersion around z=2 (the typical value estimated for SCUBA galaxies) is $\Delta z = 1.8$. To reduce this dispersion, we propose to use a radio-radio spectral index (e.g. 8.4-1.4 GHz), almost independent of z, in combination with the previous one. Indeed, with the linear combination shown Fig. 3, the Δz due to the evolution becomes 0.5 around z=2.

4. Stellar X-ray emission from galaxies

Recently we have started to include in our model the stellar contribution to the X-ray emission of galaxies. Our starting point has been the work by Van Bever & Vanbeveren (2000). They find that the dominating stellar components in the X-ray for starburst galaxies are massive binaries with a black hole or a neutron star as a primary, and an OB star as a secondary either in the H-shell burning phase or on the main sequence, and pulsars. We have included these components, and the contribution from supernova remnants, in the SSP library. The main uncertainties in the predicted output are (Fig. 4): the binary fraction, the minimum mass for BH formation and the initial spin period P_0 of pulsars (whose $L_X \propto P_0^{-2}$). A further uncertainty is the SED to assign to each component. A first attempt for the stellar X-ray SED expected from the UV-to-radio fitting model for M82, is shown in Fig. 4.

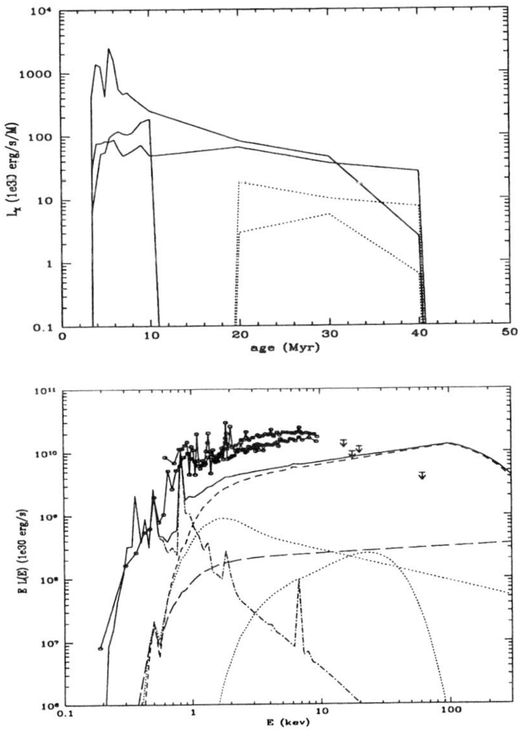

Figure 4. *Upper panel:* Predicted evolution with age of the contribution of different stellar components to the X-ray emission of Saps. The minimum mass for BH formation is $14M_\odot$ and $P_0 = 0.02$ s. *Lower panel*: Predicted stellar X-ray SED for M82 based on the UV-to-radio fitting model. Data from Moran & Lehnert (1997), Cappi et al. (1999).

References

Barger, A. J., Cowie, L. L., Smail, I., Ivison, R. J., Blain, A. W., & Kneib, J. P. 1999, AJ, 117, 2656.

Bressan, A., Granato, G. L., & Silva, L. 1998, A&A, 332, 135.

Bressan, A., Granato, G. L., & Silva, L. 2001, in preparation.

Cappi, M., Persic, M., Bassani, L., Franceschini, A., Hunt, L. K., Molendi, S. Palazzi, E., Palumbo, G. G. C., Rephaeli, Y., & Salucci, P. 1999, A&A, 350, 777.

Carilli, C. L., & Yun, M. S. 2000, ApJ, 530, 618.

Granato, G. L., Lacey, C. G., Silva, L., Bressan, A., Baugh, C. M., Cole, S., & Frenk, C. S. 2000, ApJ, 542, 710.

Moran, E. C., and Lehnert, M. D. 1997, ApJ, 478, 172.

Silva, L., Granato, G. L., Bressan, A., & Danese, L. 1998, ApJ, 509, 103.

Silva, L. 1999, PhD thesis, SISSA, Trieste, Italy.

Van Bever, J. & Vanbeveren, D. 2000, A&A, 358, 462.

POPULATION SYNTHESIS IN THE BLUE AND THE SPECTROSCOPIC AGE OF 47 TUCANAE

Ricardo P. Schiavon, S. M. Faber
UCO/Lick Observatory, University of California, Santa Cruz, CA 95064, USA

and

James A. Rose
Department of Physics and Astronomy, University of North Carolina, Chapel Hill, NC 27599, USA

Abstract We present the comparison of population synthesis predictions to the observations of the Galactic globular cluster (GC) 47 Tucanae. We show that the too large ages inferred for this and other GCs in previous work are due to a combination of uncertainties in the theoretical isochrones and stellar libraries currently in use in the literature.

Introduction

Population synthesis (PS) is the only method we have at our disposal to infer the mixture of stellar populations (SPs) in distant galaxies. The reliability of the method is crucially dependent on its ability to reproduce the observations of simple systems, like globular clusters, for a reasonable set of ages and metallicities. Recent work (Gibson et al. 1999; Vazdekis et al. 2001) pointed out important inconsistencies in SP models, in that exceedingly high ages (sometimes in excess of 20 Gyrs) are predicted for metal-rich galactic globular clusters. In this contribution, we present the results of a comparison of the integrated spectrum of the metal-rich Galactic halo globular cluster, 47 Tucanae, with predictions of our population synthesis models. In particular, we compute integrated synthetic spectra both from theoretical isochrones and from a color-magnitude diagram (CMD) of the cluster. In this way, we are able to separate effects due to uncertainties in the theoretical isochrones from the ones due to limitations of the stellar library.

M. Chávez et al. (eds.),
New Quests in Stellar Astrophysics: The Link Between Stars and Cosmology, 181–184.

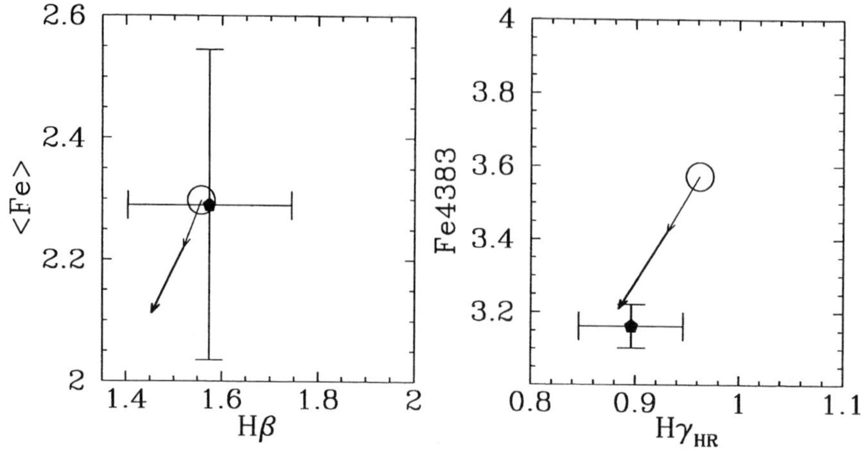

Figure 1. Comparison of CMD-based model predictions (large open circle) with the observed absorption lines indices. $< Fe >$ and $H\beta$ have larger error bars because Covino et al.'s spectrum has lower S/N than Leonardi (2001). The thin arrow indicates by how much the model predictions change if the $[Fe/H]$-scale of the giants is corrected by –0.1 dex. The thick arrow indicates by how much the model predictions change if $[Fe/H]_{47Tuc}$ is 0.1 dex lower than Carretta & Gratton (1987)'s $[Fe/H]$=–0.7. Therefore, correcting the predictions for the limitations of the spectral library bring them into agreement with the data if an uncertainty of 0.1 dex in the metallicity of 47 Tuc is considered.

1. Computations

Integrated model spectra are computed adopting either the deep HST color-magnitude diagram by Howell et al. (2000), which is complete down to 1 mag below the turn-off, or the isochrones presented by Vazdekis et al. (2001). The latter were computed for $[Fe/H] = -0.7$, $[\alpha/Fe] = +0.4$ and take He-diffusion into consideration. They do not include AGB stars. The moderately-high resolution (FWHM $\sim$ 1.8) spectral library by Jones (1996) is adopted. Further details on the computations, including the determination of photospheric parameters of the library stars and the conversion of the CMD to the (T_{eff},log g) plan are given in Schiavon et al. (2001). The observed spectra of 47 Tuc are taken from Leonardi (2001, in preparation) and Covino et al. (1995).

2. Results and Discussion

In this contribution, the comparison of model predictions with observations is focused on a few absorption line indices. A detailed discussion of colors, spectrophotometry and other indices is presented elsewhere (Schiavon et al. 2001). The indices adopted are defined by Worthey et al. (1994) and Jones & Worthey (1995). In Fig. 1 we compare the predictions of CMD-based models with the observations. It can be seen that both metal and Balmer lines are

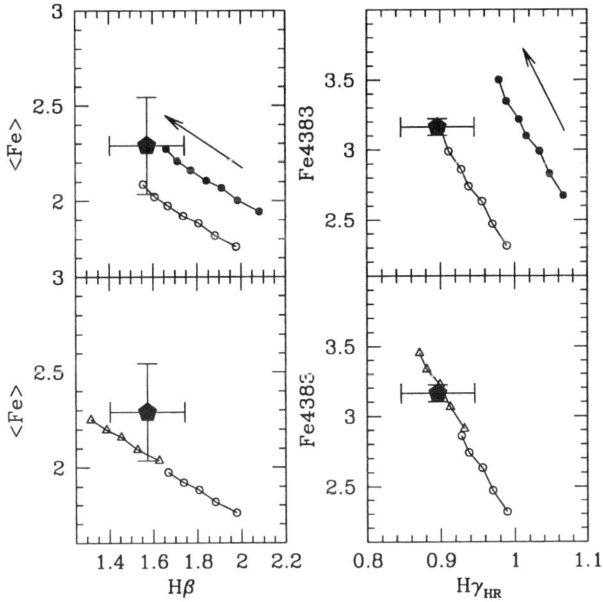

Figure 2. Comparison of isochrone-based model predictions with the observations. *Upper panels*: small filled circles are isochrone-based predictions for ages between 14 (upper left) and 8 Gyrs (lower right), in steps of 1 Gyr. Small open circles are these models corrected according to the arrows of Fig. 1. The arrow indicates how would these predictions change if we correct the theoretical value of $\log(n_{Giants})/\log(n_{Dwarfs})$ by +0.5 dex. *Lower panels*: The models from the upper panels for ages between 12 and 8 Gyrs are displayed. Small open triangles are these models corrected according to the arrows of the upper panels. In conclusion, if the isochrone-based models are corrected for the effects presented in Fig. 1 and for the luminosity function discrepancy (see text), they predict Balmer line spectroscopic ages in the range 8–11 Gyrs.

slightly over-predicted, specially in the blue region. These plots are very useful to investigate effects purely due to the spectral library, since no isochrone is involved in the computations. For instance, one limitation of the spectral library adopted is the paucity of low metallicity ($[Fe/H] < -0.6$) giants. We estimate the effect on line index predictions of correcting for this limitation. This estimate is computed with the aid of fitting functions by Worthey et al. (1994) and Concannon (2001, private communication). In Fig. 1 the thin arrow indicates that the agreement with the observations is improved when the correction is performed. Formal agreement between model and observations is achieved if we further assume that the metallicity of 47 Tuc is 0.05 – 0.1 dex below the value determined by Carretta & Gratton (1997) ([Fe/H]=–0.7).

In Fig. 2 we compare the isochrone-based model predictions to the observations. Vazdekis et al. (2001) have shown that considering the effects of He-diffusion and α-enhancement on the isochrones brings the spectroscopic ages of clusters to better agreement with the ones based on CMD-fitting techniques. In fact, as can be seen in the lower panels of Fig. 2, isochrone-based predictions are consistent with the $H\beta$ measurement for an age of 14 Gyrs. However, $H\gamma_{HR}$ is over-predicted for the same age. This is because the theoretical luminosity function (LF) underestimates the number of giants relative to turn-off stars. The latter is partly, but not entirely, due to the non-inclusion of AGB stars (see Schiavon et al. 2001 for details). A correction for this effect is shown as arrows in the lower panels. Interestingly, it runs almost perfectly parallel to the age sequence in the models.

Taking together the correction for the limitations in the spectral library and in the theoretical LF (Fig. 1 and 2), and assuming 47 Tuc has a metallicity slightly lower than the value usually adopted in the literature (yet inside the error bars), we conclude that the isochrone predictions are consistent with the observations for an age within 9–11 Gyrs, which agrees with recent age determinations based on CMD-fitting techniques (e.g., Salaris & Weiss 1998). The same results are achieved when other absorption lines, like Ca4227 and Hδ are adopted.

Acknowledgments

R.P.S. is a Gemini Fellow and acknowledges support provided by the National Science Foundation (NSF) through grant GF-1002-99 and from Association of Universities for Research in Astronomy, Inc., under NSF cooperative agreement AST 96-13615.

References

Carretta, E., & Gratton, R. G. 1997, A&AS, 121, 95.

Covino, S., Galletti, S., & Pasinetti, L. E. 1995, A&A, 303, 79.

Gibson, B. K., Madgwick, D. S., Jones, L. A., Da Costa, G. S., & Norris, J. E. 1999, AJ, 118, 1268.

Howell, J. H., Guhathakurta, P., & Gilliland, R. L. 2000, PASP, 112, 1200.

Jones, L.A. 1996, PhD Thesis, University of North Carolina.

Jones, L. A., & Worthey, G. 1995, ApJ, 446, L31.

Salaris, M., & Weiss, A. 1998, A&A, 335, 943.

Schiavon, R. P., Faber, S. M., Castilho, B. V., & Rose, J. A. 2001, ApJ, submitted.

Vazdekis, A., Salaris, M., Arimoto, N., & Rose, J. A. 2001, ApJ, 549, 274.

Worthey, G., Faber, S.M., González, J. J., & Burstein, D. 1994, ApJS, 94, 687.

ON THE PROPERTIES OF POPULATION III STARS AND "GALAXIES"

Daniel Schaerer
Observatoire Midi-Pyrénées, Laboratoire d'Astrophysique, UMR 5572,
14, Av. E. Belin, F-31400 Toulouse, France

Abstract

We present new model calculations for massive Population III stars and metal-free stellar populations. The properties of individual stars and integrated populations, including their ionizing fluxes, SEDs, and baryon production are discussed. We also study their dependence on the poorly known IMF and the star formation history. A detailed account is given in Schaerer (2001).

Introduction

Important advances have been made in recent years on the modeling of the first stars and galaxies forming out of pristine matter — so called Pop III objects — in the early Universe (cf. review of Loeb & Barkana 2001, proceedings of Weiss et al. 2000 and Umemura & Susa 2001, and other contributions in this meeting).

Tumlinson & Shull (2000, hereafter TS00) have recently pointed out the exceptionally strong He^+ ionizing flux of massive ($M \gtrsim 40$ $M_\odot$) Pop III stars, which must be a natural consequence of their compactness, i.e. high effective temperatures, and non-LTE effects in their atmospheres increasing the flux in the ionizing continua. As a consequence strong He II recombination lines such as He II λ1640 or He II λ4686 are expected; together with AGN a rather unique feature of metal-free stellar populations, as discussed by TS00 and Tumlinson et al. (2001, hereafter TGS01). Instead of assuming a "standard" Population I like Salpeter IMF up to 100 $M_\odot$ as TS00, Bromm et al. (2001, hereafter BKL01) have considered stars with masses larger than 300 $M_\odot$, which may form according to some recent hydrodynamical models. An even stronger ionizing flux and stronger H and He II emission lines was found.

Some strong simplifying assumptions are, however, made in the calculations of TS00, TGS01, and BKL01.

M. Chávez et al. (eds.),
New Quests in Stellar Astrophysics: The Link Between Stars and Cosmology, 185–188.

1) All stars are assumed to be on the zero age main sequence, i.e. stellar evolution proceeding generally to cooler temperatures is neglected

2) BKL01 take the hardness of the stellar spectrum of the hottest (1000 $M_\odot$) star as representative for all stars with masses down to 300 $M_\odot$. This leads in particular to an overestimate of the He II recombination line luminosities.

3) None of the studies include nebular continuous emission, which cannot be neglected for metal-poor objects with such strong ionizing fluxes. This process increases significantly the total continuum flux at wavelengths redward of Lyman-α and leads in turn to reduced emission line equivalent widths.

4) A single, fixed, IMF is considered only by both groups. In view of the uncertainties on this quantity it appears useful to explore a wider range of IMFs.

We here present results from new calculations which relax all the above assumptions and allow the exploration of a wide parameter space in terms of stellar tracks (including also alternate tracks with strong mass loss), IMF, and star formation history (instantaneous bursts or constant SFR). A more detailed account is given in Schaerer (2001).

1. Input physics and results for individual stars

To account for non-LTE effects, which are crucial for a proper description of the spectra of hot stars considered here we use the *TLUSTY* code of Hubeny & Lanz (1995). Several additional models were calculated with the *CMFGEN* code of Hillier & Miller (1998) to explore the possible importance of mass loss (as well known for Pop I stars, Schaerer & de Koter 1997) on the ionizing spectra of Pop III stars. For the hottest stars ($T_{\rm eff} \gtrsim$ 80 kK) dominating especially the ionizing spectrum in the He II continuum, no significant differences with plane parallel models are found.
We use stellar tracks from 1 to 500 $M_\odot$ calculated with the Geneva stellar evolution code (Desjacques 2000) or tracks from Marigo et al. (2001). Although probably unrealistic, we also explore the impact of high mass loss tracks from Klapp (1983) and El Eid et al. (1983).
The atmosphere models and tracks are included in the synthesis code of Schaerer & Vacca (1998), which also calculates nebular line and continuous emission (adapted here to include additional lines and for an appropriate electron temperature $T_e = 30000$ K).
Based on these models we have calculated the ionizing properties of ZAMS stars and their time averaged properties taking into account their evolution. The

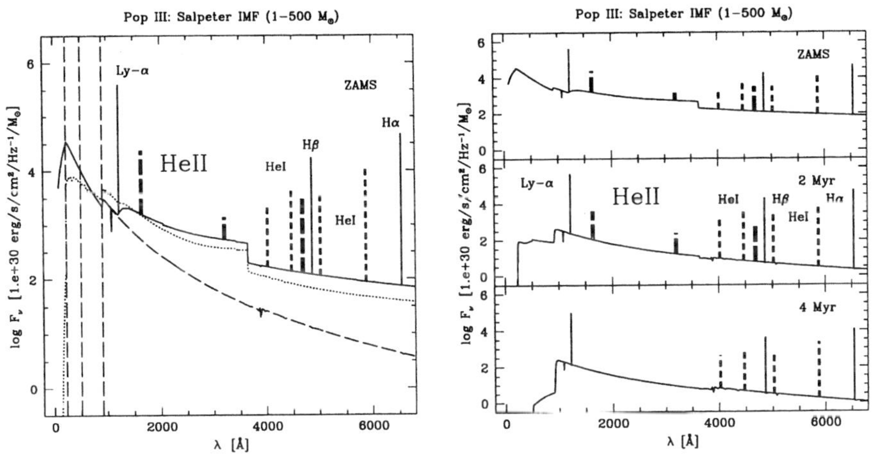

Figure 1. Spectral energy distribution (SED) including H and He recombination lines for Salpeter IMF from 1 to 500 $M_\odot$. **Left panel:** ZAMS population. The pure stellar continuum (neglecting nebular emission) is shown by the dashed line. For comparison the SED of the $Z = 1/50 Z_\odot$ population (Salpeter IMF from 1 – 150 $M_\odot$) is shown by the dotted line. The vertical dashed lines indicate the ionization potentials of H, He^0, and He^+. Note the presence of the unique He II features (shown as thick dashed lines) and the importance of nebular continuous emission. **Right panel:** temporal evolution of the spectrum for 0, 2 and 4 Myr showing the rapid change of the emission line spectrum, characterized by the disappearance of the He II lines

quantities are tabulated and given as fit formulae of use for various calculations (see Schaerer 2001).

2. Properties of Pop III "galaxies"

Integrated spectra have been calculated for a variety of IMFs. SEDs of integrated zero metallicity stellar populations are shown in Fig. 1 for the case of a Salpeter IMF from 1 to 500 $M_\odot$ and instantaneous bursts of ages 0 (ZAMS), 2, and 4 Myr. Overplotted on the continuum (including stellar + nebular emission: solid lines) are the strongest emission lines for illustration purpose. The striking differences when compared with non-zero metallicity, most importantly in the ionizing flux above the He II edge (> 54 eV), have already been discussed by TS00. The comparison of the total spectrum (solid line) with the pure stellar emission (dashed) illustrates the importance of nebular continuous emission neglected in earlier studies (TS00, BKL01), which dominates the ZAMS spectrum at $\lambda \gtrsim 1400$ Å. The nebular contribution, whose emission is proportional to $Q(\mathrm{H})$, depends rather strongly on the age, IMF, and star formation history. For the parameter space explored (see Schaerer 2001), we find that nebular continuous emission is not negligible for bursts with ages $\lesssim 2$ Myr and for constant star formation models.

The spectra in Fig. 1 show in addition to the H and He I recombination lines the presence of the strong He II $\lambda\lambda$ 1640, 3203, and 4686 recombination lines, which — due to the exceptional hardness of the ionizing spectrum — represent a unique feature of Pop III starbursts compared to metal enriched populations (cf. TGS01, Oh et al. 2001, BKL01). Another effect highlighted by this Figure is the rapid temporal evolution of the recombination line spectrum. Indeed, already $\gtrsim$ 3 Myr after the burst, the high excitation lines are absent, for the reasons discussed before.

The detailed behavior of the relative line intensities and equivalent widths and their dependence on the IMF and the star formation history are discussed in Schaerer (2001). For the case of constant star formation, the He II or H recombination lines can be used as a measure of the star formation rate (SFR). For obvious reasons, strong variations of the various SFR indicators on the poorly known IMF are obtained.

Last, but not least, we have also calculated the metal production of Pop III "galaxies", taking into account SNII and pair creation SN, based on recent calculations of Woosley & Weaver (1995) and Heger & Woosley (2001). The photon conversion efficiency of rest mass to ionizing continuum, η (cf. Madau & Shull 1996), of metal-poor and metal-free populations is increased by factors of $\sim$ 3 - 18 compared to solar metallicity. If stars in the range of $\sim$ 130 to 260 $M_\odot$ leading to pair creation SN are present, high overabundances of Si/C and O/C are obtained. See Schaerer (2001) for more details.

References

Bromm, V. , Kudritzki, R.P., & Loeb, A., 2001, ApJ, 552, 464 (BKL01).

Desjacques, J., 2000, Diploma thesis, Geneva Observatory.

El Eid, M.F., Fricke, K.J., & Ober, W.W. 1983, A&A, 119, 54.

Heger, A., & Woosley, S.E. 2001, ApJ, in press (astro-ph/0107037).

Hillier, D.J., & Miller, D.L. 1998, ApJ, 496, 407.

Hubeny, I., & Lanz, T. 1995, ApJ, 439, 875.

Klapp, J. 1983, ApSS, 93, 313.

Loeb, A., & Barkana, R. 2001, ARA&A, 39, 19.

Madau, P., & Shull, J.M. 1996, ApJ, 457, 551.

Marigo, P., Girardi, L., Chiosi, C., & Wood, R. 2001, A&A, 371, 152.

Oh, S.P., Haiman, Z., & Rees, M.J. 2001, ApJ, 553, 73.

Schaerer, D. 2001, A&A, submitted.

Schaerer, D., & de Koter, A. 1997, A&A, 322, 598.

Schaerer, D., & Vacca, W.D. 1998, ApJ, 497, 618.

Tumlinson, J., & Shull, J.M. 2000, ApJ, 528, L65 (TS00).

Tumlinson, J., Giroux, M.L., & Shull, J.M. 2001, ApJ, 550, L1 (TGS01).

Umemura, M., & Susa, H. eds. 2001, in The physics of galaxy formation, (San Francisco: ASP).

Weiss, A., Abel, T., & Hill, V., eds. 2000, in The First Stars (Heidelberg: Springer Verlag).

Woosley, S.E., & Weaver, T.A. 1995, ApJS, 101, 181.

POPULATION SYNTHESIS AND THE DIAGNOSTICS OF HIGH-REDSHIFT GALAXIES

Alberto Buzzoni

Telescopio Nazionale Galileo, A.P. 565, 38700 S/Cruz de La Palma (Tf), Spain, and

Osservatorio Astronomico di Brera, Milano Italy

Abstract The effect of redshift on the observation of distant galaxies is briefly discussed emphasizing the possible sources of bias in the interpretation of high-z data. A general energetic criterion to assess physical self-consistency of evolutionary population synthesis models is also proposed, for a more appropriate use of this important tool to investigate distinctive properties of primeval galaxies.

Introduction

The recent major advances in the observation of the deep Universe have substantially improved our chance to reach the redshift of galaxy formation. Objects at $z > 5$ begin in fact to emerge from HST observations and other deep surveys with the new-generation telescopes (Cohen et al.2000).

As far as we move to larger distances, however, optical (and infrared) observations probe galaxy spectral energy distribution (SED) at much shorter restframe wavelength, in the ultraviolet domain. If neglected, this change in our spectral vantage point could lead to an important bias in the interpretation of high-redshift data. In this note, we will briefly analyze some aspects of this problem dealing both with the observational side and the appropriate use of theoretical tools such as population synthesis models to match primeval galaxy evolution.

1. Redshift constraints and galaxy spectral properties

One important consequence of redshift (besides the first and most obvious effect of moving spectral features to longer wavelength) is that the wavelength interval $\Delta\lambda_{obs}$ sampled by any fixed set of photometric bands will span a narrower portion of the restframe SED when observing galaxies at increasing z, such as $\Delta\lambda_{rest} = \Delta\lambda_{obs}/(1+z)$.

M. Chávez et al. (eds.),
New Quests in Stellar Astrophysics: The Link Between Stars and Cosmology, 189–196.

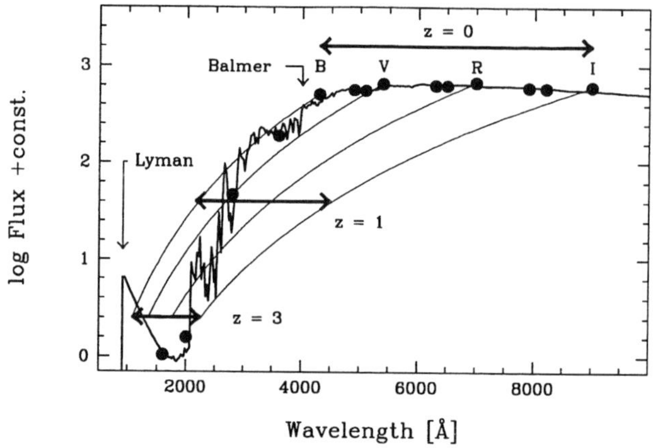

Figure 1. Restframe portion of galaxy SED explored at different redshift by the Johnson four-color photometry through the B, V, R, and I bands. The wide wavelength range, between 4000 and 9000 Å, attained for local galaxies, "shrinks" to a 1000–2200 Å interval when observing objects at $z = 3$. The synthetic SED of a model elliptical is displayed, for reference.

This effect, by itself, can be of paramount importance as far as we try to gain information on galaxy stellar content by means of multicolor photometry. Figure 1 gives an illuminating example in this sense.

For a $z = 0$ galaxy, for instance, Johnson four-color photometry in the *BVRI* bands would span a wide wavelength range, from 4000 to 9000 Å, and collect a substantial contribution from the galaxy stellar population. On the contrary, when observing at $z = 3$ with the same photometric system, we would be covering a scarce ±600 Å spectral window centered about 1600 Å in the galaxy restframe. Quite deceivingly, our "multicolor" photometry at $z = 0$ would translate into a nearly "monochromatic" estimate of the galaxy flux at $z = 3$.

1.1 Sampling stellar populations of high-z galaxies

According to the overall color-magnitude distribution, each star in a galaxy aggregate contributes with a different weight to the integrated luminosity with varying wavelength. In Buzzoni (1993) we approached this problem in a statistical way introducing the concept of "effective" contributors in a stellar population. This quantity directly relates to galaxy integrated properties, such as total luminosity or surface brightness fluctuations (as further explored, for instance, by Tonry and Schneider 1988).

We showed, in particular, that by sampling at a given wavelength a total luminosity L_{tot} provided by N stars, the expected Poissonian fluctuation results:

$$\frac{\sigma(L)}{L_{tot}} = \frac{(\sum^N l_i^2)^{1/2}}{\sum^N l_i}, \quad (1)$$

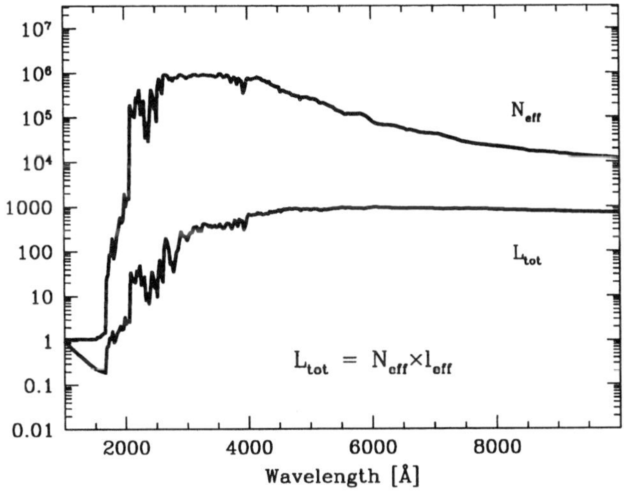

Figure 2. Relative number of "effective" stellar contributors (N_{eff}) and SED (L_{tot}) for a 5 Gyr SSP model with Salpeter IMF and [Fe/H] $= +0.3$ dex according to Buzzoni (1993). Both quantities are in logarithm scale and arbitrarily normalized to their values at 1000 Å.

where l_i is the luminosity of each composing star.
If all stars have the same luminosity, then simply $\sigma(L)/L_{tot} = 1/\sqrt{N}$. More generally, however, a quantity

$$N_{\mathit{eff}} = \frac{(\sum^N l_i)^2}{\sum^N l_i^2} \leq N \tag{2}$$

can be defined, as a function of λ. Its value gives a statistical measure of the effective number of stars that provide L_{tot} at a given wavelength. Following Buzzoni (1993), a "mean" luminosity (l_{eff}) for the composing stars can also be computed as

$$l_{\mathit{eff}} = \frac{\sum^N l_i^2}{\sum^N l_i} \tag{3}$$

so that, at every wavelength,

$$L_{tot} = N_{\mathit{eff}} \times l_{\mathit{eff}}. \tag{4}$$

In Fig. 2 we show the expected value of N_{eff} for the illustrative case of a 5 Gyr simple stellar population (SSP) with Salpeter IMF and [Fe/H] $= +0.3$ along with the integrated SED of the model. This should closely resemble the real case of primeval elliptical galaxies (Buzzoni 1995).
As a striking feature of the model, note that below 2000 Å the effective number of "luminous" contributors dramatically drops by over six (!) orders of

magnitude compared for example with the B band value at 4000 Å. On the contrary, luminosity over the same wavelength range only dims by a factor of $L_B/L_{2000} \sim 10^3$. In terms of surface brightness fluctuations this means, for instance, that the ultraviolet image of a galaxy consisting of 10^{11} stars should fairly compare with the B observations of a (10^6 star-poorer) Galactic globular cluster.

Such a severe undersampling of the galaxy stellar population in the ultraviolet range could have an important impact when studying morphology of $z > 1$ ellipticals at optical wavelength (but the same argument holds also for late-type systems). In particular, high-redshift objects would more likely appear as coarser systems with increasing z (e.g. Puerari 2001, this conference).

1.2 Redshift and age bias

The selective sampling of galaxy stellar population with changing wavelength has even more pervasive consequences, as far as we track evolution of star-forming systems at different redshift.

More generally, we could describe a composite stellar population in terms of a suitable convolution of SSPs according to the star formation rate (SFR) at the different epochs. The total luminosity at age "t" results therefore:

$$L_{tot}(t) = \int_0^t L_{SSP}(\tau)\ SFR(t-\tau)\ d\tau \tag{5}$$

A mean luminosity-weighted age of the composing SSPs can also be derived such as

$$\bar{t}_* = \frac{\int_0^t \tau\ L_{SSP}(\tau) SFR(t-\tau)\ d\tau}{L_{tot}(t)} \tag{6}$$

This could be regarded as the "representative" age of the stars that contribute to galaxy luminosity. We know (Buzzoni 1995; Tinsley and Gunn 1976) that SSP luminosity evolution can be pretty well described by a simple power law such as $L_{SSP} \propto t^{-\alpha}$ with the power index α that is a function of wavelength. As an instructive example, if we consider for the SFR a power-law time decay such as SFR $\propto t^{-\eta}$,[1] eq. (6) becomes

$$\bar{t}_* = \frac{\int_0^t \tau^{1-\alpha}\ (t-\tau)^{-\eta}\ d\tau}{\int_0^t \tau^{-\alpha}\ (t-\tau)^{-\eta}\ d\tau}. \tag{7}$$

[1]The major advantage of this parameterization is that the full range of galaxy morphologies can simply be accounted for by an age-independent distinctive birthrate: $b = \mathrm{SFR}/ < \mathrm{SFR} >= (1-\eta)$. This easily derives recalling that, at age t, $\mathrm{SFR} = C\, t^{-\eta}$ and its time average is $< \mathrm{SFR} >= C\, t^{-1} \int_0^t \tau^{-\eta}\, d\tau$. Providing that $\eta < 1$, we always have $< \mathrm{SFR} >= \mathrm{SFR}/(1-\eta)$, from which the value of b directly follows, by definition.

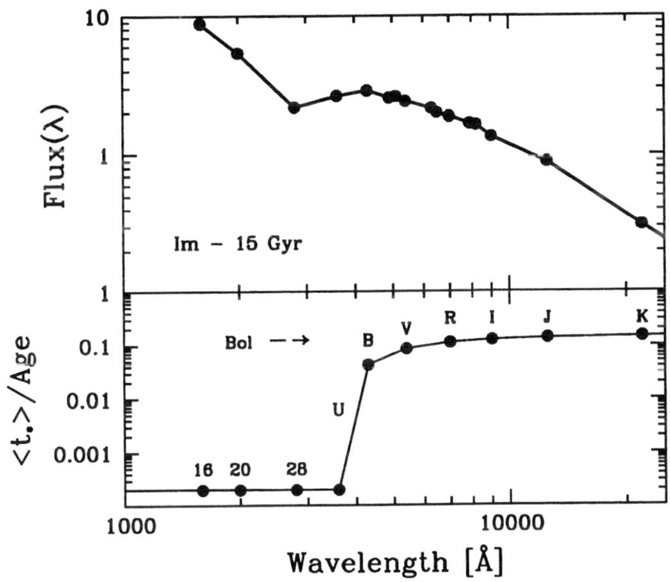

Figure 3. Theoretical SED (*upper panel*), and "mean" age (relative to galaxy age) of stars at different photometric bands (and in bolometric as well), according to eq. (7) (*lower panel*), for a 15 Gyr Im model galaxy from Buzzoni (2001).

The integral has an analytical solution such as

$$\bar{t}_* = \frac{1-\alpha}{2-\alpha-\eta}\, t. \tag{8}$$

In bolometric, $\alpha = 0.76$ (Buzzoni 1989) and, for a constant SFR (i.e. $\eta = 0$), $\bar{t}_* = 0.2\,t$. At 15 Gyr, bright stars are therefore in average 3 Gyr old, but this value changes according to the $\alpha(\lambda)$ function. For a SSP with Salpeter IMF and $Z_\odot$, the luminosity power index smoothly increases from $\alpha = 0.71$ in the K band to 0.94 in the B band, and exceeds unity below 3500 Å. A value of $\alpha \geq 1$ implies that $\bar{t}_* \to 0$, that is only the youngest and more massive stars are representative contributors to galaxy ultraviolet luminosity, as discussed in the previous section.

The mean age of stars contributing to galaxy luminosity at different wavelength for a synthesis model of a 15 Gyr Magellanic irregular (see Buzzoni 2001 for details) is reported in Fig. 3. Note from the figure that the value of $\bar{t}_*$ smoothly decreases with decreasing wavelength. As a result, high-redshift objects forcedly appear to be younger than local homologues at $z = 0$ *in spite of any intrinsic evolution.*

2. Energetic self-consistency of population synthesis models

Evolutionary population synthesis models are extensively used as a reference tool to reproduce galaxy colors and investigate galaxy spectrophotometric properties at the different cosmic epochs (Buzzoni 1989, 1995, 2001; Bruzual and Charlot 1993; Worthey 1994; Bressan et al.1994; Fioc and Rocca-Volmerange 1997).

Besides any difference in the computational details and input physics, a crucial constraint that needs to be consistently fulfilled in the synthesis procedure concerns a suitable match of the MS and Post-MS luminosity contribution in the SSP models. The so-called "Fuel consumption theorem" of Renzini and Buzzoni (1986) provides a simple and very powerful tool in this sense. According to the theorem, the Post-MS bolometric luminosity in a SSP of total luminosity (L_{bol}) is

$$L_{PMS} = \mathcal{B} \times L_{bol} \times \text{Fuel}. \tag{9}$$

In the equation, the nuclear fuel refers to the amount consumed by one star leaving the MS turn-off (TO) point along the whole Post-MS evolution. This is a natural output of the theoretical stellar tracks and is therefore known with high accuracy. The scaling factor $\mathcal{B}$ is the "specific evolutionary flux" of the SSP, and directly deals with the stellar clock, that is the relationship between MS lifetime and TO stellar mass (see Renzini and Buzzoni 1986). In particular, we have that

$$\mathcal{B} \propto M_{TO}^{-s} \times \dot{M}_{TO} \tag{10}$$

assuming that a power-law IMF holds with $dN_* \propto M_*^{-s}\, dM_*$ ($s = 2.35$ for the Salpeter case). Again, the r.h. term of eq. (10) is a direct output of stellar evolution theory, and $\mathcal{B}$ should therefore be regarded as a "universal" function, that is nearly insensitive to the specific input physics of the SSP models.

A comparison of the $\mathcal{B}$ scaling factor, as derived from the synthesis codes of Bruzual and Charlot (1993), Buzzoni (1989), and the Padova group (Bressan et al.1994) for a SSP with $Z_\odot$ and Salpeter IMF is displayed in Fig. 4. While both the Buzzoni and Padova codes consistently match the fuel consumption theorem, the Bruzual and Charlot models show a much higher value for the specific evolutionary flux. The effect is a macroscopic one, and reaches about a 100% discrepancy at intermediate age. Such higher scale factor leads, in eq. (9), to a corresponding overestimate of Post-MS luminosity at every age. The effect is quite shifty as it will not explicitly appear when considering luminosity evolution (cf. Fig. 5, left panel), but would strongly affect the theoretical prediction of galaxy M/L ratio and apparent color evolution. In the latter case, enhanced Post-MS contribution would lead to a larger number of red giant stars and therefore a systematically redder color for the stellar aggregate as a whole. This is shown in Fig. 5 (right panel) when comparing the Bruzual and Charlot

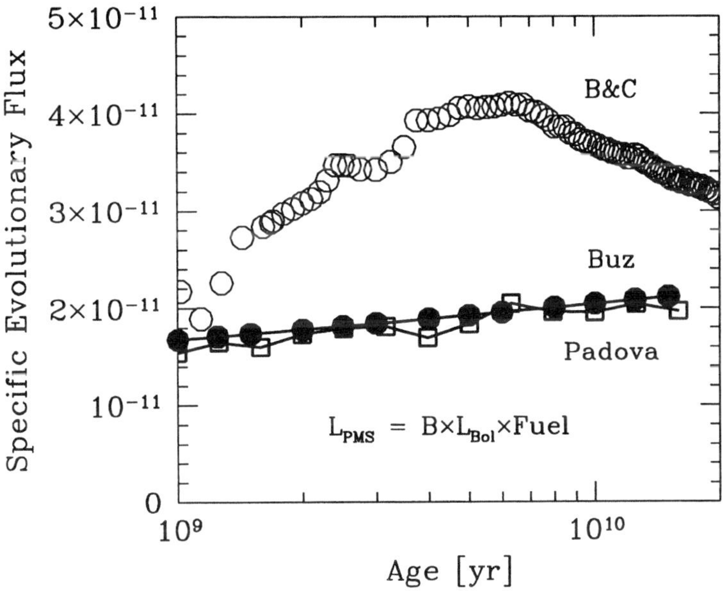

Figure 4. The "specific evolutionary flux", $\mathcal{B}$, for SSP models with $Z_\odot$ and Salpeter IMF according to Buzzoni (1989; "•" markers), Bressan et al.(1994; "*square*") and Bruzual and Charlot (1993; "o"). The value of $\mathcal{B}$ is given in unit of $yr^{-1}\ L_\odot^{-1}$. It directly scales Post-MS contribution to SSP luminosity through eq. (9).

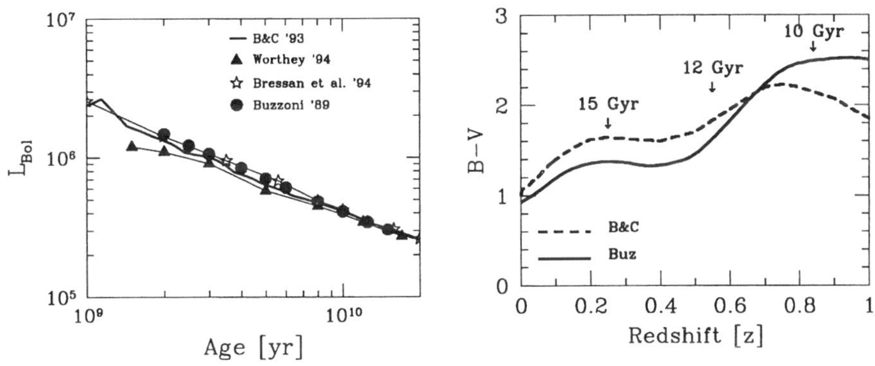

Figure 5. *Left panel:* bolometric evolution for a SSP with $Z_\odot$ and Salpeter IMF according to different synthesis codes, as labeled. Luminosity is normalized at 12 Gyr. *Right panel:* apparent $B - V$ evolution of elliptical galaxies according to Buzzoni (1995; "Buz") and Bruzual and Charlot (1993; "B&C"). A cosmological model with $(H_o, q_o, z_f) = (50\ \text{km s}^{-1}\ \text{Mpc}^{-1}, 0, 30)$ is assumed. The absolute age of galaxies at the different redshift is also labeled on the plot.

(1993) results with the expected evolution of elliptical galaxies according to Buzzoni (1995).

Acknowledgments

It is a pleasure to thank the INAOE for the kind invitation and the excellent organization of this meeting. This work received partial financial support from the Italian MURST under COFIN'00 02-016 grant.

References

Bressan, A., Chiosi, C., & Fagotto, F. 1994, ApJS, 94, 63.

Bruzual, G., & Charlot, S. 1993, ApJ, 405, 538.

Buzzoni, A. 1989, ApJS, 71, 817.

Buzzoni, A. 1993, A&A, 275, 433.

Buzzoni, A. 1995, ApJS, 98, 69.

Buzzoni, A. 2001, AJ, in press.

Cohen, J.G., Hogg, D.W., Blandford, R., Cowie, L.L., Hu, E., Songaila, A., Shopbell, P., & Richberg, K. 2000, ApJ, 538, 29.

Fioc, M., & Rocca-Volmerange, B. 1997, A&A, 326, 950.

Renzini, A., & Buzzoni, A. 1986, in Spectral Evolution of Galaxies, eds. A. Renzini & C. Chiosi (Dordrecht: Reidel), p. 195.

Tinsley, B. M., & Gunn, J. E. 1976, ApJ, 203, 52.

Tonry, J., & Schneider, D.P. 1988, AJ, 96, 807.

Worthey, G. 1994, ApJS, 95, 107.

VI

HIGH–MASS STELLAR PRODUCTS AND COSMOLOGY

EXPLOSIVE YIELDS OF METAL-FREE SNII AND THE ABUNDANCES IN EXTREMELY METAL-POOR STARS

Marco Limongi
Osservatorio Astronomico di Roma, Italy

Alessandro Chieffi
Istituto di Astrofisica Spaziale (CNR), Roma, Italy

Oscar Straniero
Osservatorio Astronomico di Collurania, Teramo, Italy

and

Inma Domínguez
Universidad de Granada, Spain

Abstract The presupernova evolutions and the explosive yields of massive stars in the range $13-80\ M_{\odot}$ with metallicity Z=0 and $Z = 10^{-3}$ are presented. Differences in the final elemental yields between the two sets of models are addressed and discussed in terms of differences in the presupernova evolutionary properties. Implications on the enrichment of the interstellar medium for the various generations of stars are also discussed.

Introduction

Current theories of Galaxy formation would imply that extremely metal poor stars may retain information on the ejecta of a single, or at most, few zero metallicity type II supernovae (Argast et al. 2000; Auduze & Silk 1995). Assuming that the formation of each extremely metal poor star is triggered by the explosion of a single zero metallicity type II SN, taking the observed abundances for these extremely metal poor stars and the latest theoretical yields

M. Chávez et al. (eds.),
New Quests in Stellar Astrophysics: The Link Between Stars and Cosmology, 199–206.

for zero metallicity massive stars we intend to verify if this can be considered a self-consistent scenario.

One of the latest high resolution abundance survey of extremely metal poor stars is the one of McWilliam et al. (1995) (MPSS hereinafter). In this study they found the following results for the trends of the various elements in the metallicity range between –4 and –2.5. [Co/Fe] increases with declining the metallicity; [Cr,Mn/Fe] decrease with decreasing the metallicity; [Ni,Sc] seem flat around the solar value; [α/Fe] are constant down the lower metallicities and enhanced with respect to the solar value by 0.44 dex, for Mg, Si, and Ca, and 0.31 dex for Ti; [Al/Fe] decreases while [C,Na] are almost constant with decreasing the metallicity.

Many groups have tried to interpret these data but looking only at some observed element abundance ratios, i.e., Nakamura et al. (1999) studied only the observed trends of [Cr,Mn,Co,Ni/Fe] with [Fe/H], while Shigeyama & Tsujimoto (1998) analyzed only the trends of [C,Si,Ca/Mg] with [Mg/H].

Our approach is completely different, in the sense that we intend to compare the abundance pattern of all the elements of each extremely metal poor star with the theoretical yields of each supernova model as a function of the initial mass, and not only some of the observed element abundance ratios. We think that in this way we can better constrain both the theoretical models and the nucleosynthesis, moreover in this approach we do not need to assume any relation between the exploding mass and the metallicity ([Fe/H]) or the dilution factor.

1. Observed element abundance ratios and theoretical yields of zero metallicity massive stars.

We have selected 4 stars from the MPSS database, namely, CD-38°245 ([Fe/H] = –4.01), CS22891-200 ([Fe/H] = –3.49), CS22892-052 ([Fe/H] = –3.04), CS22947-187 ([Fe/H] = –2.5). These stars should sample the typical abundance pattern for any [Fe/H] in the range between –4.0 and –2.5. A more extended discussion on the differences in the abundance pattern among stars having a similar metallicity will be addressed in a forthcoming paper (Chieffi, Limongi & Straniero 2001).

The theoretical yields have been computed by means of a simulated explosion of a grid of zero metallicity presupernova models in the mass range between 15 and 100 $M_{\odot}$. Due to the present uncertainties on the computation of the explosion of a SN II (Mezzacappa et al. 1998), the location of the mass cut, i.e. the mass coordinate between the remnant and the ejecta, is a free parameter and is chosen by imposing the amount of ^{56}Ni ejected. Both the presupernova models and the explosive yields are presented and extensively discussed in (Limongi, Straniero & Chieffi 2000; Limongi, Chieffi & Straniero 2001; Chieffi, Limongi & Straniero 2001), however it is important to remind here that the various

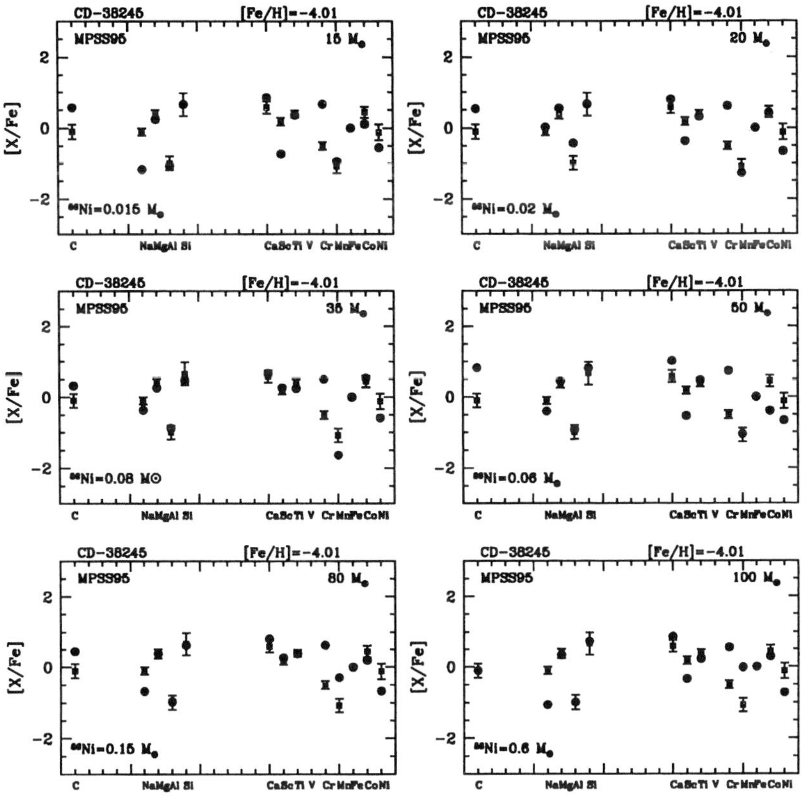

Figure 1. Theoretical yields of supernova models for star CD-38°245. In each panel The observed element abundance ratios ($[X/Fe] = \mathrm{Log}(X/Fe)_{star} - \mathrm{Log}(X/Fe)_{\odot}$) are plotted in each panel according to the different mass of the supernova progenitor: 15 $M_{\odot}$ (upper left panel), 20 $M_{\odot}$ (upper right), 35 $M_{\odot}$ (middle left panel), 50 $M_{\odot}$ (middle right), 80 $M_{\odot}$ (lower left panel) and 100 $M_{\odot}$ (lower right).

elements can be divided into different groups depending on their production site (see also Chieffi et al. this meeting). In particular we have the following groups: 1) C, Na, Mg and Al produced during the hydrostatic evolution of the star within the convective carbon shell and then partially modified by the explosion by the explosive Ne and C burnings; 2) Si and Ca produced only by both the explosive O and explosive incomplete Si burnings; 3) Cr and Mn produced only by the explosive incomplete Si burning; 4) Ti and Fe produced by a combination of both the explosive complete and explosive incomplete Si burnings and finally 5) Sc, Co and Ni produced only by the explosive complete Si burning. Among these groups there are some elements (C, Na, Mg, Al, Si and Ca) that do not depend on the location of the mass cut, since they are produced in a rather external zones,

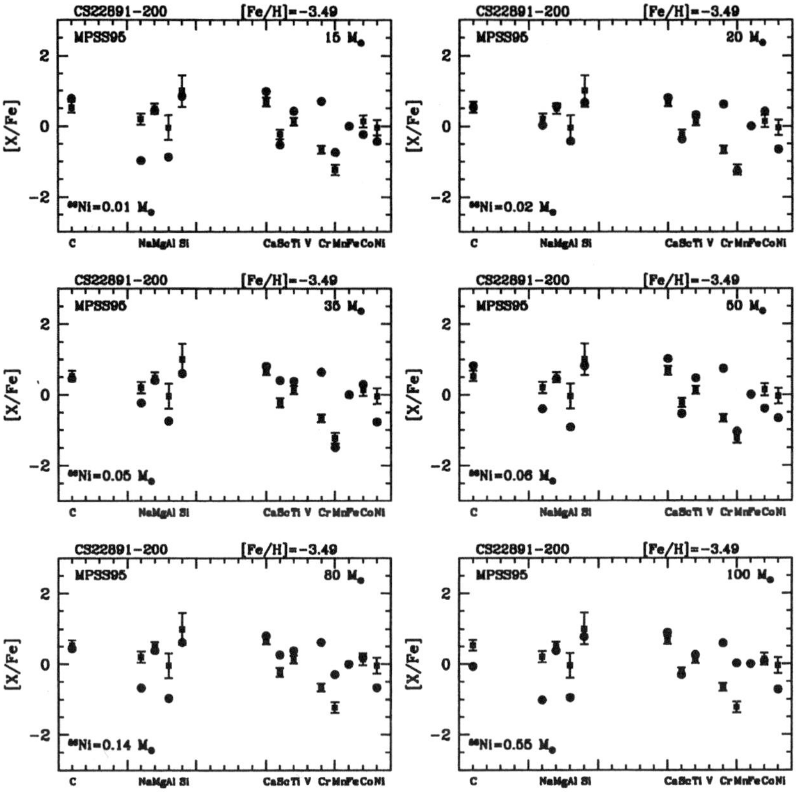

Figure 2. Same as Fig. 1, but for star CS22891-200.

while some other (Sc, Cr, Mn, Ti, Fe, Co and Ni), produced in the innermost layers of the exploding envelope, whose final abundance significantly depends on this parameter. As a consequence if we study the element abundance ratios relative to Fe ([X/Fe]) a changing of the mass cut location has two different effects: [C,Na,Mg,Al,Si,Ca/Fe] are simply vertically shifted maintaining their relative scaling, while, on the contrary, [Sc,Cr,Mn,Ti,Co,Ni/Fe] may change, even significantly, their relative proportions.

2. Comparison between observations and theoretical yields

Figure 1 shows the comparison between the most metal poor star in the MPSS database, namely the CD-38°245, and the theoretical yields of each supernova model. In particular in each panel of Fig. 1 we plot the observed element abundance ratios ($[X/Fe] = \mathrm{Log}(X/Fe)_{star} - \mathrm{Log}(X/Fe)_{\odot}$) and the theoretical

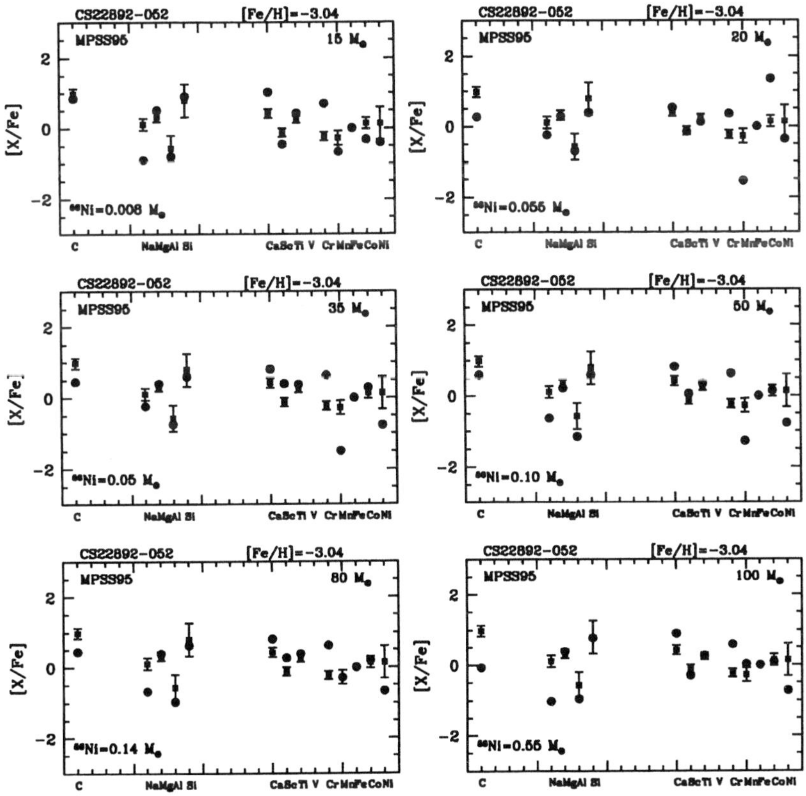

Figure 3. Same as Fig. 1, but for star CS22892-052

yields of any given mass of the supernova progenitor: 15 $M_\odot$ in the upper left panel, 20 $M_\odot$ in the upper right panel, 35 $M_\odot$ in the middle left panel, 50 $M_\odot$ in the middle right panel, 80 $M_\odot$ in the lower left panel and 100 $M_\odot$ in the lower right panel. The amount of ^{56}Ni ejected is chosen in order to better reproduce the elements from C to Ca, i.e. all the elements whose relative scaling does not depend on the location of the mass cut; once the amount of ^{56}Ni ejected is chosen the relative scaling of the iron peak elements is fixed. If we look to the relative scaling of all the element abundance ratios, and not only to some of these, we find that the 35 $M_\odot$ is the model which is in better agreement with the observations, a part for few elements. In particular [Cr/Fe] and [Mn/Fe] are overestimated and underestimated respectively my the model, although they are produced in the same region (i.e., the explosive incomplete Si burning); [Ni/Fe] is underestimated by the model, in spite of the fact that both [Sc/Fe] and [Co/Fe], produced in the same region (i.e., the complete explosive Si burning), are in excellent agreement with the observations.

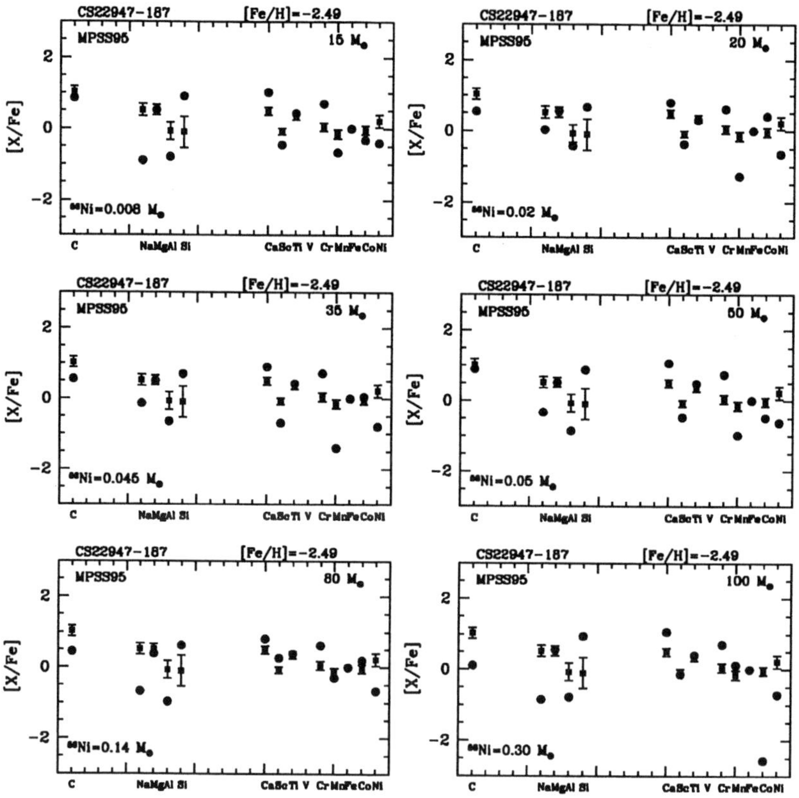

Figure 4. Same as Fig. 1, but for star CS22947-187

Following a similar analysis for the other three stars (cf. Fig. 2, 3, and 4) we find that: 1) the CS22891-200 is compatible with a 20 $M_\odot$ model, although [Cr/Fe] and [Ni/Fe] are overestimated and underestimated respectively by the model; 2) any model reproduces simultaneously the observed element abundance ratios of the light and intermediate mass elements (from C to Ca) of both the CS22892-052 and the CS22947-187 for any choice of the mass cut.

3. Conclusions

The comparison between the observed element abundance ratios and the theoretical yields of zero metallicity massive stars shows that at very low metallicities ($[Fe/H] \leq -3.5$) there exists at least one model which is in good agreement with the observations while at higher metallicities none of the models reproduce the observed abundance pattern. These preliminary results and their implications will be addressed in a future paper (Chieffi, Limongi & Straniero 2001).

References

Argast, D., Samland, M., Gerhard, O. E., & Thielemann, F. K. 2000, A&A, 356, 873.

Chieffi, M., Limongi, M., & Straniero, O. 2001, in preparation.

Limongi, M., Straniero, O., & Chieffi, A. 2000, ApJ, 129, 625.

Limongi, M., Chieffi, A., & Straniero, O. 2001, in preparation.

Mezzacappa, A., Calder, A. C., Bruenn, S. W., Blondin, J. M., Guidry, M. W., Strayer, M. R., & Umar, A. S. 1998, ApJ, 493, 848.

McWilliam, A., Preston, G. W., Sneden, C., & Searle, L. 1995, AJ, 109, 2757.

Nakamura, T., Umeda, H., Nomoto, K., Thielemann, F. K., & Burrows, A. 1999, ApJ, 517, 193.

Shigeyama, T., & Tsujimoto, T. 1998, ApJ, 508, L151.

THE FIRST DETECTION OF COBALT IN A DAMPED LYMAN ALPHA SYSTEM

Sara L. Ellison
European Southern Observatory, Chile

Sean G. Ryan
Open University, UK

and

Jason X. Prochaska
Carnegie Observatories, USA

Abstract

We present the first ever detection of Cobalt in a Damped Lyman Alpha system (DLA) at $z \sim 2$. In addition to providing important clues to the star formation history of these high redshift galaxies, we discuss how studying the Co abundance in DLAs may also help to constrain models of stellar nucleosynthesis in a regime not probed by Galactic stars.

Introduction

The study of elemental abundances in Damped Lyman Alpha systems (DLAs) at high redshift represents one of our best opportunities to probe galaxy formation and chemical evolution at early times. By coupling measurements made in high z DLAs with our knowledge of abundances determined locally and with nucleosynthetic models, we can start to piece together the star formation histories of these galaxies. To this end, certain key elements are generally targeted, for example Zn and Cr are traditionally used as a measure of the overall metallicity and gas-to-dust ratio in DLAs, whereas Si and S are used to determine whether there is any evidence for α element overabundances, indicative of SN Type II enrichment. Here, we propose that Co, which has never before been featured in DLA studies, yet is well-studied in Galactic stars, may be an additional element that can provide useful clues to the nature of star formation at high redshift.

M. Chávez et al. (eds.),
New Quests in Stellar Astrophysics: The Link Between Stars and Cosmology, 207–210.

1. Why Study Cobalt?

The intriguing trends of [Co/Fe] as a function of [Fe/H] in the various Galactic stellar populations indicate that cobalt may be an element that can provide important clues to chemical evolution, see Figure 1. In sub-solar metallicity thin disk stars, Co is found to be slightly underabundant with respect to Fe ([Co/Fe] ~ -0.1, Gratton and Sneden 1991). This trend continues in moderately metal-poor halo stars down to metallicities of [Fe/H] ~ -2.5. For more metal-poor halo stars, a considerable overabundance of Co with respect to Fe is observed, $0 <$ [Co/Fe] < 0.8, (McWilliam et al. 1995; Ryan, Norris and Beers 1996). Interestingly, this coincides with a downturn in relative Cr to Fe abundances, such that Co and Cr are closely anti-correlated in the metal poor population. Similar [Co/Fe] overabundances are observed in the bulge ([Co/Fe] $\sim +0.3$, McWilliam & Rich 1994), which, we recall, is as old as the halo, and to a lesser degree in the thick disk ([Co/Fe] $\sim +0.1$, Prochaska et al. 2000).
The detection of Co II transitions is a challenging possibility for ISM absorption line spectroscopy, both locally and at high z. Although there are numerous transitions that lie at convenient rest wavelengths for both local and high redshift studies, only two interstellar measurements currently exist (e.g. Mullman et al 1998). One of the advantages of moving to higher redshifts is that the relatively strong UV lines become shifted into the optical where one has access to more efficient detectors and larger telescopes. In addition, the observed equivalent width is increased by a factor of $(z + 1)$. However, the fundamental problem remains that the Co II lines are all intrinsically weak and that the solar abundance of Co is 8×10^{-8} that of hydrogen.

2. Cobalt in DLAs

In order to maximize the likelihood of detecting Co, we have selected two DLAs with high N(Fe) that lie in front of relatively bright background QSOs. The former criterion is to enable us to more easily reach our target detection limit of [Co/Fe] ≤ 0, which will be sufficient to distinguish between most of the populations plotted in Figure 1. Selecting bright QSOs obviously allows us to reach the requisite S/N in feasible amounts of time. The two systems chosen for this pilot study are Q2206$-$199 ($B = 17.8, z_{em} = 2.559$, $z_{abs} = 1.92$) and Q1223+17 ($B = 18.5, z_{em} = 2.936$, $z_{abs} = 2.466$). Observations were obtained with UVES (on the VLT) and HIRES (on Keck) the details of which (along with data reduction steps and complete abundance analysis) are described in full by Ellison, Ryan and Prochaska (2001). Here, we limit ourselves to reporting the final determined Co abundances which are [Co/H] $= -0.51$ for Q2206$-$199 and [Co/H] < -1.76 for Q1223+17. These points are plotted in Figure 1 along with Galactic stellar values and additional DLA limits derived from the literature.

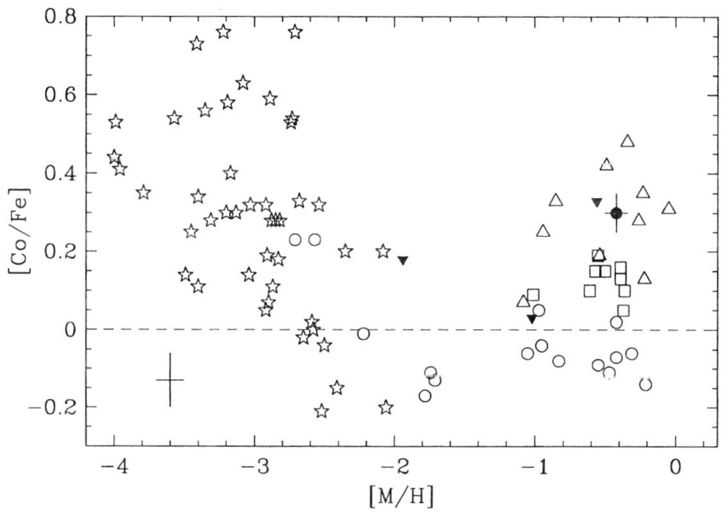

Figure 1. Relative cobalt abundances as a function of metallicity for different Galactic populations and DLAs: open stars — metal-poor halo stars from McWilliam et al (1995) and Ryan, Norris, & Beers (1996); open circles — moderately metal-poor thin disk and halo stars from Sneden & Gratton (1991); open triangles — bulge stars from McWilliam & Rich (1994); open squares — thick-disk stars from Prochaska et al. (2000). The solid circle represents Q2206-199 and the down-turned filled triangles are upper limits for three other DLAs (Q1223+17 [this work], HE1104−1805 [Lopez et al 1999] and Q302−223 [Pettini et al 2000]). The cross in the lower left corner of the figure shows the typical error bar for stellar points.

3. Results

We have shown that the detection of Co in DLAs is feasible with echelle spectrographs on 8-m class telescopes. The first detection of Co in a DLA appears to be consistent with the abundances measured in the Galactic bulge and although this doesn't necessarily equate DLAs with high redshift bulges, it may indicate that the two share some similarities in their star formation histories. However, the interpretation of these results is hindered by our lack of understanding of the nucleosynthetic processes that produce Co. Only recently have models been formulated that can successfully reproduce Galactic [Co/Fe] trends by forcing a deeper mass cut or higher explosion energies (Nakamura et al 1999, 2000). However, matching the [Co/Fe] has come at the expense of high Ni abundances which are not seen in stars. In the case of Q2206−199, however, we *do* find that high [Co/Fe] is accompanied by high [Ni/Fe]. Therefore, although the results here represent a somewhat preliminary study of Co with a clear need for more data points, we have demonstrated the interest of studying this element both in the context of understanding better the nature of DLAs and for providing a regime for the testing of nucleosynthetic models.

Acknowledgments

The authors would like to express their gratitude to their colleagues Jacqueline Bergeron, Patrick Petitjean and Max Pettini for their consent to use some of the data presented here.

References

Ellison, S. L., Ryan, S. G., & Prochaska, J. X. 2001, MNRAS, 326, 628.

Gratton, R., & Sneden, C. 1991, A&A, 241, 541.

Lopez, S., Reimers, D., Rauch, M., Sargent, W., & Smette, A. 1999, ApJ, 513, 598.

McWilliam, A., & Rich, M. 1994, ApJS, 91, 749.

McWilliam, A., Preston, G. W., Sneden, C., & Searle, L. 1995, AJ, 109, 2757.

Mullman, K., Lawler, J., Zsargo, J., & Federman, S. 1998, ApJ, 500, 1064.

Nakamura, T., Umeda, H., Iwamoto, K., Nomoto, K., Hashimoto, M., Hix, R., & Thielemann, F. 2000, ApJ, 555, 880.

Nakamura, T., Umeda, H., Nomoto, K., Thielemann, F., & Burrows, A. 1999, ApJ, 517, 193.

Pettini, M., Ellison, S. L., Steidel, C. C., Shapley, A. E., & Bowen, D. V. 2000, ApJ, 532, 65.

Prochaska J. X, Naumov S. O., Carney B. W., McWilliam A., & Wolfe A. M. 2000, AJ, 120, 2513.

Ryan, S. G., Norris, J. E., & Beers, T. C. 1996, ApJ, 471, 254.

TYPE IB-IC SUPERNOVAE: TIME CLOCKS OF SEYFERT GALAXIES?

Alessandro Bressan, P. Marziani
Osservatorio Astronomico, Padova, Italia

and

M. Della Valle
Osservatorio Astrofisico di Arcetri, Italia

Abstract

Type Ib-c supernovæ appear to be more frequent with respect to type II supernovæ in Seyfert than in normal galaxies by a factor ≈ 3. We suggest that this statistically significant difference may be the signature of young star formation in Seyferts, with average age $\tau \leq 20$ Myr.

Introduction

We consider here whether there is a difference in the rate of core-collapse supernovæ between Seyfert galaxies and "normal" galaxies. Core collapse supernovæ have been subdivided into supernovæ of type Ib and type Ic (hereafter referred to as Ib-c) and supernovæ of type II. As we will briefly discuss below, differences in the ratio between the two types of supernovæ, f = N(Ib-c)/N(II), may hint at metallicity, age, or at Initial Mass Function (IMF) slope effects.

1. Methodology & Results

From the Asiago Supernova Catalogue we extracted all galaxies with $v_r \leq 8000$ km s^{-1}, classified as Seyfert 1 and Seyfert 2 in the 9th edition of *A Catalogue of Quasars and Active Galactic Nuclei* by M.- P. Véron-Cetty and P. Véron with supernovæ revealed after 1985 (22 supernovæ). Seyfert nuclei in our sample are predominantly of type 2-like (19 host galaxies: 4 Seyfert 1-1.5; 15 Seyfert 1.8–1.9–2). After eliminating supernovæ in objects classified as S3 (LINER) or H2 (HII regions), we kept all other supernovæ as the ones produced by normal galaxies (365 events).

M. Chávez et al. (eds.),
New Quests in Stellar Astrophysics: The Link Between Stars and Cosmology, 211–214.

We found the ratio $\overline{f}_{Seyf}$ to be ≈ 0.83 for our Seyfert galaxy sample, and $\overline{f}_{norm} \approx 0.27$ for normal galaxies. It is important to stress that this difference should be almost independent from the control time, because we are considering relative rates of supernovæ whose average magnitudes at maximum fall within a narrow range.

The excess of type Ib-c supernovæ in the Seyfert with respect to the "normal" galaxy sample is found to be significant at a 2σ confidence level from binomial statistics as well as from a more appropriate approach based on the "bootstrap" technique. The bootstrap method consists in the generation of a large number ($\sim 10^4$) of synthetic data sets matching the Seyfert galaxy sample i.e., with the same number of objects and statistically indistinguishable distribution of any parameter influencing either the probability of discovery of supernovæ as well as the intrinsic rate of supernova production (galaxy inclination, standard isophotal diameter, morphological type, radial velocity; Cappellaro et al. 1993). An additional result is that the distribution of supernova distance from the galaxy center (normalized to the standard diameter D_{25}) is more centrally peaked in Seyfert than in normal galaxies. This result has been found in our sample as well as in previous works.

2. Implications: A young circumnuclear Starburst in Seyfert Galaxies?

The ratio $\overline{f}_{norm} \approx 0.27$ found for non-active galaxies is close to the value expected for constant star formation rate and a Salpeter Initial Mass Function (IMF), assuming that the minimum progenitor mass of a type II supernova is ≈ 8 $M_\odot$. So, as far as normal galaxies are concerned, we do not see anything extraordinary.

On the contrary, something unusual occurs in Seyfert galaxies if $\overline{f}_{Seyf} \approx 0.83$. In the case of constant star formation rate, the slope of the IMF cannot be $\alpha \approx 2.35$ (the one of Salpeter law) unless the minimum mass needed to give rise to a type Ib-c supernova is 15 $M_\odot$. We must note in this respect that the median distances from the galactic center of Ib-c supernovæ in Seyfert and in normal galaxies are 0.05 D_{25} and 0.25 D_{25} respectively. This implies that Ib-c supernovæ could be typically produced in 2 $Z_\odot$ regions in Seyfert and in $Z_\odot$ regions in normal galaxies, assuming a representative abundance gradient for early type-spirals (Vila-Costas & Edmunds 1992). Such increase in metallicity may lower the minimum mass for Ib-c production from 30 to 25 $M_\odot$ (Bressan 1994) and yield a ratio $\overline{f}_{Seyf}/\overline{f}_{norm} \approx 1.4$, but not $\overline{f}_{Seyf} \approx 0.83$.

If the minimum mass of the type Ib-c progenitors is ≥ 20 $M_\odot$, then $\alpha \leq 1.6$. The necessity of an IMF different from the one by Salpeter disappears if, in Seyferts, we are observing a very young burst of star formation. A Salpeter IMF is consistent with the observed $\overline{f}_{Seyf}$ if the main sequence turnover mass is $M(\tau) \approx 15\ M_{\odot}$, which corresponds to a starburst age $\tau \approx 20$ Myr. An upper limit to the starburst age of $\tau \approx 20$ Myr is set right by the Salpeter case since, if $\alpha < 2.35$, τ should be even shorter. We conclude that a "young" burst of star formation is the most likely explanation for an excess of type Ib-c supernovæ in Seyfert galaxies.

Acknowledgments

This work has benefited from research grant Cofin 00-02-007 of Italian MURST.

References

Bressan, A., 1994, Space Sci. Rev., 66, 373.
Cappellaro, E., et al. 1993, A&A, 268, 472.
Vila-Costas, M. B., & Edmunds, M. G. 1992, MNRAS, 259, 121.

ON THE EVOLUTION OF COSMOLOGICAL TYPE IA SUPERNOVAE AND THE GRAVITATIONAL CONSTANT

E. García-Berro
IEEC/UPC, Física Aplicada, Mòdul B–5, Campus Nord, 08034 Barcelona, Spain

E. Gaztañaga
INAOE, Astrofisica, Tonantzintla, Apdo Postal 216 y 51, Puebla 7200, Mexico

J. Isern
IEEC/CSIC, Barcelona, Spain

O. Benvenuto, and L. Althaus
Facultad de Ciencias Astronómicas y Geofísicas, La Plata, Argentina

Abstract A varying gravitational constant G could affect the interpretation of the recent high-redshift Type Ia supernovae results. If the local value of G at the space-time location of distant supernovae is different, it would change both the thermonuclear energy release and the time scale of the supernova outburst. In both cases the effect is related to a change in the Chandrasekhar mass $M_{\rm Ch} \propto G^{-3/2}$. In addition the integrated variation of G with time would also affect cosmic evolution and therefore the luminosity distance relation. Here we investigate in a consistent way how these different effects of a varying G could change the current interpretation of the Hubble diagram of Type Ia supernovae and use the results to set upper bounds on an hypothetical variation of G.

Introduction

Type Ia supernovae (SNeIa) are supposed to be one of the best examples of standard candles. This is because, although the nature of their progenitors and the detailed mechanism of explosion are still the subject of a strong debate, their observational light curves are relatively well understood and, consequently, their individual intrinsic differences can be easily accounted for. Therefore, ther-

M. Chávez et al. (eds.),
New Quests in Stellar Astrophysics: The Link Between Stars and Cosmology, 215–222.

monuclear supernovae are well suited objects to study the Universe at large, especially at high redshifts ($z \sim 0.5$), where the rest of standard candles fail in deriving reliable distances, thus providing an unique tool for determining cosmological parameters or discriminating among different alternative cosmological theories.

Using the observations of 42 high redshift Type Ia supernovae and 18 low redshift supernovae (Riess et al. 1998; Perlmutter et al. 1999), both the Supernova Cosmology Project and the High-z Supernova Search Team found that the peak luminosities of distant supernovae appear to be ~ 0.2 magnitude fainter than predicted by a standard decelerating universe ($q_0 > 0$). Based on this, the Supernova Cosmology Project derived $\Omega_M = 0.28^{+0.14}_{-0.12}$ at 1σ, for a flat universe, thus forcing a non-vanishing cosmological constant. However this conclusion is based on the assumption that there is no mechanism likely to produce an evolution of the observed light curves over cosmological distances. In other words: both teams assumed that the intrinsic peak luminosity and the time scales of the light curve were exactly the same for both the low-z and the high-z supernovae. More recently Riess et al. (1999a,b) have found evidences of evolution between the samples of nearby supernovae and those observed at high redshifts by comparing their respective risetimes, thus casting some doubts about the derived cosmological parameters. In particular Riess et al. (1999a,b) find that the sample of low-z supernovae has an average risetime of 19.98 ± 0.15 days whereas the sample of high-z supernovae has an average risetime of 17.50 ± 0.40 days. The statistical likelihood that the two samples are different is high (5.8σ). Riess et al. (1999b) also analyze several potential alternatives to produce, within a family of theoretical models, an evolution with the observed properties: distant supernovae should be intrinsically fainter and at the same time should have smaller risetimes. All the families of models studied so far have the inverse trend: decreasing peak luminosities correspond to longer risetimes.

On the other hand, and from the theoretical point of view, it is easy to show that a time variation of the gravitational constant, in the framework of a Scalar-Tensor cosmological theory, can reconcile the observational Hubble diagram of SNeIa with an open $\Omega_\Lambda = 0$ universe. The purpose of this paper is to analyze the effect of varying G in the current interpretation of the Hubble diagram of distant SNeIa and to use this analysis to set up upper bounds on its rate of change. The paper is organized as follows: in section Ÿ2 we describe the effects of a varying G on the physics of supernovae; in section Ÿ3 we analyze the effects of a varying G on the luminosity distance of distant supernovae and, finally in section Ÿ4 we discuss our results and draw our conclusions.

1. The effects of a varying G on the physics of supernovae

Our starting point is simple: assume that all thermonuclear supernovae release the same amount of energy (E). In a simple model of light curve (Arnett 1982) the peak luminosity is proportional to the mass of nickel synthesized, which in turn, to a good approximation, is a fixed fraction of the Chandrasekhar mass ($M_{\rm Ni} \propto M_{\rm Ch}$), which depends on the value of gravitational constant: $M_{\rm Ch} \propto G^{-3/2}$. Thus we have $E \propto G^{-3/2}$, and if one assumes a slow decrease of G with time, distant supernovae should be dimmer. Therefore, we have:

$$M - M_0 = \frac{15}{4} \log\left(\frac{G}{G_0}\right) \tag{1}$$

where M stands, as usual, for the absolute magnitude, G for the the precise value of the gravitational constant at a given redshift and the subscript 0 denotes their local values. From this equation we can see that in order to reduce the apparent luminosity of distant supernovae by $\Delta m \simeq 0.2$ a dramatic change of G is required (of the order of $\sim$13%).

However, the time scales of supernovae depend as well on the Chandrasekhar mass. Let us elaborate on this last point. According to the analytic model of light curve of Arnett (1982), the width of the peak of the light curve of SNeIa is given by:

$$\tau \propto \left(\frac{M_{\rm ej}^3}{M_{\rm inc}}\right)^{1/4} \tag{2}$$

where $M_{\rm ej}$ is the ejected mass and $M_{\rm inc}$ is the incinerated mass. Within our current knowledge of the mechanisms of explosion of SNeIa both masses can be considered proportional to the Chandrasekhar mass, and therefore we have $\tau \propto M_{\rm Ch}^{1/2}$ or, equivalently, $\tau \propto G^{-3/4}$. Since the risetime for distant supernovae is obtained from semi-empirical models, that is a template light curve which takes into account the decline rate and the width of the peak, one can then also assume this dependence on G for the risetime. This expression has the right properties since distant supernovae have smaller peak luminosities and, at the rise time, smaller risetimes, as required by observations

$$\left\langle\frac{\tau}{\tau_0}\right\rangle \simeq \left\langle\frac{G}{G_0}\right\rangle^{-3/4} \tag{3}$$

To make some quantitative estimates, we can use the mean evolution found by Riess et al. (1999a,b). From their Fig. 1 we obtain the following widths of the light curve when the supernova is 2.5 magnitudes fainter than the peak luminosity: $\tau_0 = 45.0 \pm 0.15$ (at $z \simeq 0$) and $\tau = 43.8 \pm 0.40$ (at $z \simeq 0.5$), were the errors in the widths have been ascribed solely to the errors in the risetimes.

Thus, from Eq. (3) we obtain $G/G_0 \leq 1.037 \pm 0.017$ (2σ errors) at $z \simeq 0.5$. Therefore, a very small variation of the gravitational constant could account for the reported differences in the SNeIa time scales. On one hand these limits on the variation of G could be considered as weak, in the sense that since most SNeIa are discovered close to its peak luminosity the width of the light curve is poorly determined. However, on the other hand, the upper limit derived here can be regarded as very strong since: i) the statistical significance of the observational data is very strong, and ii) the value of this upper limit *does not depend* on the adopted cosmological model but on the well calibrated relation between the duration of the supernova outburst and its peak magnitude. We would like to stress here that this upper bound is a *maximum* estimate of the rate of change of G since we are attributing the reported inhomogeneity of SNeIa exclusively to a potential variation of G. Should new supernova models ultimately describe the reported difference in the risetimes and peak luminosities of distant supernovae this upper bound would become even tighter.

2. The Hubble diagram

The next question we want to address here is how an hypothetical variation of the gravitational constant (of the amount of $\sim 3.7\%$ derived in the previous section) translates in the Hubble diagram of distant supernovae. To be consistent we need to derive a luminosity distance relation which also includes the possible effects of a varying G on the cosmological evolution (see Luke & Szamosi 1972). The possibility that G could vary in space and/or time is motivated in the framework of Scalar-Tensor theories of gravity (STTs) such as Jordan-Brans-Dicke (JBD) theory or its extensions. JBD scalar fields have been also used in the early Universe in the context of the so-called extended inflation (La & Steinhardt 1989) and in quintessence models (Wang et al. 1999). To make quantitative predictions we will consider cosmic evolution in STTs, where G is derived from a scalar field ϕ which is characterized by a function $\omega = \omega(\phi)$ determining the strength of the coupling between the scalar field and gravity. In the simplest JBD models, ω is just a constant and $G \simeq \phi^{-1}$, however if ω varies then it can increase with cosmic time so that $\omega = \omega(z)$. The Hubble rate H in these models is given by:

$$H^2 \equiv \left(\frac{\dot{a}}{a}\right)^2 = \frac{8\pi\rho}{3\phi} + \frac{1}{a^2R^2} + \frac{\Lambda}{3} + \frac{\omega}{6}\frac{\dot{\phi}^2}{\phi^2} - H\frac{\dot{\phi}}{\phi}, \tag{4}$$

this equation has to be complemented with the acceleration equations for a and ϕ, and with the equation of state for a perfect fluid: $p = (\gamma - 1)\rho$ and $\dot{\rho} + 3\gamma H\rho = 0$. The structure of the solutions to this set of equations is quite rich and depends crucially on the coupling function $\omega(\phi)$ (see Barrow & Parsons

1996). Here we are only interested in the matter dominated regime: $\gamma = 1$. In the weak field limit and a flat universe the exact solution is given by:

$$G = \frac{4+2\omega}{3+2\omega}\phi^{-1} = G_0(1+z)^{1/(1+\omega)}. \tag{5}$$

In this case we also have that $a = (t/t_0)^{(2\omega+2)/(3\omega+4)}$. This solution for the flat universe is recovered in a general case in the limit $t \to \infty$ and also arises as an exact solution of Newtonian gravity with a power law $G \propto t^n$ (Barrow 1996). For non-flat models, $a(t)$ is not a simple power-law and the solutions get far more complicated. To illustrate the effects of a non-flat cosmology we will consider general solutions that can be parameterized as Eq. (5) but which are not simple power-laws in $a(t)$. In this case, it is easy to check that the new Hubble law given by Eq. (4) becomes:

$$H^2(z) = H_0^2\left[\hat{\Omega}_M(1+z)^{3+1/(1+\omega)} + \hat{\Omega}_R(1+z)^2 + \hat{\Omega}_\Lambda\right] \tag{6}$$

where $\hat{\Omega}_M$,$\hat{\Omega}_R$ and $\hat{\Omega}_\Lambda$ follow the usual relation: $\hat{\Omega}_M + \hat{\Omega}_R + \hat{\Omega}_\Lambda = 1$ and are related to the familiar local ratios ($z \to 0$): $\Omega_M \equiv 8\pi G_0\rho_0/(3H_0^2)$, $\Omega_R = 1/(a_0RH_0)^2$ and $\Omega_\Lambda = \Lambda/(3H_0^2)$ by:

$$\hat{\Omega}_M = \frac{\Omega_M}{g}\left(\frac{3+2\omega}{4+2\omega}\right) \; ; \; \hat{\Omega}_\Lambda = \frac{\Omega_\Lambda}{g} \; ; \; \hat{\Omega}_R = \frac{\Omega_R}{g} \tag{7}$$

$$g \equiv 1 + \frac{1}{(1+\omega)} - \frac{1}{6}\frac{\omega}{(1+\omega)^2} \tag{8}$$

Thus the general relativity limit is recovered as $\omega \to \infty$. The luminosity distance $d_L = d_L(z, \Omega_M, \Omega_\Lambda, \omega)$ is obtained as usual from the (line-of-sight) comoving coordinate distance: $r(z) = \int dz'/H(z')$, with the trigonometric or the hyperbolic sinus to account for curvature (Peebles 1993). In the limit of small z we recover the usual Hubble relation: $y = H_0 r = z - (1+\hat{q}_0)z^2/2$ where a new deceleration $\hat{q}_0$ parameter is related to the standard one by:

$$\hat{q}_0 = \frac{q_0}{g} + \frac{\hat{\Omega}_M}{2(1+\omega)}. \tag{9}$$

One can see from these equations that even for relative small values of ω the effect of a varying G on d_L is small. For example for $\Omega_R = 0$ and $\Omega_M = 1$ at $z = 0.5$ we have $d_L = 0.5513$ for $\omega = 10$ and $d_L = 0.5505$ for $\omega = 10^3$. Therefore, the cosmological effect in the Hubble diagram of SNeIa is one order of magnitude smaller than the effect produced by a varying G on the Chandrasekhar mass. Nevertheless, it should be noted that the former tends to decrease the acceleration and therefore it partially compensates the later effect.

In summary, Eq. (5) parameterizes the change in G as a function of ω while Eqs. (6-8) parameterize the corresponding cosmic evolution. Thus, using Eq. (5), we have:

$$m(z) = M_0 + 5\ \log d_L(z, \Omega_M, \Omega_\Lambda, \omega) + \frac{15}{4(1+\omega)} \log{(1+z)} + 25 \quad (10)$$

which reduces to the standard relation as $\omega \to \infty$.

3. Discussion and Conclusions

In astrophysics and cosmology the laws of physics (and in particular the simplest version of general relativity) are extrapolated outside its observational range of validity. It is therefore important to test for deviations of these laws at increasing cosmological scales and times (redshifts). SNeIa provide us with a new tool to test how the laws of gravity and cosmology were in faraway galaxies ($z \simeq 0.5$). The observational limits on $\dot{G}/G$ come from quite different times and scales (Barrow & Parsons 1996; Will 1993), but mostly in the local and nearby environments at $z \simeq 0$ (solar system, binary pulsars, white dwarf cooling, neutron stars). Typical upper bounds give $\dot{G}/G \lesssim 10^{-11} - 10^{-12}\ \mathrm{yr}^{-1}$, or $G/G_0 \lesssim 1.10 - 1.01$, which in the JBD theory of gravitation translates into $\omega \gtrsim 10 - 100$. Here we have found from the analysis of the risetimes of SNeIa that $G/G_0 \lesssim 1.04$ at $z \simeq 0.5$, *independently of the adopted cosmological model*, which is roughly consistent with the local upper bounds. In the context of JBD or STT models these limits on $\dot{G}$ correspond to $\omega \gtrsim 10$ and are therefore less restrictive than the solar system limits $\omega \gtrsim 500$. However, STTs predict $\omega = \omega(\phi)$. To be precise, ω is not required to be a constant, so that ω can increase with cosmic time, $\omega = \omega(z)$, in such a way that it could approach the general relativity predictions ($\omega \to \infty$) at present time and still give significant deviations at earlier cosmological times. Furthermore, it has been shown (Damour & Nordtvedt 1993) that the cosmological evolution makes STTs practically indistinguishable from General Relativity at the present epoch. Our results set strong constraints at cosmological distances. It is however beyond the scope of this paper to study STTs in detail since we have shown that for the upper limit to the rate of change of G derived so far, the dominant contribution of a varying G to the Hubble diagram at $z \sim 0.5$ comes from the change in the Chandrasekhar mass rather than from changes in the theory of gravitation. In this sense, cosmological nucleosynthesis also offers some limit on the amount of variation of G. Roughly speaking, the bounds from nucleosynthesis arise from the sensitivity of the abundances of light elements produced at high temperature to the expansion rate of the Universe at those temperatures, especially ^{4}He. There is a range of opinions, but there is also

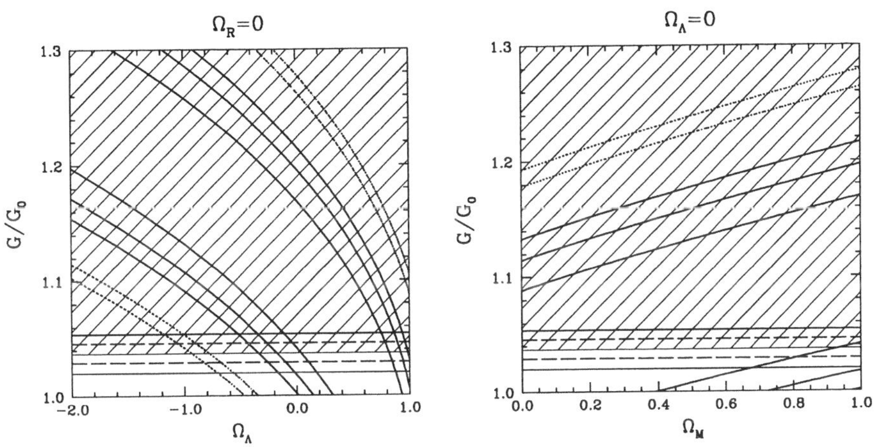

Figure 1. Confidence contours of G/G_0 for $\Omega_R = 0$ (left) and $\Omega_\Lambda = 0$ (right).

the widespread agreement that the expansion rate must have been well within a factor of two of the standard model. Some might even push for more stringent limits that would exclude changes by even as little as $\sim$ 10%, which would be marginally consistent with our analysis — see the most recent analysis by Santiago et al. (1997) for a detailed discussion.

Finally, we have re-done the likelihood analysis of the Supernova Cosmology Project allowing for a varying G. Thus, we use the same observational data but with the magnitude-distance relation given by Eq. (10). This analysis is shown in Fig. 1 where we show the confidence contours (at the 99%, 90%, 68% — solid lines — 5% and 1% confidence level — dotted lines) obtained from the analysis of the Hubble diagram of SNeIa (at $z \sim 0.5$) with the procedure outlined in the previous sections in the $(G/G_0, \Omega_\Lambda)$ plane for a flat $\Omega_R = 0$ universe and in the $(G/G_0, \Omega_M)$ plane for the case $\Omega_\Lambda = 0$ (right panel). The region excluded by the analysis of the rise times is the shaded region shown in both panels and the corresponding errors in the upper bound are shown as horizontal dashed (1σ) and continuous (2σ) lines. The confidence intervals in Fig. 1 show that an accelerating universe (for instance, one with $\Omega_M = 1$ and $\Omega_\Lambda = 0$) is still marginally compatible with the observations, in contrast with the standard analysis in general relativity (that is, with a constant G). Regarding these diagrams it should be stressed that since the dependence of the luminosity distance on the gravitation theory is small for the rate of change of G derived in Ÿ1, the constraints in the cosmological parameters shown in Fig.1 do not rely strongly on the assumed STT model. In this sense, our approach is quite different from the "quintessence" solution to the SNeIa results (see Wang et al. 1999, and references therein).

Obviously more work is needed both regarding other observational consequences of a varying G and on the physics of supernovae. In particular, an improvement of our knowledge of the physics of thermonuclear supernovae would provide us with an unique tool to test fundamental laws of physics over cosmological distances. In addition it should be stressed that new observations of distant supernovae, or other standard candles, at higher redshifts ($z > 1$) could constrain even more the current limits on the variation of the fundamental constants.[1]

Acknowledgments

This work was supported by the DGES grants PB97–0983–C03–02 and PB96–0925 and by the CIRIT (Generalitat de Catalunya) grant 1995SGR-0602. EGB acknowledges support from Sun MicroSystems under the Academic Equipment Grant AEG-7824-990325-SP.

References

Arnett, D. 1982, ApJ, 253, 785.

Barrow, J.D. 1996, MNRAS, 282, 1397.

Barrow, J.D., & Parsons, P. 1997, Phys. Rev. D, 55, 1906.

Damour, T., & Nortdvedt, K. 1993, Phys. Rev. D, 48, 3436.

La, D., & Steinhardt, P.J. 1989, Phys. Rev. Lett., 62, 376.

Luke, S.K., & Szamosi, G. 1972, A&A, 20, 397.

Peebles, J. 1993, Principles of Physical Cosmology, (Princeton: Princenton University Press).

Perlmutter, S., et al. (The Supernova Cosmology Project) 1999, ApJ, 517, 565.

Riess, A.G., et al. 1998, AJ, 116, 1009.

Riess, A.G., et al. 1999a, AJ, 118, 2675.

Riess, A.G., et al. 1999b, AJ, 118, 2668.

Santiago, D.I., Kalligas, D., & Wagoner, R.V. 1997, Phys. Rev. D, 56, 7627.

Wang, L., Caldwell, R.R., Ostriker, J.P., & Steinhard, P.J. 1999, ApJ, 530, 17.

Will, C.M. 1993, Theory and experiment in Gravitational Physics, 2nd edition, (Cambridge: Cambridge University Press).

[1]After this work was completed, and submitted to astro-ph/9907440, Aldering et al. (2000, AJ 119, 2110) demonstrated that the risetime uncertainties were underestimated in Riess et al. (1999a,b), being revised upwards to ± 1.2 days statistical and $^{+3.6}_{-1.9}$ days systematical. Gaztañaga, García-Berro, Isern, Bravo and Domínguez (2001, submitted) present a revision of the implications for G/G_0 and $\dot{G}$.

LIFTING THE VEIL: γ-RAY BURSTS AS BEACONS FOR COSMIC STAR FORMATION

Enrico Ramirez-Ruiz, Neil Trentham, Andrew W. Blain
Institute of Astronomy, Madingley Road, Cambridge, CB3 OHA, UK

Abstract

How much of the star formation that has happened over the age of the Universe is obscured? We attempt to address this question by studying the extinction towards γ-ray bursts (GRBs) and their afterglows. Several arguments suggest such an approach. First, GRBs are thought to be associated with the deaths of massive stars and so their formation rate ought to trace that of the massive stars responsible for the observed UV and far-infrared luminosities of both the Lyman break and SCUBA galaxies. Secondly, GRBs are the final phase of the evolution of these massive stars. These stars do not live long enough to travel far from their place of birth, and so GRBs should be located where the stars formed. Thirdly, GRB afterglows are seen in both the optical and X-ray wave-bands, and so we can hope to exploit the differential effects of extinction between the two wave-bands to test whether large amounts of dust and gas are present in the galaxies hosting the GRBs. We can then infer how many of the GRBs occurred in galaxies with both moderate and considerable amounts of dust, likely to be Lyman break and SCUBA galaxies respectively.

1. γ-Ray Bursts: a New Cosmic Window

The γ-ray bursts (GRBs) for which redshifts are known are bright enough to be detectable, in principle, out to much larger distances than those of the most luminous quasars or galaxies detected at present (Lamb & Reichart 2000). Within the first minutes to hours after the burst, the optical light from afterglows is known to have a range of visual magnitudes $m_v \sim 10 - 15$, far brighter than quasars. Thus, promptly localized GRBs could serve as beacons which, shining through the pregalactic gas, provide information about much earlier epochs in the history of the Universe. For the long burst afterglows localized so far, the host galaxies show signs of ongoing star formation activity necessary for the presence of young, massive progenitor stars. The physical properties of the afterglows, their locations at a few kpc from the center of the host galaxies, the presence of iron line features, and the evidence of a supernova component

M. Chávez et al. (eds.),
New Quests in Stellar Astrophysics: The Link Between Stars and Cosmology, 223–226.

detected several weeks after the burst (see Mészáros 2001 for a review), give strong support to the idea that the most common GRBs are linked to the cataclysmic collapse of massive stars into black holes (MacFadyen & Woosley 1999). In this context, high-redshift GRBs could become invaluable probes of the history of high-mass star formation and galaxy evolution (Blain & Natarajan 2000; Lamb & Reichart 2000). Because γ-rays are not attenuated by intervening dust and gas, the selection of cosmic sites of massive star formation by this method is unlikely to be affected by the biases associated with dust absorption.

2. Star Formation: Obscured vs. Unobscured

One of the major recent achievements in extra-galactic astronomy has been an estimate of the history of the cosmic star formation rate, using two types of complementary measurements. First, optical and near-infrared observations can be made of distant galaxies. The rest-frame ultraviolet (UV) or optical luminosities of these galaxies can then be converted into star formation rates and summed over a representative sample of galaxies in each redshift interval to derive the form of the comoving star formation rate of the Universe as a function of redshift, as is conventionally presented in the Madau plot (Madau et al. 1996). Secondly, submillimeter (submm) and far-infrared measurements of source counts and the extra-galactic background light can be used to infer the amount of dust-enshrouded high-redshift star-formation activity. The redshift distribution of sources contributing to these source counts and backgrounds is poorly known; nevertheless the constraints are tight enough that reasonable progress has been made in constructing a far-infrared/submm Madau plot (Blain 2001).

A compilation of measurements of the UV star-formation rate history is presented in Figure 1 (Somerville, Primack & Faber 2000), along with a model curve that shows the total (UV and submm) star formation rate history that has been derived from a fit to the far-infrared and submm counts and backgrounds (Blain 2001). Since the curve is so much higher than the points it is immediately apparent that most star formation has happened in the submm mode. High-redshift submm-selected galaxies are thought to have properties similar to local ultraluminous infrared galaxies (ULIGs) like Arp 220, based on the concordance between their luminosities and spectral energy distributions. If this is the case, we expect the star formation to occur in either relatively diffuse and unobscured UV-bright star-forming galaxies or in very compact and heavily obscured submm-bright ULIGs. Given that a significant fraction of the submm-wave background originates in luminous SCUBA galaxies (and subject to the caveats discussed in Ramirez-Ruiz, Trentham & Blain 2001), we estimate about **three quarters** of the star formation in the Universe occurred in the

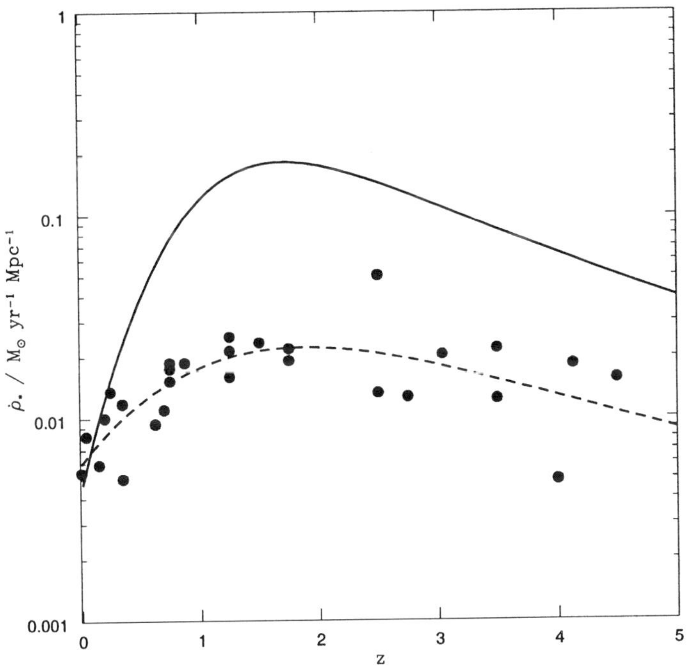

Figure 1. The comoving star-formation rate of the Universe as a function of redshift. The points represent the optical/UV data compiled by Somerville, Primack & Faber (2000) with no corrections for dust extinction – the UV–optical Madau plot. The solid line represents the total (UV and submm) Madau plot, from the models of Blain (2001), assuming no AGN contribution to the far-infrared and submm backgrounds and counts.

submm-bright mode. This suggests that most of the high-mass star-formation activity in the Universe is heavily obscured.

3. Extinction of γ-Ray Burst Afterglows

Because GRBs represent the final phase of the evolution of very massive stars that do not live long enough to leave their birth place, a large fraction are likely to be still enshrouded in their placental clouds of molecular gas and dust. This environment will attenuate the GRB afterglow radiation at X-ray and optical wavelengths. Since we expect the molecular clouds and dust to be more prevalent in submm galaxies than in UV ones, we expect the afterglows originating in the submm galaxies to be more attenuated.

About three quarters of the star formation in the Universe appears to be heavily obscured. If the SCUBA galaxies are high-redshift counterparts of the ULIGs, and are powered mainly by star-formation, then one would expect three quarters of GRBs to have no optical afterglows, although almost all would have an X-ray afterglow. The observations of GRB afterglows are consistent with this picture. First, among the 31 X-ray afterglows detected by the WFC of *BeppoSAX*, $60 \pm 15\%$ had no optical counterparts. Secondly, X-ray afterglows were observed in almost all the bursts detected by the WFC of *BeppoSAX*. Even for extreme submm galaxies like Arp 220, the obscuring column densities are not high enough to completely extinguish the hard X-ray fluxes. Column densities of the order of $10^{24}\,\mathrm{cm}^{-2}$ or higher are required for noticeable effects on the X-ray spectrum. Based on the *BeppoSAX* Hardness Ratio (defined as the ratio of the fluence at 0.2–6 keV and the fluence at 2–10 keV) derived for the GRB with non-detected afterglows, we estimate that the line-of-sight column densities for these bursts are characteristically below $10^{23.5}\,\mathrm{cm}^{-2}$. These results imply that even if GRBs are located in the most absorbed regions of SCUBA galaxies, then the corresponding absorption is not enough to hide their X-ray afterglows.

4. Conclusions

In summary, we expect about three quarters of the star formation in the Universe to have occurred in submm-bright as opposed to UV-bright galaxies. GRBs in submm-bright galaxies should have their optical afterglows extinguished by internal dust absorption, but only very few should have their 2–10 keV X-ray afterglows obscured. The value of three quarters that we quote is marginally consistent with observations of GRBs: 60 ± 15 per cent of GRBs have no detected optical afterglow, whereas almost all have an X-ray afterglow. Therefore, GRBs can be used to detect a putative population of dust-enshrouded forming galaxies at high redshifts.

References

Blain, A. W. 2001, in the Ringberg Conference, in press.

Blain, A. W., & Natarajan, P. 2000, MNRAS, 312, L35.

Lamb, D. Q. & Reichart, D. E. 2000, ApJ, 538, 1.

Lazzati D. et al. 2001, MNRAS, submitted.

MacFadyen, A. I. & Woosley, S. E. 1999, ApJ, 524, 262.

Madau, P., Ferguson, H. C., Dickinson, M. E., Giavalisco, M., Steidel, C. C., & Fruchter, A. 1996, MNRAS, 283, 1388.

Mészáros, P. 2001, Science, 291, 79.

Ramirez-Ruiz, E., Trentham, N., & Blain, A. W. 2001, MNRAS, submitted.

Somerville, R. S, Primack, J. R., & Faber, S. M. 2000, MNRAS, 320, 504.

VII

STELLAR ENVIRONMENTS IN HIGH-REDSHIFT GALAXIES

STARS AND GAS IN HIGH REDSHIFT GALAXIES

Max Pettini
Institute of Astronomy, Cambridge, UK

Abstract Recent advances in instrumentation and observing techniques have made it possible to begin to study in detail the stellar populations and the interstellar media of galaxies at redshift $z = 3$, when the universe was still in its 'teen years'. In keeping with the theme of this conference, I show how our knowledge of local star-forming regions can be applied directly to these distant galaxies to deduce their ages, metallicities, initial mass function, and masses. I also discuss areas where current limitations in stellar astrophysics have a direct bearing on the interpretation of the data being gathered, at an ever increasing rate, on the high redshift universe.

Introduction

This meeting comes at a most opportune time. Thanks to major technological advances in the last few years, we are now in the lucky position of not only being able to identify large numbers of galaxies at redshift $z \simeq 3$ (corresponding to look-back times of 85% of the age of the universe for an $\Omega_{\rm M} = 0.3$, $\Omega_\Lambda = 0.7$ cosmology), but also to begin to study their physical properties in some detail. The limiting factors to our understanding of the nature of the high redshift galaxy population, and what it tells us about the process of galaxy formation in general, are no longer lack of direct observations but rather the tools at our disposal for interpreting the increasing body of data. The fundamental link between stellar astrophysics and observational cosmology has never been more relevant. I shall illustrate the strong interplay between these two areas of research with a few topical examples, concentrating in particular on (a) the stellar populations of high redshift galaxies; (b) abundance determinations in stars and gas; and (c) the impact of star formation at high z on the intergalactic medium (IGM).

1. Stellar Populations

One of the turning points in extragalactic astronomy in the 1990s was the realization that high redshift galaxies can be found in large numbers using a highly efficient photometric selection technique based on the passage of the Lyman

M. Chávez et al. (eds.),
New Quests in Stellar Astrophysics: The Link Between Stars and Cosmology, 229–240.

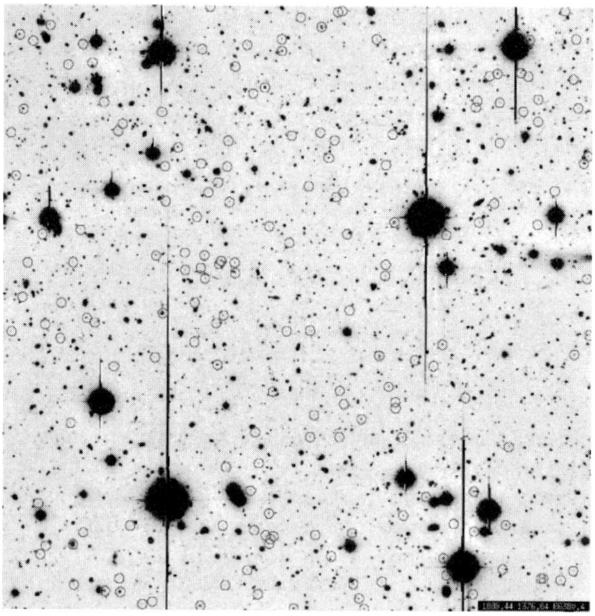

Figure 1. A typical deep CCD frame recorded at the prime focus of a 4m-class telescope. This particular image (of the field designated DSF2237a) was obtained with the COSMIC camera of the Palomar Hale telescope, by exposing for a total of two hours through our custom made $\mathcal{R}$ filter. In the 9×9 arcmin field of view (corresponding to co-moving linear dimensions of $11.6 \times 11.6\,h^{-1}$ Mpc at $z = 3$) there are more than 4000 galaxies brighter than $\mathcal{R} = 25.5$; of these about 100–150 (circled) show Lyman breaks which place them at redshifts between $z = 2.5$ and 3.5 .

edge—at the rest wavelength of 912 Å—through the U-band. After many years of fruitless searches (targeted mainly at Lyα emission which turned out to be a less reliable marker than anticipated), we have witnessed a veritable explosion of data from the *Hubble Deep Fields* and ground-based surveys (see Fig. 1). Samples of these Lyman break galaxies (LBGs) now include more than a thousand objects with spectroscopic redshifts (e.g. Steidel 2000).

1.1 The Upper End of the Initial Mass Function

At $z \simeq 3$ optical wavelengths correspond to the rest-frame far-UV, where we see the integrated light of short-lived O and early B stars. Such spectra are most effectively analyzed with population synthesis models, the most sophisticated of which is *Starburst99* developed by the Baltimore group (Leitherer et al. 1999). In Fig. 2 we compare *Starburst99* model predictions for different IMFs with the observed spectrum of a particularly well-studied LBG, the gravitationally lensed galaxy MS 1512-cB58 at $z = 2.7268$, in the wavelength region encompassing the C IV $\lambda\lambda$1548, 1550 doublet.

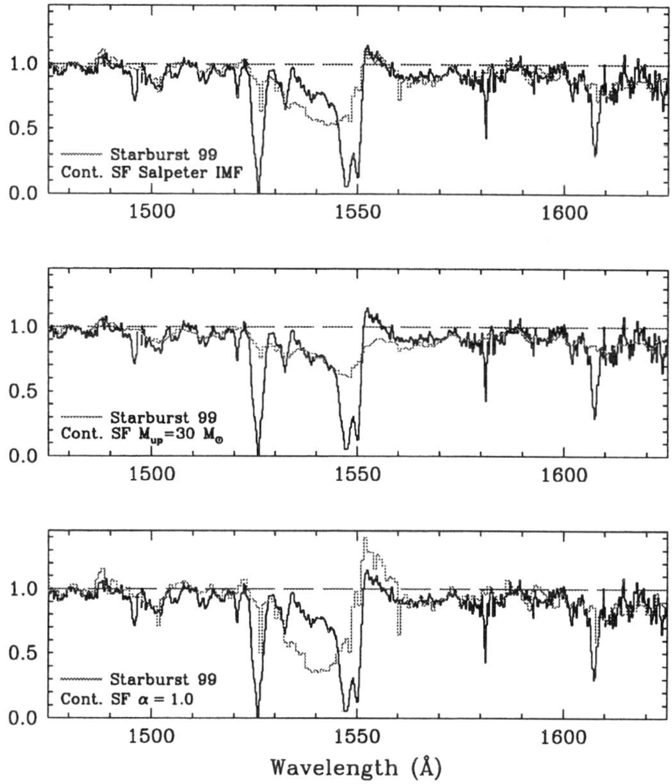

Figure 2. Comparisons between *Starburst99* (Leitherer et al. 1999) population synthesis models with different IMFs (green lines) and the Keck spectrum of MS 1512-cB58 analyzed by Pettini et al. (2000) in the region near the C IV doublet (black histogram). The y-axis is residual intensity.

It is important to realize that the comparison only refers to *stellar* spectral features and does not include the *interstellar* lines, readily recognizable by their narrower widths (these IS lines are much stronger in cB58, where we sample the whole ISM of the galaxy, than in the models which are based on libraries of nearby Galactic O and B stars). With this clarification, it is evident from Fig. 2 that the spectral properties of at least this Lyman break galaxy are remarkably similar to those of present-day starbursts—a continuous star formation model with a Salpeter IMF provides a very good fit to the observations. In particular, the P-Cygni profiles of C IV, Si IV and N V are sensitive to the slope and upper mass limit of the IMF; the best fit in cB58 is obtained with a standard Salpeter IMF with slope $\alpha = 2.35$ and $M_{\rm up} = 100 M_\odot$ (top panel of Fig. 2). IMFs either lacking in the most massive stars or, conversely, top-heavy seem to be excluded by the data (middle and bottom panels of Fig. 2 respectively).

The only significant difference between the observed and synthesized spectra is in the optical depth of the P-Cygni absorption trough which is lower than predicted (top panel of Fig. 2). A possible explanation is that this is an abundance effect, since an analogous weakening of the absorption is seen in OB stars in the Magellanic Clouds (e.g. Lennon 1999) and is also predicted by stellar wind theory (e.g. Kudritzki 1998; Vink, de Koter, & Lamers 2001). However, the recent upgrade of the *Starburst99* code, which now includes spectral libraries of stars in the Magellanic Clouds (Leitherer et al. 2001), has shown that this is only part of the story.

The best fit to the UV spectrum of cB58 is actually obtained with a *composite* stellar population based on *Hubble Space Telescope* observations of the nearby starburst galaxy NGC 5253 by Tremonti et al. (2001). Their long-slit STIS spectra show differences between the young star clusters, less than 20 Myr old, and the more uniform background in this galaxy. While the former require the presence of stars up to 100 $M_\odot$ to reproduce the details of their UV spectra, the field lacks these most massive stars, perhaps because it consists of a slightly older population. Possibly the clusters dissolve over timescales of 10-20 Myr and their surviving stars disperse into the field. In any case, an equal mixture of the field and cluster UV spectra observed in NGC 5253 provides an excellent fit to the integrated (and unresolved) spectrum of cB58 (see Fig. 3).

Before we leave this topic, the point I should like to emphasize is that with a modest amount of effort (and luck in the form of gravitational lensing) it is now possible to obtain spectra of high redshift galaxies of sufficient quality to distinguish between the OB stellar populations of the Milky Way, the Magellanic Clouds, and local starburst galaxies such as NGC 5253!

1.2 Dating the Star Formation Activity

Realistically, the detailed analysis described above can only be applied to a few objects, at least until 30-m telescopes come on line. However, the coarser spectral energy distribution (SED) of Lyman break galaxies still holds important information on the star formation episodes. Broad-band photometry in the optical and near-infrared, spanning the wavelength interval 900–5500 Å in the rest-frame, is now available for more than one hundred galaxies at $z \simeq 3$ (Papovich, Dickinson, & Ferguson 2001; Shapley et al. 2001). The colors over this range (typically four colors are used in the analysis) depend on the degree of dust reddening, $E(B-V)$, and on the age of the stellar population, $t_{\rm sf}$. The two can be decoupled with some degree of confidence provided that the SED includes the age-sensitive Balmer break near 3650 Å which at $z = 3$ falls between the H and K bands—hence the need for accurate near-IR photometry.

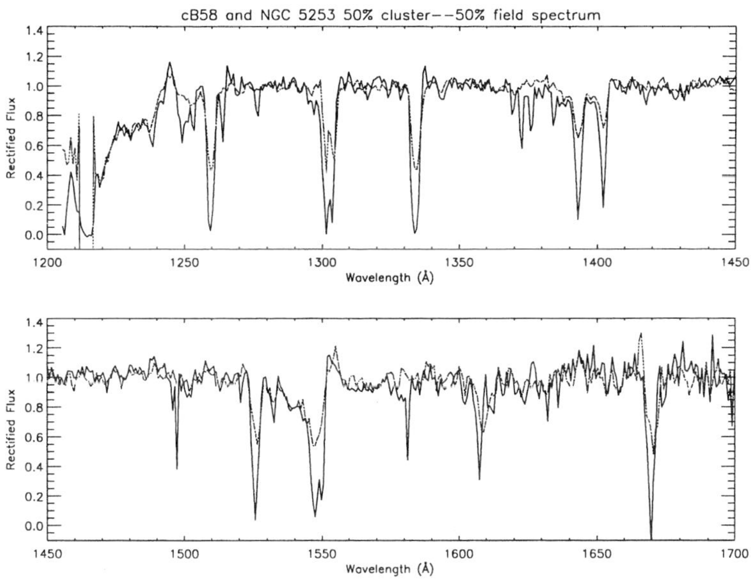

Figure 3. Comparison between the spectrum of MS 1512-cB58 (black) and a composite based on the observed UV spectra of cluster and field stars in the nearby starburst galaxy NGC 5253 (red, or light grey). Such a population mix can reproduce remarkably well the integrated stellar light of this high redshift galaxy (reproduced from Tremonti et al. 2001).

A third parameter, the instantaneous star formation rate, $\Psi(t_{\rm sf})$, determines the normalization (rather than the shape) of the SED. The analyses by Papovich et al. (2001) and Shapley et al. (2001) deduced the best-fitting values of $E(B-V)$, $t_{\rm sf}$, and $\Psi(t_{\rm sf})$ by χ^2 minimization of the differences between the observed SEDs and those predicted by the widely used population synthesis code of Bruzual & Charlot. The results have turned out to be very interesting—some would say surprising (see Fig. 4 and 5).

Evidently, Lyman break galaxies span a wide range of ages. One fifth of the sample considered by Shapley et al. (2001) consists of objects which seem to have just collapsed and are forming stars on a dynamical timescale ($\sim$ 35 Myr). At the other end of the scale, some 20% of the galaxies at $z=3$ have been forming stars for more than 1 Gyr, placing the onset of star formation at much higher redshifts ($z > 5-10$). Furthermore, there appears to be a correlation between age and star formation rate, with the younger objects typically forming stars at about ten times the rate of the older ones and being on average significantly more reddened. The mean SFR is $\langle\Psi(t_{\rm sf})\rangle = 210h^{-2}\,M_\odot\ {\rm yr}^{-1}$ for galaxies with $t_{\rm sf} < 35$ Myr while, for the 20% of the sample with $t_{\rm sf} > 1$ Gyr, $\langle\Psi(t_{\rm sf})\rangle = 25h^{-2}\,M_\odot\ {\rm yr}^{-1}$. This range of properties is further reflected in the total formed stellar masses $m_{\rm star}$ obtained by integrating $\Psi(t_{\rm sf})$ over $t_{\rm sf}$ (a

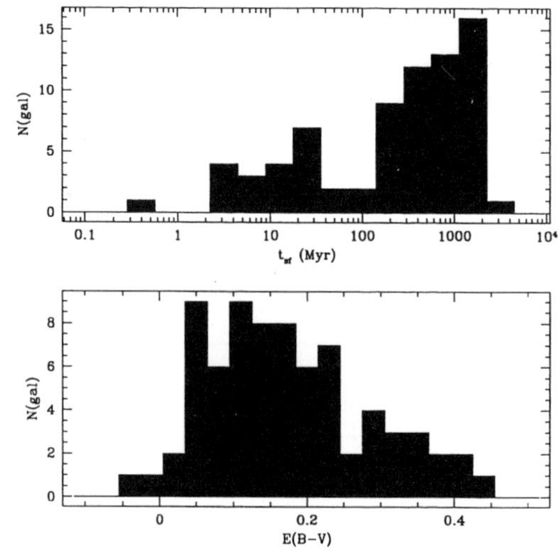

Figure 4. Histograms of best-fitting ages and reddening for the sample of 81 $z \simeq 3$ LBGs analyzed by Shapley et al. (2001). There is a large spread of ages in the population; the median age is 320 Myr and 20% of the objects are older than 1 Gyr. The median $E(B - V) = 0.155$ for the sample corresponds to attenuations by factors of ~ 4.5 and ~ 2 at 1600 Å and 5500 Å respectively.

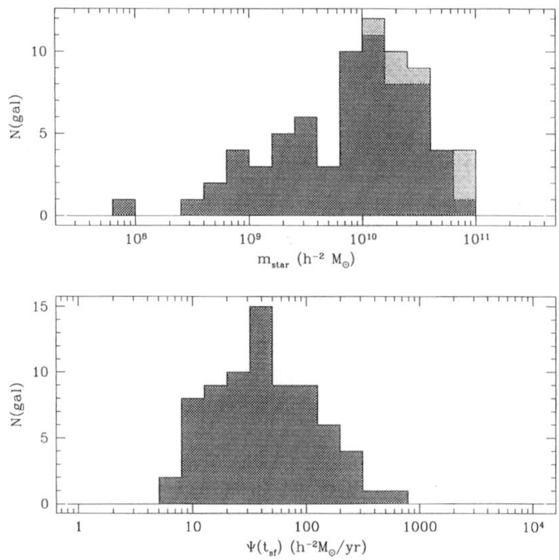

Figure 5. Histograms of assembled stellar mass and star formation rates from Shapley et al. (2001). By redshift $z \simeq 3$ a significant fraction of LBGs seem to be approaching the stellar mass of today's L^* galaxies, $m_{\rm star} \simeq 4 \times 10^{10} M_\odot$.

variety of star formation histories was considered but $m_{\rm star}$ does not depend sensitively on this choice). As can be seen from Fig. 5, by redshift $z \simeq 3$ some galaxies had apparently already assembled a stellar mass comparable to that of an L^* galaxy today, $m_{\rm star} \simeq 4 \times 10^{10} M_\odot$, while 20% of the sample have values of $m_{\rm star}$ one order of magnitude smaller. These findings led Shapley et al. (2001) to spcculatc that we may beginning to discern an evolutionary sequence in Lyman break galaxies, with the younger, dustier, more actively star-forming objects evolving to the older, less reddened, and more quiescent phase. It remains to be seen how this scenario stands up to the scrutiny of future observations, as we try to link the properties of the stellar populations of *individual* galaxies to other parameters, such as dynamical mass and metallicity.

2. Element Abundances

Both the overall level of metal enrichment and the relative abundance pattern of different elements are key pieces of information which would throw light on the nature of high redshift galaxies, their past history of star formation, and their descendants today (e.g. Kobulnicky & Zaritsky 1999). Unfortunately, it has not been easy up to now to measure element abundances from the UV spectral region of LBGs, where data are abundant. Here the strongest—and therefore most easily recorded—interstellar absorption lines are usually saturated and tell us more about the kinematics of the gas than its chemical composition. Stellar abundance indicators in the UV are still very approximate or non-existent (Leitherer 1999). One of the messages from the cosmologists to the stellar astronomers at this meeting is the urgent need for some abundance diagnostics based on stellar UV features. What would be particularly useful are absorption indexes based on the strengths of *broad* blends of stellar photospheric lines, such as those near 1370, 1425 and 1940 Å (Leitherer et al. 2001). Such indexes, which presumably could be measured even from spectra of only moderate resolution, would make the large body of data already in hand accessible to abundance determinations. The development of these tools should probably be attempted theoretically at first, since the acquisition of the relevant stellar libraries is likely to require space observations over a period of many years.

In the nearby universe, element abundances in star forming regions have traditionally been measured from the ratios of optical emission lines from ionized gas. At $z = 3$ these features move to near-infrared wavelengths and have only become accessible in the last eighteen months with the commissioning of high resolution spectrographs on the VLT (ISAAC) and Keck telescopes (NIRSPEC). Using these facilities, we have recently completed the first spectroscopic survey of Lyman break galaxies in the near-IR, bringing together data for 19 objects; the galaxies are drawn from the bright end of the luminosity function, from $\sim L^*$ to $\sim 4\,L^*$ (Pettini et al. 2001). Figure 6 shows an example of the quality

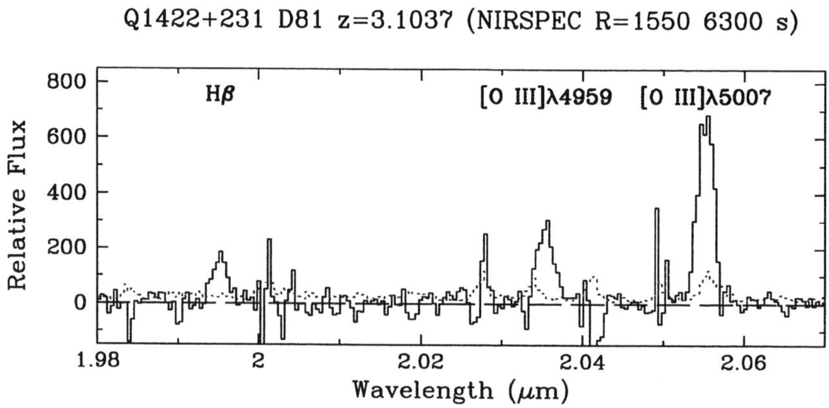

Figure 6. Example of a NIRSPEC K-band spectrum of a Lyman break galaxy. The objects targeted in our survey typically have $K = 21$ (on the Vega scale) and remain undetected in the continuum. However, the nebular emission lines of [O III] $\lambda\lambda$4859, 5007, [O II] 3727 (not shown), and Hβ usually show up clearly with exposures of 2–3 hours. The dotted line is the 1σ error spectrum.

of spectra which can be secured with a 2–3 hour integration. In five cases we attempted to deduce values of the abundance of oxygen by applying the familiar R_{23} = ([O II] + [O III])/Hβ method first proposed by Pagel et al. (1979). We found that generally there remains a significant uncertainty, by up to 1 dex, in the value of (O/H) because of the double-valued nature of the R_{23} calibrator (see Fig. 7). Thus, in the galaxies observed, oxygen could be as abundant as in the interstellar medium near the Sun (Meyer, Jura, & Cardelli 1998; Esteban et al. 1998), or as low as $\sim 1/10$ solar. This degeneracy can in principle be resolved by measuring the [N II]/Hα ratio (Terlevich, Denicolo, & Terlevich 2001); in the one case where this has proved possible so far (Teplitz et al. 2000) values of (O/H) near the upper end of the range are indicated. It will soon be feasible to carry out abundance studies based on the full complement of nebular emission lines, from [O II] λ3727 to Hα, on galaxies at more easily accessible redshifts near $z = 2.2$, where all the lines of interest fall in near-IR atmospheric transmission windows. However, this will remain a time consuming task until multi-object spectrographs operating at near-IR wavelengths become available on large telescopes.

3. Galactic-Scale Outflows

When the redshifts of the interstellar absorption lines, of the nebular emission lines, and of the resonantly scattered Lyα emission line are compared within

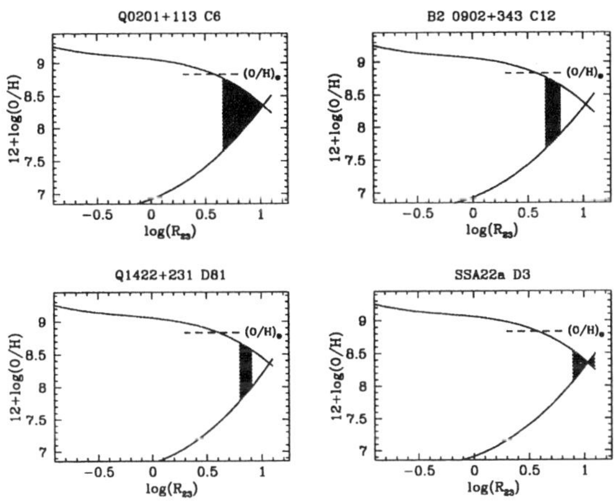

Figure 7. Oxygen abundance from the R_{23} = ([O II]+[O III])/Hβ ratio. In each panel the continuous lines are the calibration by McGaugh (1991) for the ionization index O_{32} = [O III]/[O II] appropriate to that object. The shaded area shows the values allowed by the measured R_{23} and its statistical 1σ error. The broken horizontal line gives for reference the solar abundance $12 + \log(\mathrm{O/H}) = 8.83$ (Grevesse & Sauval 1998).

the same galaxy, a systematic pattern of velocity differences emerges in all LBGs observed up to now (see Fig. 8). We interpret this effect as indicative of galaxy-wide outflows, presumably driven by the supernova activity associated with the star-formation episodes. Such 'superwinds' appear to be a common characteristic of galaxies with large rates of star formation per unit area at high, as well as low, redshifts (e.g. Heckman 2000). They involve comparable amounts of matter as is being turned into stars (the mass outflow rate is of the same order as the star formation rate) and about 10% of the total kinetic energy delivered by the starburst (Pettini et al. 2000). These outflows have a number of important astrophysical consequences.

First, they provide self-regulation to the star formation process (Efstathiou 2000). Galactic winds may well be the key factor at the root of the 'evolutionary sequence' for LBGs discussed in Ÿ2.2 above, and have been invoked to resolve the dark matter problem in disk galaxies (Binney, Gerhard, & Silk 2000).

Second, they can distribute the products of stellar nucleosynthesis over large volumes of the intergalactic medium (the outflow speeds often exceed the escape velocities—Ferrara, Pettini, & Shchekinov 2000). As we have seen, many LBGs are already metal-enriched at $z = 3$ and have by then been forming stars for much of the Hubble time. There is therefore at least the potential for widespread pollution of the IGM with metals, thereby explaining the common occurrence of C IV in the Lyα forest (Ellison et al. 2000; Aguirre et al. 2001).

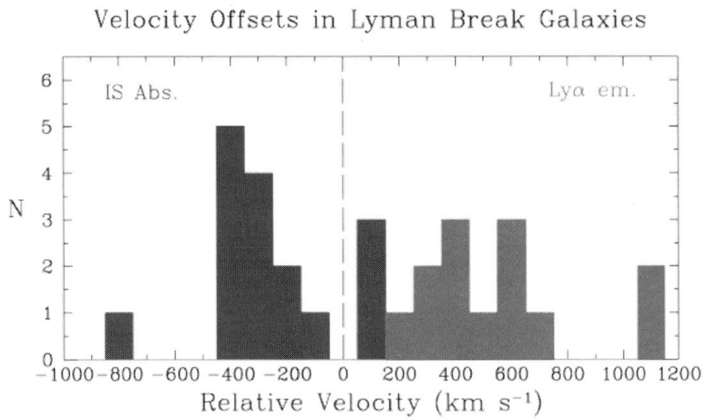

Figure 8. Velocity offsets of the interstellar absorption lines (blue or dark grey) and of the Lyα emission line (red or light grey) relative to [O III] and Hβ. Large scale motions of the order of several hundred km s^{-1} are indicated by the systematic tendency for the former to be blueshifted and the latter redshifted relative to the nebular emission lines.

Third, the outflowing hot gas is likely to 'punch' through the neutral interstellar medium of the galaxies and provide a route through which Lyman continuum photons can leak out of the galaxies, easing the problem of how the universe came to be re-ionized (Steidel, Pettini, & Adelberger 2001). Indeed it now appears (Adelberger et al. 2002, in preparation) that LBGs have a substantial impact on the nearby IGM, and that shock-ionization by their winds leads to a pronounced 'proximity effect'—the Lyα forest is essentially cleared out by these outflows over radii of $\sim 200h^{-1}$ kpc, as also discussed by Rupert Croft at this meeting.

4. Conclusions

In summary, a clearer picture of the Lyman break galaxy population is emerging from the combination of optical and near-IR observations. At redshifts $z \simeq 3$, many of these galaxies already resemble closely what today we call Population I stars in the Milky Way, with a Salpeter slope at the upper end of the IMF and near-solar metallicities. Generally, the onset of star formation seems to have occurred well before the time we observe the galaxies; in some cases perhaps at redshifts as high as $z \simeq 10$. Galactic superwinds, driven by the kinetic energy deposited by the star formation activity, are a common feature and probably play a key role in the evolution of LBGs. These outflows of metal-enriched, hot material are also likely to have a significant impact on the intergalactic medium, with far-reaching consequences which we are only now beginning to appreciate fully. In particular, they have the potential for explaining the widespread distribution of metals in the IGM, its re-ionization, and its structure over scales of hundreds of kpc. As we learn more about LBGs, the similarity with today's starburst galaxies is as striking as it appeared when they were first discovered (Steidel et al. 1996). Thus the link between observational cosmology and stellar astrophysics remains a strong one. In particular, it is clear from this brief review that future advances in the modeling of young stellar populations and in the tools at our disposal for interpreting their spectra will feed directly into a much improved understanding of some of the earliest galaxies in the universe.

Acknowledgments

The work described in this review is part of an on-going collaboration with Chuck Steidel, Kurt Adelberger, Mark Dickinson, Mauro Giavalisco, and Alice Shapley whose major contributions it is a pleasure to acknowledge. I should like to thank the organizers of this enjoyable meeting for their generous invitation, the editors of the proceedings for their patience, and my Mexican colleagues for their warm hospitality.

References

Aguirre, A., Hernquist, L., Schaye, J., Katz,N., Weinberg, D. H., & Gardner, J. 2001, ApJ, 561, 521.

Binney, J., Gerhard, O., & Silk, J. 2001, MNRAS, 321, 471.

Efstathiou, G. 2000, MNRAS, 317, 697.

Ellison, S. L., Songaila, A., Schaye, J., Cowie, L. L., & Pettini, M. 2000, AJ, 120, 1175.

Esteban, C., Peimbert, M., Torres-Peimbert, S., & Escalante, V. 1998, MNRAS, 295, 401.

Ferrara, A., Pettini, M., & Shchekinov, Y. 2000, MNRAS, 319, 539.

Grevesse, N., & Sauval, A. J. 1998, Space Sci. Rev., 85, 161.

Heckman, T. M. 2000, in ASP Conf. Ser. 240, Gas and Galaxy Evolution, eds. J. E. Hibbard, M. P. Rupen, & J. H. van Gorkom, (San Francisco: ASP), p. 345.

Kobulnicky, H. A., & Zaritsky, D. 1999, ApJ, 511, 118.

Kudritzki, R. P. 1998, in Stellar Astrophysics for the Local Group, eds. A. Aparicio, A. Herrero, & F. Sánchez (Cambridge: Cambridge Univ. Press), p. 149.

Leitherer, C. 1999, in Chemical Evolution from Zero to High Redshift, eds. J. R. Walsh & M. R. Rosa (Berlin: Springer Verlag), p. 204.

Leitherer, C., et al. 1999, ApJS, 123, 3.

Leitherer, C., Leão, J. R. S., Heckman, T. M., Lennon, D. J., Pettini, M., & Robert, C. 2001, ApJ, 550, 724

Lennon, D. J. 1999, Rev. Mex. Astron. Astrof. Conf. Series, 8, 21.

McGaugh, S. 1991, ApJ, 380, 140.

Meyer, D. M., Jura, M., & Cardelli, J. A. 1998, ApJ, 493, 222.

Pagel, B. E. J., Edmunds, M. G., Blackwell, D. E., Chun, M. S., & Smith, G. 1979, MNRAS, 189, 95.

Papovich, C., Dickinson, M., & Ferguson, H. C. 2001, ApJ, 559, 620.

Pettini, M., Shapley, A. E., Steidel, C. C., Cuby, J. G., Dickinson, M., Moorwood, A. F. M., Adelberger, K. L., & Giavalisco, M. 2001, ApJ, 554, 981.

Pettini, M., Steidel, C. C., Adelberger, K. L., Dickinson, M., & Giavalisco, M. 2000, ApJ, 528, 96.

Shapley, A. E., Steidel, C. C., Adelberger, K. L., Dickinson, M., Giavalisco, M., & Pettini, M. 2001, ApJ, 562, 95.

Steidel, C.C. 2000, Proc. SPIE, 4005, 22.

Steidel, C.C., Giavalisco, M., Pettini, M., Dickinson, M., & Adelberger, K. L. 1996, ApJ, 462, L17.

Steidel, C. C., Pettini, M., & Adelberger, K. L. 2001, ApJ, 546, 665.

Teplitz, H. I., et al. 2000, ApJ, 533, L65.

Terlevich, R. J., Denicolo, G., & Terlevich, E. 2001, Rev. Mex. Astron. Astrof., in press.

Tremonti, C. A., Calzetti, D., Leitherer, C., & Heckman, T. M. 2001, ApJ, 555, 322.

Vink, J. S., de Koter, A., & Lamers, H. J. G. L. M. 2001, A&A, 369, 574.

ABUNDANCES IN DAMPED LYα SYSTEMS

Paolo Molaro
Osservatorio Astronomico di Trieste, Italy

Abstract Absorption due to clouds along the line of sight to quasars with hydrogen column density large enough to be shielded from intergalactic ionizing photons probes the chemistry of the cosmic gas with unique accuracy. The status of available chemical abundances in Damped Lyα systems is reviewed and compared with those of other astrophysical sites. Some implications for the very early stellar formation and the nature of the high redshift gas giving rise to the Damped Lyα systems are also discussed.

Introduction

Damped Lyα (hereinafter DLAs) are absorption systems toward QSO characterized by the largest hydrogen column density, $\log N(HI) \geq 20.2$ cm^{-2} according to a rather conservative definition introduced by A. Wolfe. They are observed in the whole redshift range from zero up to z= 4.46 (Dessegues-Zavadsky et al. 2001). Consensus regarding DLAs as progenitors of the present day galaxies is general, however, their specific nature has still to be clearly identified. The notion of DLAs as tracers of already formed rotating disks, which was rather popular only few years ago, is now being revised to include other galaxian morphologies. At low redshift some DLA galaxies have been imaged showing a heterogeneous population which comprises dwarfs galaxies, low surface brightness galaxies, irregulars and also spirals (Le Brun et al. 1997, Rao & Turnshek 2000). At high redshift no imaging is feasible and any association is even more uncertain. We cannot exclude here that we are sampling the building blocks of galaxies or even some pregalactic clumps as the lack of large HI column density systems for z$\geq$ 4 seems to suggest (Peroux et al. 2001).

1. Abundances, dust & ionization

At present there are something like 180 known DLAs, out of which about 70 give some sort of chemical information. The most frequently measured ions are SiII, FeII, CrII, ZnII, NiII, AlII, MnII, TiII, SII and NI. New entries are ArI, PII, CoI (cfr Sara Ellison, this conference) and OI. For the interpretation of the

M. Chávez et al. (eds.),
New Quests in Stellar Astrophysics: The Link Between Stars and Cosmology, 241–248.

chemical pattern it is rather important to have access to a large set of chemical elements, in particular elements with different nucleosynthetic origin, different sensitivities to the radiation field or different affinity with dust, since all these act to modify the observed abundances.

In general, column densities in the DLAs can be rather accurate, depending from the complexity of the absorption system. Typical errors in the absolute abundances are of the order of 25%, with the error budget dominated by the uncertainty of the hydrogen column density, while relative abundances have errors of 10% or even better in case of optically thin lines and of simple velocity structure of the absorber. There are no other techniques applied to the high redshift universe where chemical abundance can be obtained with comparable precision. Somewhat paradoxically the precision is often even higher than to that obtained in the Galactic ISM, due to the fact that the most relevant UV resonance lines are redshifted into the optical and resonance lines are de-saturated by the abundance decrease of DLAs.

The large column densities of the DLA shield efficiently the clouds from the surrounding IGM radiation field. Thus DLAs are HI regions with lower ionization elements in their dominant state with negligible ionization corrections (Viegas 1995). C^{3+} and Si^{3+} are often observed but show a different velocity structure and likely originate somewhere outside the HI region. Recently it has been pointed out that Al^{++} and Al^{+} have similar profiles and their ratio anticorrelates with HI suggesting the presence of HII interfaces. However, even when such interfaces are accounted for, the corrections remain negligible for all elements but possibly Al (Vladilo et al. 2001). Inspired by observations of the local metal poor blue compact galaxy IZw18, Izotov et al. (2001) elaborated an alternative model for the DLA composed by two or more components with different metallicities where ionization corrections are significant. New observations of Ar have become available although only in a couple of DLAs, with abundances which are in line with the other elemental abundances (Molaro et al. 2001, Levshakov et al. 2001). Argon is very sensitive to ionization effects and the Izotov et al. correction would result in large Ar overabundunces, which are very unlikely.

The presence of dust in the DLA is rather crucial for the interpretation of the abundance of refractory elements. The refractory elements are depleted from the diffuse phase and locked into dust grains in the DLA as it is revealed by the general deficiency of Fe and Cr compared to Zn (Pettini et al. 1997). Zn is not depleted in the ISM and tracks Fe closely according to the observations in the metal poor stars of the Galactic halo. This argument has been often criticized since Zn has a more complex nucleosynthesis than Fe. Further evidence of the presence of dust has been given by the excess of reddening of quasars showing a DLA in the foreground, by the absence of Lyα emission which has been interpreted as being due to dust grain absorption before escaping, and by

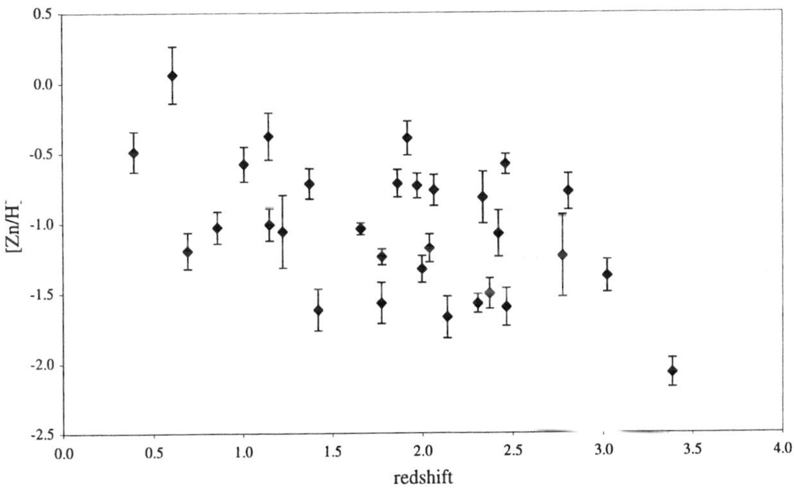

Figure 1. [Zn/H] abundance vs. redshift distribution

the anti-correlations of abundances of refractory elements with the elemental column density (Hou, Boissier & Prantzos 2001) and of molecular hydrogen with the [Cr/Zn] ratio (Levshakov et al. 2001).

2. Cosmological chemical evolution

Considering the uncertainties which affect the refractory elements due to the presence of dust, the evolution of DLA abundances with redshift is safely traced by Zn. In Fig. 1 the [Zn/H] abundances are plotted as a function of redshift. The data are from the compilation of Vladilo et al (2000) with the inclusion of a new measurement for Q0347-3819 at z_{abs}=3.025 obtained by Levshakov et al. (2001). As it was first recognized by Pettini et al. the [Zn/H] abundances show an intrinsic scatter by a factor of 15-30 at all redshifts. Unfortunately, this interpretation is not univocal since the scatter can be equally due to different object ages, different line paths as well as to different galaxian morphologies. The second notable feature, which was quite a surprise, is the reduced evolution shown by the DLAs as a whole with only a modest decrease of the mean metallicity along the four units of redshift spanned by the observations. At $z \approx 1$ where the bulk of star formation occurred and our own Galaxy had already achieved almost solar metallicity the abundances in DLAs are significantly below solar. The rate of the increase is of 0.3 $\pm$0.1 in metallicity per unit of redshift (Vladilo et al. 2000). The same rate is found by Savaglio (2000) who considered a variety of elements after having elaborated corrections for dust depletion. The important fact to note here is that the observed variation is much lower than what predicted by theoretical models (cfr for instance Prantzos &

Boissier 2000). According to Jimenez et al. (1999) such a modest evolution is more consistent with what expected from a population of galaxies characterized by low SFR such as the LSB galaxies.
The moderate evolution shown by DLAs vanishes when the column density-weighted mean is considered (Pettini et al. 1999, Vladilo et al. 2000). As Pettini et al. pointed out, it looks as if the DLAs did not trace the galaxy population responsible for the bulk of star formation, or that for some reasons DLAs were sampling only the low end of the mass-luminosity function for galaxies. The column density-weighted mean is controlled by only few systems with the largest HI column densities, which have also the lowest metallicity. An anti-correlation between [Zn/H] and $\log N(HI)$ was noted by Boisse et al. (1998) who suggested the possibility of an observational bias. When metallicity indicators other than Zn are considered, it is clear that low $\log N(HI)$ systems with low metallicities do exist, but the lack of high $\log N(HI)$ systems with high metallicity is confirmed. This lack suggests the possibility of a significant dust obscuration of QSO behind a metal rich DLA which therefore would tend to be underrepresented in optically selected quasars. There are ongoing projects for studies of the chemical abundances in DLA towards radio-selected QSOs, which hopefully will clarify the problem.
From Fig. 1 it is also possible to see that DLA metallicities are [Zn/H] $\geq$ -2.0 over the whole redshift spanned by observations. Lower metallicities would be easily detected if present, thus revealing that processed material is spread out over all the observable universe. At redshift higher than that shown in Fig. 1 the metallicity is better inferred from FeII or from other elements, despite the dust uncertainty, since the Zn UV lines are redshifted into the infrared. In recent years metallicities have been measured in DLAs with ever increasing redshifts. The current record is the DLA measured at z_{abs}=4.466 towards APM BR J0307-4945 with [Fe/H]=-1.97$\pm$0.19 (Dessagues-Zavadsky et al. 2000), and several DLAs with redshift greater than 4 also show rather high metallicity, thus disappointing naive expectation.
According to the current ΛCDM cosmology with Ω_λ=0.7, Ω_m=0.3, H_0= 70 km s^{-1} Mpc^{-1} at z=4.5 the age is $\approx$ 1.3 Gyrs, which is something about 800 Myr after the formation of the first objects at z$\approx$ 10. Thus the DLAs show evidence of the presence of a prompt enrichment in the metallicity within the first Gyr, followed by essentially no evolution for the following 5 or 6 Gyrs, which is a rather surprising behavior.

3. The chemical pattern

The chemical pattern shown by different elements in a galaxy is of particular interest for recovering its chemical history. A key feature is the ratio between α elements and the iron-peak elements which is related to the relative contribution of Type II and Type I supernovae. This varies with the type of galaxy so that from the [α/Fe] ratio it may be possible to obtain clues on the nature of the galaxies producing the DLAs.

Most of the early works focused on the [Si/Fe] ratios since Si is the most accessible α element. The [Si/Fe] ratio is now available in 38 DLA systems and the straight mean gives [Si/Fe] = 0.39 $\pm$ 0.17, which is about the value observed in the Galactic halo stars with comparable metallicity. This has been originally interpreted as evidence of the signature of TypeII SN yields, which lended support to the hypothesis of DLAs being already formed rotating disks (Lu et al. 1996, Prochaska & Wolfe 1999). However, a very similar ratio of [Si/Fe] = 0.35 $\pm$ 0.19 is observed in the warm phase gas, and $\approx$ 0.80 in the cool disk phase (Savage and Sembach 1996). Therefore, the former interpretation is strictly valid only in complete absence of dust, which, as we mentioned, is rather unlikely.

Different approaches have been followed to resolve this degeneracy in the interpretation of the [Si/Fe] ratio. Vladilo (1998) corrected for the dust contribution by assuming that dust is the same as that in the Milky Way and by scaling down from observations of weakly and strongly depleted elements. When abundances are corrected for the presence of dust the [Si/Fe] ratios become about solar. Similarly, DLAs characterized by low dust content, namely those with [Fe/Zn] $\approx$ 0, show a mixture of low and moderately enhanced [Si/Fe] (Pettini et al. 2000).

To avoid dust complication the best way is to follow non-refractory elements such as the [S/Zn] ratio. S is found enhanced in halo Galactic stars and both S and Zn elements show little affinity with dust grains. The SII 1250 Å multiplet generally lies in the Lyα forest, but the presence of three lines allows to evaluate the contribution from HI interlopers. The first measures of the [S/Zn] ratios as pointed out by Molaro et al. (1996) were in fact about solar. In Fig. 2 the data points from Centurion et al. (2000), up-dated with recent literature, are shown. It is clear that the behavior of the majority of the [S/Zn] in the DLA is different from the [S/Fe] of the halo stars. All but two of the data points have about solar ratios with a trend of the values to become negative with the increase of metallicity. To be noted is the rather extremely negative value found by Ge et al. (2001).

Recently, UVES-VLT observations of high redshift DLAs allowed to enlarge the set of abundances by including elements with transitions falling in the forest which were overlooked in previous studies. For the first time abundances of O, Ar and P have been derived. The O and Ar are both α elements and not depleted into dust grains, too. In the DLA at z=3.39 towards QSO 0000-2621 the mean ratio involving four α species and 3 Fe-peak elements is [O,S,Si,Ar/Fe,Zn,Cr] = 0.11$\pm$ 0.014 (Molaro et al. 2001). The refractory elements Si, Fe, and Cr can be considered here since this DLA is also dust-free in the sense that [Fe/H] $\approx$

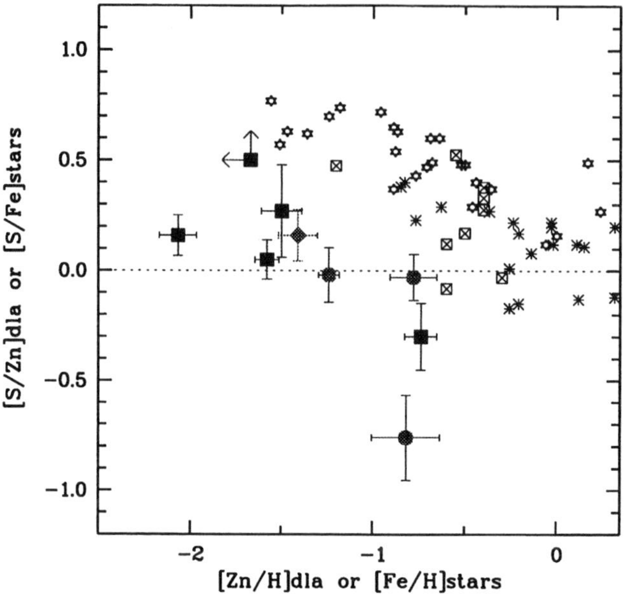

Figure 2. [S/Zn] ratio in the DLA systems. Squares are from Centurion et al. (2000) and Levshakov (2001). The diamond is a [S/Fe] value from Ellison et al. (2001). Circles are from literature (cfr Centurion et al. 2000) and Ge et al. (2001). Stellar [S/Fe] values are: stars from Francois (1988), asterisks from Clegg et al. (1981), crossed squares from Prochaska et al. (2000).

[Zn/H]. This analysis shows that there exists a DLA with modest enhancement of α elements compared to the Fe-peak ones.

On the other hand a similar analysis conducted on the DLA system at z_{abs} = 3.025 towards QSO 0347- has shown O, S, Ar enhanced by $\approx$ 0.5 dex in comparison to Zn (Levshakov et al. 2001). Thus the picture that is emerging is that a variety of chemical patterns with both halo-like and solar chemical patterns are present in DLAs, with the latter found more frequently. If enhancement of α elements is naturally expected in young objects when the TypeII SNe yields are dominating, solar ratio at low metallicity requires that the yields of TypeI SN have already become effective in determining the global metallicity of the system. This can be achieved either with a slow SFR or from a burst of SF followed by a long quiescent period in which TypeI supernovae evolve. A first paper addressing the possibility of reproducing solar ratios already at low metallicity is the one by Matteucci et al. (1997) where a single burst of star formation was assumed. Certainly, the new evidence that DLAs can have solar ratios is a new important constraint which has to be considered in detailed DLA galaxies chemical evolution models.

Acknowledgments

I warmly acknowledge Miriam Centurion, Piercarlo Bonifacio, Sergio Levshakov, Giovanni Vladilo, Sandro D'Odorico, and Miroslava Dessaugues-Zavadsky for enlightening discussions and for allowing me to present common work.

References

Boisse, P., Le Brun, V., Bergeron, J., & Deharveng, J. M. 1998, A&A, 333, 841.

Centurión, M., Bonifacio, P., Molaro, P., & Vladilo, G. 2000, ApJ 536, 540.

Clegg, R. E. S., Lambert, D. L., & Tomkin, J. 1981 Apj 250, 262.

Dessauges-Zavadsky, M., D'Odorico, S., McMahon, R. G., Molaro, P., Ledoux, C., Peroux, C., & Storrie-Lombardi, L. J. 2001, A&A, 370, 426.

Ellison S., Pettini, M., Steidel C. C., & Shapley A., E., 2001, ApJ 549, 770.

Francois, P. 1988, A&A 195, 226.

Ge, J., Bechtold, J., & Kulkarni V. P., 2001, ApJ, 547, L1.

Hou, J. L., Boissier, S., & Prantzos, N. 2001, A&A, 370, 23.

Izotov, Y.,Schaerer, D., & Charbonnel C., 2001, ApJ 549, 878.

Jimenez R., Padoan, P., Matteucci F., & Heavens A. F., 1998, MNRAS, 299, 123.

Le Brun, V., Bergeron, J., Boisse, P., & Deharveng, J. M. 1997, A&A, 321, 733.

Levshakov, S. A., Dessauges-Zavadsky, M., D'Odorico, S., & Molaro P. 2001, ApJ in press (see also astro-ph/0105529).

Lu, L., Sargent, W.L.W., Barlow, T.A., Churchill, C.W., & Vogt, S. 1996, ApJS, 107, 475, 1996.

Matteucci F., Molaro, P., & Vladilo G. 1997, A&A 321, 45.

Molaro, P., D'Odorico, S., Fontana, A., Savaglio, S., & Vladilo, G. 1996, A&A, 308, 1.

Molaro, P., Bonifacio, P., Centurión, M., D'Odorico, S., Vladilo, G., Santin, P., & Di Marcantonio, P. 2000, ApJ, 541, 54.

Molaro, P., Levshakov, S. A., D'Odorico, S., Bonifacio, P., & Centurión, M., 2001, ApJ, 549, 90.

Peroux, C., Storrie-Lombardi, L. J., McMahon, R. G., Irwin, M., & Hook, I. M. 2001, AJ, 121, 1799.

Pettini, M., King D., L., Smith L., J., & Hunstead R. W. 1997, ApJ, 478, 536.

Pettini, M., Ellison, S.L., Steidel, C.C., & Bowen, D.V. 1999, ApJ, 510, 576.

Pettini, M., Ellison S., Steidel C., Shapley A. E., & Bowen D. V. 2000 ApJ, 532, 65.

Prantzos N., & Boisser S. 2000, MNRAS, 315, 82.

Prochaska, J. X., & Wolfe, A. M. 1999, ApJS, 121, 369.

Prochaska J. X, Naumov S. O., Carney B. W., McWilliam A., & Wolfe A. M. 2000, AJ, 120, 2513.

Rao, S. M., & Turnshek, D. A. 2000, ApJS, 130, 1.

Savage B. D., & Sembach K.R., 1996, ARA&A, 34, 279.

Savaglio, S. 2000, in IAU Symp. 204, The Extragalactic Infrared Background and its Cosmological Implications, in press (see also astro-ph/0011473).

Viegas S. M. 1995, MNRAS, 276, 268.

Vladilo G., 1998, ApJ, 493, 583.

Vladilo, G., Bonifacio, P., Centurión, M., & Molaro, P. 2000, ApJ, 543, 24.

Vladilo, G., Centurión, M., Bonifacio, P., & Howk J. C. 2001, ApJ, 557, 1007.

HIGH REDSHIFT GALAXIES AND THE INTER-GALACTIC MEDIUM

Rupert A. C. Croft, Volker Springel, Lars Hernquist, and Martin White
Harvard-Smithsonian Center for Astrophysics, Cambridge, MA, USA.

Abstract

We present results from a high resolution cosmological simulation of galaxy formation in a cold dark matter universe, focusing on a redshift of 3. At this early time, star formation (which we follow using a multiphase prescription for the ISM) is occurring vigorously, but most of the baryons in the universe are still diffuse, and have yet to form into galaxies. Absorption lines due to neutral hydrogen in the spectra of nearby quasars provide a useful way to probe physical conditions in the gas around the galaxies. We make simulated spectra, compare them with observational data, and draw some conclusions about the relationship between galaxies, star formation, and the IGM.

Introduction

In the context of modern cosmological structure formation theories, it is possible to look at galaxy formation starting from well motivated initial conditions. Simulation codes include processes such as gravitational collapse, gas dynamics, and radiative cooling, as well as heuristic prescriptions for star formation. New observational data on galaxies (e.g., Steidel et al. 1999) and the intergalactic medium is rapidly becoming available for us to compare to the predictions of these models.

1. Simulation and Analysis

We have simulated a 33.5 h^{-1}Mpc cube in the currently popular ΛCDM cosmology, using the smoothed particle hydrodynamics (SPH) code GADGET (Springel et al. 2001). The simulation itself and its parameters are described more fully in White et al. (2001). We use 300^3 gas and 300^3 dark matter particles, giving a baryonic mass resolution of 1.5×10^{10} h^{-1}Mpc. The force resolution is $6h^{-1}$kpc comoving.

The star formation is followed using a multiphase prescription for the interstellar medium (details are given in Springel 2001). Each gas particle represents a cold

M. Chávez et al. (eds.),
New Quests in Stellar Astrophysics: The Link Between Stars and Cosmology, 249–252.

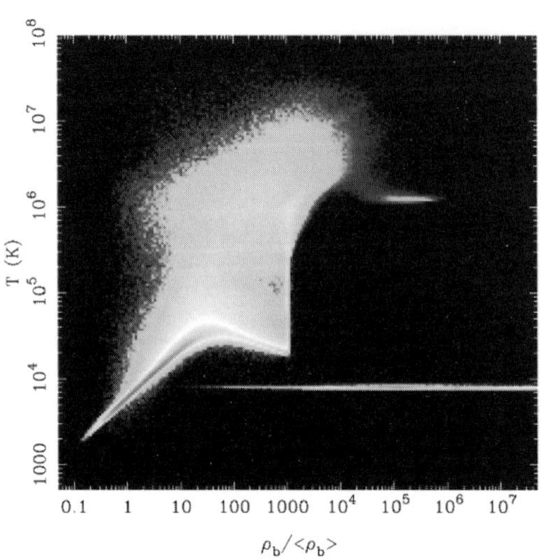

Figure 1. Phase diagram of temperature vs. density for particles in the simulation.

molecular cloud component, a stellar component, and a hot ISM. The equations of mass and energy transfer are solved internal to each particle. An important physical process which we model is supernova heating of the ISM. A phase diagram showing the temperature and density of the particles at redshift $z = 3$ is shown in Fig. 1, where the hot ISM ($T \sim 10^6$K) and cold cloud component ($T \sim 10^4$K) of galaxies is clearly visible on the right.

If we look at the star formation rate as a function of redshift (see Fig. 2), we can see that the simulation gives results which are in reasonable agreement with observations. Without the multiphase formulation, simpler prescriptions for star formation do not appear to converge readily with increasing resolution (see results taken from Weinberg 1999 in the figure). Energy from feedback processes is added to denser and denser gas as the resolution is increased, which cools rapidly and has no lasting effect. With the multiphase particles, this is no longer a problem, as the feedback energy is (more realistically) recycled into the hot ISM.

The relationship between the sites of star formation and the surrounding IGM can be probed by looking at absorption spectra of quasars whose sightlines pass nearby. In Fig. 3 we show small regions taken from the simulation surrounding two galaxies with different masses. We also show segments of Lyα forest spectra close by. We can see that even though the galaxies are surrounded by clouds of relatively hot ($T \sim 10^6$) gas, which is collisionally ionized, there is still substantial neutral hydrogen absorption. We can quantify this by looking at the mean absorption level in pixels as a function of their distance from each galaxy. This is shown in Fig. 4(a) for different mass galaxies. We also show

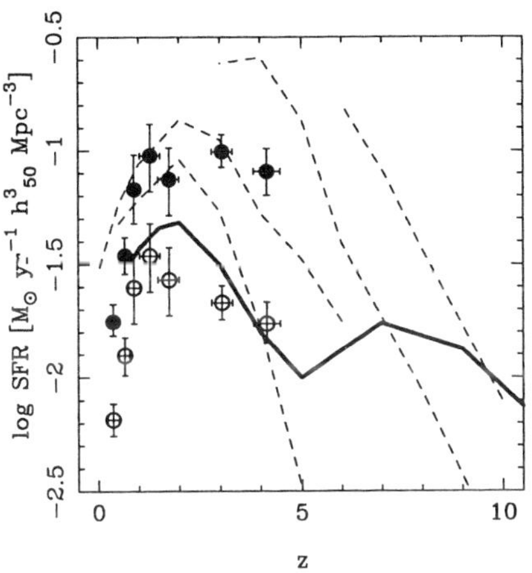

Figure 2. Star formation rate in the simulation as a function of redshift (solid line). Observations from the compilation of Steidel et al. (1999) are shown as open points (uncorrected) and solid points (extinction corrected). Results from SPH simulations of an LCDM model at different resolutions with the single-phase star formation prescription of Katz et al. (1996) are shown as dashed lines (data taken from Weinberg et al. 1999). The multiphase simulation reported here has a resolution (mass and force) closest to the second from top of the dashed lines.

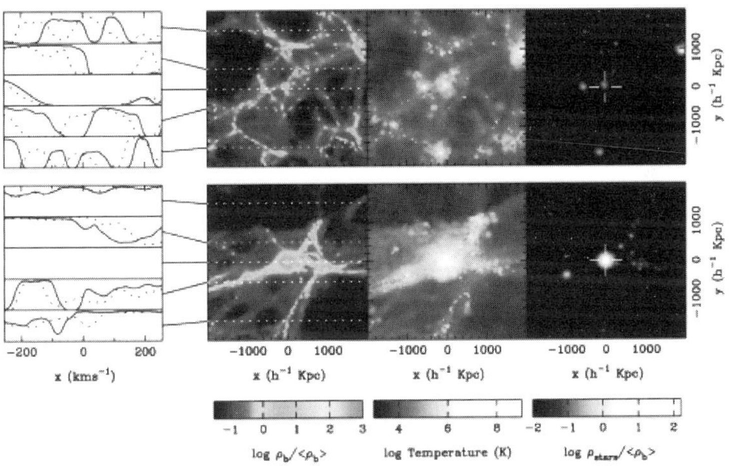

Figure 3. Rightmost three panels in each row: the gas density, gas temperature, and stellar density in small regions (slices 2 h^{-1}Mpc thick) centered on galaxies in the simulation. The leftmost panel shows Lyα absorption spectra extracted along lines of sight shown as dashed lines in the density panel. The solid spectra are the full result and the dotted spectra have been calculated neglecting the effect of peculiar velocities. The bottom row is for a massive galaxy (baryonic mass $\sim 10^{11}M_\odot$) and the top row shows results for a much smaller galaxy (baryonic mass $\sim 10^9 M_\odot$).

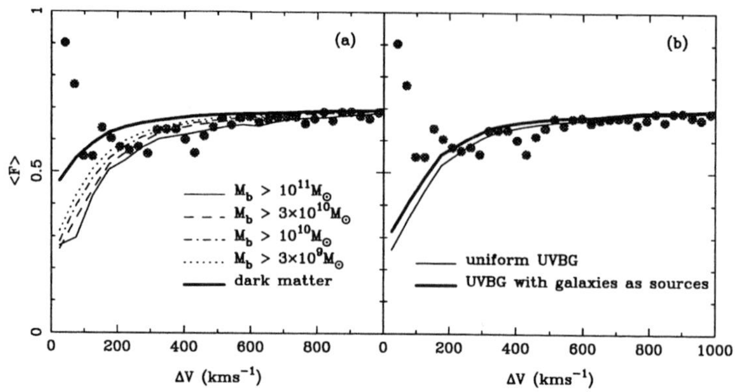

Figure 4. Mean QSO Lyα forest flux in pixels at a given distance (in km/s) from a galaxy. The points in both panels are the observational results of Adelberger et al. (2001). The lines in (a) are for galaxies in the simulation with different lower baryonic mass limits. The lines in (b) are both for galaxies with baryonic mass greater than $3 \times 10^{10} M_\odot$. The difference between them is that in one we have assumed that the ultra-violet ionizing background is spatially uniform, and in the other we have assumed that it is entirely generated by the galaxies themselves.

observational results for Lyman-break galaxies from Adelberger et al. (2001). The results are similar over most of the range, except for the two bins closest to galaxies. This cannot be explained by a an increase in the photoionizing flux close to galaxies, as even if we assume that the entirety of the ultra-violet background is produced by the galaxies, the effect is too small [see Fig. 4(b)]. It is more likely that galactic winds, or maybe a sample selection effect are responsible.

Acknowledgments

We thank Kurt Adelberger for permission to show some of his results in advance of publication. RACC thanks the organizers for support and for a very enjoyable and interesting conference.

References

Adelberger, K., et al. 2001, in preparation

Katz, N., Hernquist, L., & Weinberg, D. H. 1996, ApJS, 105, 19.

Springel, V., 2001, in preparation.

Springel, V., Yoshida, N., & White, S.D. M. 2001, New Astronomy, 6, 79.

Steidel, C.C., Adelberger, K.L., Giavalisco, M., Dickinson, M., & Pettini, M. 1999, ApJ, 519, 1.

Weinberg, D. H., Davé, R., Gardner, J. P., Hernquist, L., & Katz, N. 1999, in ASP Conf. Series 191, Photometric Redshifts and High Redshift Galaxies, eds. R. Weymann, L. Storrie-Lombardi, M. Sawicki, & R. Brunner (San Francisco: ASP), p. 341.

White, M., Henrquist, L., & Springel, V. 2001, ApJ, 2001, 550, L129.

DUST PENETRATED MORPHOLOGY IN THE HIGH REDSHIFT UNIVERSE

D. L. Block
University of Witwatersrand, South Africa

I. Puerari
Instituto Nacional de Astrofísica, Optica y Electrónica, Mexico

M. Takamiya
Gemini Observatory, USA

R. Abraham
University of Toronto, Canada

A. Stockton
University of Hawaii, USA

I. Robson, and W. Holland
Joint Astronomy Center, USA

Abstract Images from the Hubble Deep Field (HDF) North and South show a large percentage of dusty, high redshift galaxies whose appearance falls outside traditional classification systems. The nature of these objects is not yet fully understood. Since the HDF preferentially samples restframe UV light, HDF morphologies are not dust or 'mask' penetrated. The appearance of high redshift galaxies at near-infrared restframes remains a challenge for the New Millennium. The Next Generation Space Telescope (NGST) could routinely provide us with such images. In this contribution, we quantitatively determine the dust-penetrated structures of high redshift galaxies such as NGC 922 in their near-infrared restframes. We show that such optically peculiar objects may readily be classified using the dust penetrated $z \sim 0$ templates of Block and Puerari (1999), Block and Puerari (1999) and Buta and Block (2001).

M. Chávez et al. (eds.),
New Quests in Stellar Astrophysics: The Link Between Stars and Cosmology, 253–256.

Introduction

Several studies have recently been conducted to elucidate the morphology of high redshift objects (e.g. Glazebrook et al. 1995; Driver et al. 1998; Abraham et al. 1996a, 1996b, Bouwens et al. 1998; Giavalisco et al. 1996).
Near-infrared studies ($\sim$ 2.1 μm) probe underlying stellar masses far more accurately than do optical morphologies. In order to 'penetrate the mask' and effectively delineate the stellar mass distribution of the older population, optical detectors do not suffice.
To dust-penetrate galaxies at high redshifts in their K$'$ restframes with the NGST necessitates the design of detectors sensitive to light whose wavelengths are longward of K$'$ (2.1 μm). A preliminary description of the NGST facilities are given in Gillett & Mountain (1998). Bally and Morse (1999) describe a detector which will operate in the wavelength range 0.6 to 5.3 μm. This detector can observe K$'$ restframe light of objects out to redshifts of $z \approx 1.5$.

1. Analysis

The images analyzed here were generated following the simulation methodology of Takamiya (1999). The bidimensional Fourier method was used to secure information about the morphology (see e.g. Puerari & Dottori 1992; Block & Puerari 1999, The simulations recreate a given image when moved out to higher redshifts, always in a pre-selected restframe. In this contribution, the images of NGC 922 are simulated at redshifts $z = 0.7$ and $z = 1.2$ in the K$'$ restframe. The original image of NGC 922 was obtained at the NASA Infrared Telescope Facility with NSFCam (more details can be found in Block et al. 2001).
The bidimensional Fourier analysis shows that the near-infrared morphology of NGC 922 remains the same when the galaxy is moved out to redshifts $z = 0.7$ and $z = 1.2$ (see Fig. 1). The analysis also confirms that only two low-order m components ($m = 1, 2$) are sufficient to describe the dust-penetrated morphology of NGC 922 (see Fig. 2).

2. Conclusions

In this paper, we have simulated images of NGC 922 when it is systematically moved out in redshift space, always in restframe K$'$. Our analysis confirms that the morphology of such optically peculiar objects may successfully be described by low-order m modes, as for galaxies at $z \sim 0$ in our local Universe.

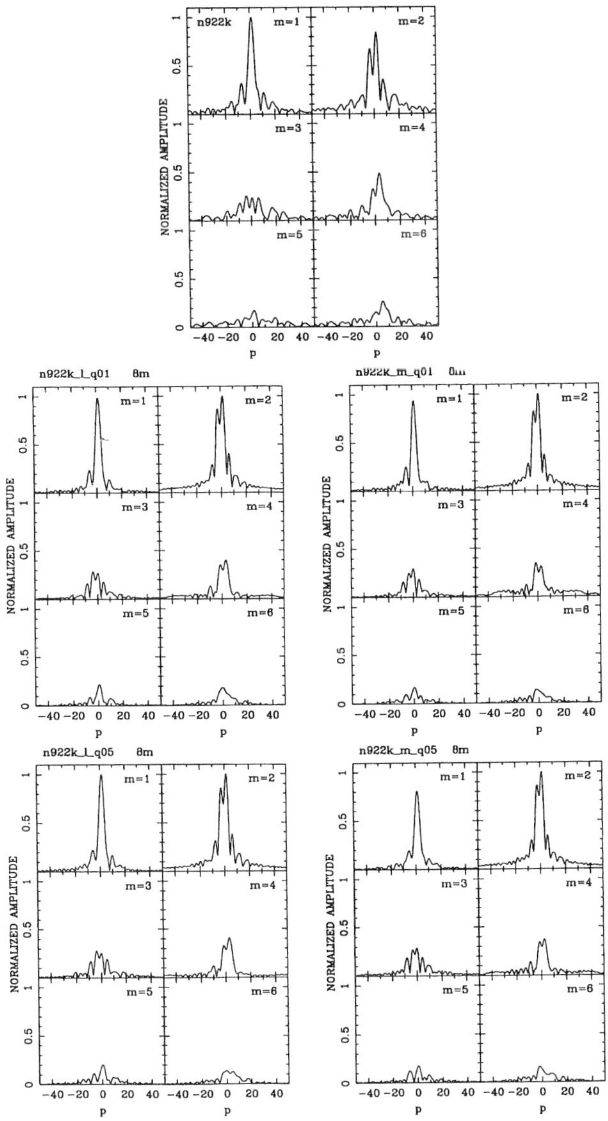

Figure 1. The Fourier spectra of the K′ image of NGC 922 at its original distance (top). The middle panels show the Fourier spectra when NGC 922 is moved to redshifts $z = 0.7$ (L-band) and $z = 1.2$ (M-band), where a deceleration parameter $q_0 = 0.1$ is used. The bottom panels are similar to the middle ones, except that a deceleration parameter of $q_0 = 0.5$ is adopted. A remarkable similarity in the restframe K′ images of NGC 922 is found, independent of redshift and of the deceleration parameter assumed.

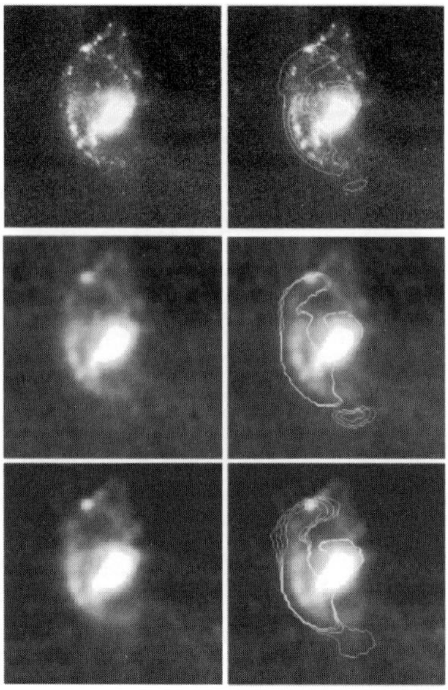

Figure 2. Dust penetrated images of NGC 922. Upper left: Groundbased K′ image. Middle left: simulated 1h exposure (assuming an 8m NGST), at redshift 0.7 (L band). Bottom left: NGC 922 simulated at redshift $z = 1.2$ (M band) with an 8m NGST. The right hand panels show contour overlays of the low order $m = 1$ and 2 inverse Fourier transforms.

References

Abraham, R. G., Tanvir, N. R., Santiago, B. X., Ellis, R. S., Glazebrook, K., & van den Bergh, S. 1996a, MNRAS, 279, L47.

Abraham, R. G., van den Bergh, S., Glazebrook, K., Ellis, R. S., Santiago, B. X., Surma, P., & Griffiths, R. E. 1996b, ApJS, 107, 1.

Bally, J., & Morse J. A. 1999, AAS, 195, 86.

Block, D. L., & Puerari, I. 1999, A&A, 342, 627.

Block, D. L., Puerari, I., Takamiya, M., Abraham, R., Stockton, A., Robson, I., Holland, W. 2001, A&A, 371, 393.

Bouwens, R., Broadhurst, T. & Silk, J. 1998, ApJ, 506, 557.

Buta, R. J., & Block, D. L. 2001, ApJ, 550, 243.

Driver, S. P., Fernandez-Soto, A., Couch, W. J., Odewahn, S. C., Windhorst, R. A., Phillips, S., Lanzetta, K., & Yahil, A. 1998, ApJ, 496, L93.

Giavalisco, M., Livio, M., Bohlin, R. C., Macchetto, F. D., & Stecher, T. P. 1996, ApJ, 112, 369.

Gillett, F., & Mountain, M. 1998, in ASP Conf. Series 133, Science With The Next Generation of Space Telescope, eds. E. P. Smith & A. Koratkar (San Francisco: ASP), p. 42.

Glazebrook, K., Ellis, R., Colless, M., Broadhurst, T., Allington-Smith, J., & Tanvir, N. 1995, MNRAS, 273, 157.

Puerari, I., & Dottori, H. 1992, A&ASS, 93, 469.

Takamiya, M. 1999, ApJS, 122, 109.

THE DUSTY STAR FORMATION HISTORY OF HIGH-Z GALAXIES, MODELING TOOLS AND FUTURE PROSPECTS

Gian Luigi Granato
Osservatorio Astronomico di Padova, Italy and SISSA, Trieste, Italy

and

Laura Silva
Osservatorio Astronomico di Trieste and SISSA, Trieste, Italy

Abstract

We summarize recent advances in the determination of the cosmic history of star formation and other properties of high-z galaxies, and the relevance of this information in our understanding of the formation of structures. We emphasize the importance of dust reprocessing in the high–z universe, as demonstrated in particular by IR and sub-mm data. This demand a panchromatic approach to observations and suitable modeling tools. We spend also some words on expectations from future instruments.

Introduction

In the last few years, a huge number of studies of the high-z universe have been devoted to the determination of the cosmic history of star formation $SFR(z)$. The main motivation for these efforts is that baryons are the only observational tracers of the evolution of large scale structures, which are believed to be driven by the gravitational collapse of dark matter (DM). The determination of $SFR(z)$ can in principle help to discriminate between different proposed scenarios for the formation and evolution of galaxies.

A fundamental issue connected to the determination of $SFR(z)$ is the fraction of light produced by stars which has been reprocessed by dust into IR photons. We know that in the local universe surveyed by IRAS this fraction is $\sim 30\%$. Though this is a significative percentage, it is not dramatic, in the sense that if the same were true also at high–z, optical–UV observations would determine $SFR(z)$ with a small uncertainty. But IRAS observations demonstrated also

M. Chávez et al. (eds.),
New Quests in Stellar Astrophysics: The Link Between Stars and Cosmology, 257–264.

that dust reprocessing in local galaxies is a strong increasing function of their star formation activity, and can't be reliably determined by UV and optical data alone. This is vividly illustrated for instance by Fig. 2 of Sanders & Mirabel (1996), which shows that infrared-selected galaxies range over 3 orders of magnitude in L_{IR}, while the optical luminosity changes only by a factor of 3-4, with minor differences in the shape of the optical-UV SED. A more model dependent example of this is shown in Fig. 1. The possibility to measure the amount of dust hidden SF from the optical-UV SED has been discussed quite a lot, for instance using the Meuer's correlation between the UV spectral index and the ratio L_{UV}/L_{IR}. This relationship is found to be quite narrow for some classes of starburst galaxies, however it has recently been shown that it does not hold for the most luminous starburst galaxies at low redshift (Meuer et al. 1999; Meuer et al. 2000; see also Panuzzo et al., these proceedings).
It is therefore quite natural to suspect that in the younger, more active universe an higher fraction of star luminosity was reprocessed by dust, and that in this situation the global SF activity can be reliably measured only by means of FIR and sub-mm data.

1. The dusty SF history of galaxies

Indeed, several pieces of evidence in the paste few years have shown that *most of the SF in the high–z universe is dust obscured or dimmed to a substantial degree:*

(1) The discovery of a cosmic far-IR/sub-mm background by the COBE satellite (Puget et al. 1996; Fixsen et al. 1998; Hauser et al. 1998), whose energy density, which is at least a factor 2 larger than the optical-UV one, indicates that a large fraction of the energy radiated by stars over the history of the universe has been reprocessed by dust.

(2) The SCUBA detection of a population of bright 850μm sources. Surveys at around 1 mm are very effective in discovering high-z dust enshrouded star formation, due to the steep increase of the expected SED of dusty star forming galaxies shortward this wavelength (see Silva et al. in these proceedings). Indeed, the resulting strong positive K correction is able to counterbalance the cosmological dimming with increasing z for $1 \lesssim z \lesssim 8$ (details depend of course on the precise λ of the survey and on the cosmological parameters). As a result, the expected mm flux of a given luminosity source remains almost constant in this redshift range. Since the comoving cosmic volume element increases up to $z \sim 2$, the conclusion is that mm or sub-mm surveys are strongly biased in favor of dusty high redshift objects.

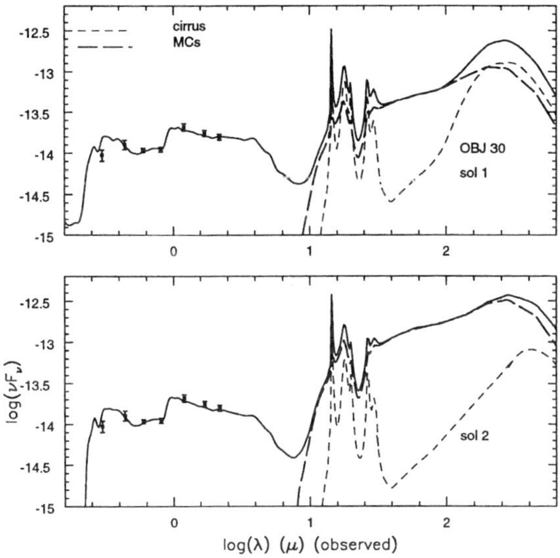

Figure 1. The observed optical-NIR SED of a late–type galaxy in HDF-N at $z = 1.3$ is reproduced with plausible models differing by a factor 3 in SFR. Adapted from Rodighiero et al. (2000).

In the last few years this fact has been exploited in particular by surveys at 850μm performed with the SCUBA camera at JCMT. These surveys led to the discovery of a population of sub-mm sources at high redshift ($z \gtrsim 1.5$), whose luminosities, if powered by star formation in dust-enshrouded galaxies, imply very large star formation rates ($\gtrsim 10^2 M_\odot \mathrm{yr}^{-1}$), and a total star formation density probably greater than that inferred from the UV luminosities of the Lyman-break galaxies (Smail et al. 1997; Hughes et al. 1998). Also, this bright population is originating a significant fraction of the IR background (e.g. Blain 2000)

(3) The ISO detection of a population of strong IR sources; 15 μm ISOCAM (Oliver et al. 1997) and 175 μm ISOPHOT surveys (Kawara et al. 1998; Puget et al. 1999) show a population of actively star forming galaxies at $0.4 < z < 1.3$, mostly disk/interacting galaxies with K typical of a L* galaxy. This population boosts the cosmic star formation density by a factor ~ 3 with respect to that estimated in the optical from the CFRS in the same redshift range.

For (1) and (2), there is the caveat that the contribution from dust-enshrouded AGNs to the sub-mm counts and background is currently uncertain, but probably the AGNs do not dominate (e.g. Granato, Danese & Franceschini 1997).

A summary of the present status of the determination of $SFR(z)$ is given for instance by figure 17 of Genzel & Cesarsky (2000). The main point is that while a few years ago it was claimed, based on optical observations, that this function had a peak at $z \simeq 1$ and declined at higher redshifts, it is now clear instead that $SFR(z)$ steeply increases from $z = 0$ to $z = 1$, but than stays essentially flat to at least $z \simeq 4$.

2. Interpretative tools

These observations should be framed in the hierarchical structure formation paradigm. In this scenario, structures result from the gravitational amplification of small, primordial density fluctuations, possibly quantum ripples boosted to macroscopic scales by inflation. The subsequent formation and merging of DM haloes is driven by gravitation and fully determined by the initial density fluctuations and by the cosmology.

While the simulation of the behaviour of the DM component of the universe is a relatively simple task, a much more difficult problem is to predict the evolution of the baryonic component, subject to a much more complex physics. As a matter of fact, to date it is not possible to follow by direct numerical simulations these processes with the dynamical range relevant for galaxy formation. The most employed technique is instead that of semi-analytical models, in which simplified analytical descriptions of baryon processes (e.g. gas cooling and collapse, star formation, supernovae feedback and galaxy merging) are adopted (see Cole et al. 2000). The predicted SF histories are then combined with stellar population models including dust reprocessing, in order to predict galaxy observational properties.

Present semi-analytical models compare favorably with most observations, including many IR constraints (Granato et al. 2000 and references therein; Silva 1999) but are seriously challenged by sub-mm counts, which turn out to be underpredicted by about one order of magnitude (see Fig. 2). This, coupled with the fact that these models are instead able to reproduce the observed background from the FUV to the sub-mm, leads to the conclusion that the main problem is that they spread the total SF activity in too many but too faint episodes, with respect to what sub-mm counts indicate. To fix this problem, the simplified prescriptions used in semi-analytical models to describe the behavior of baryons in DM halos need some substantial revision, on which work is in progress (Lacey et al., in preparation). It is clear that, in some sense, semi-analytical models, based on the hierarchical clustering paradigm, should be modified in such a way that the formation of a subset of galaxies resemble the prediction of the so called "monolithic scenario" (Eggen, Lynden-Bell & Sandage, 1962; Larson, 1975)

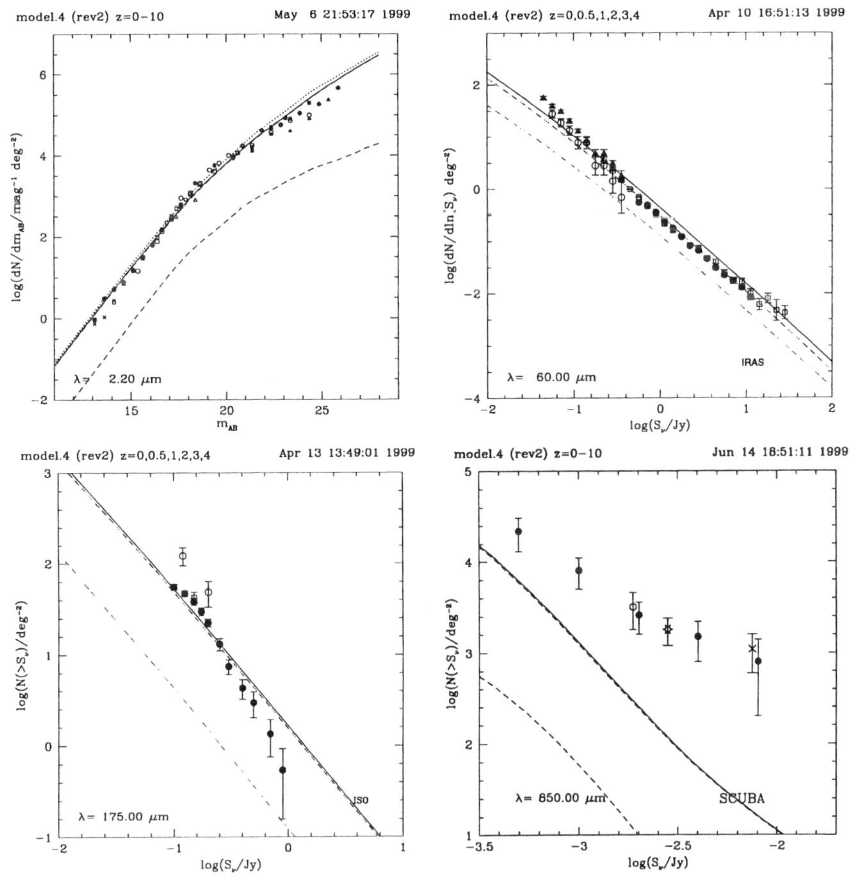

Figure 2. Number counts at 2.2, 60, 175 and 850 μm. Standard semi-analytical models fail to reproduce the 850μm SCUBA counts. Some substantial revision of the prescriptions used for star formation at high redshift seems unavoidable (Silva 1999; see also Lacey et al. 2001, in preparation)

Further indications of this can be gained taking advantage of the observational evidence suggesting that QSOs did shine in the core of early-type protogalaxies, during their main episode of star formation. Granato et al. (2001) used this information to derive the formation rate of spheroids as a function of redshift, essentially by means of a deconvolution of the rather well known evolution of the QSO luminosity function. Then they used GRASIL (Silva et al. 1998) to predict their SED evolution in the FIR sub-mm region. In this way they have shown that SCUBA counts are well reproduced by early-type protogalaxies (see Fig. 3), in particular if the active formation phase was quicker in more massive objects, as suggested by chemical enrichment studies (Thomas, Greggio & Bender 1999 and references therein).

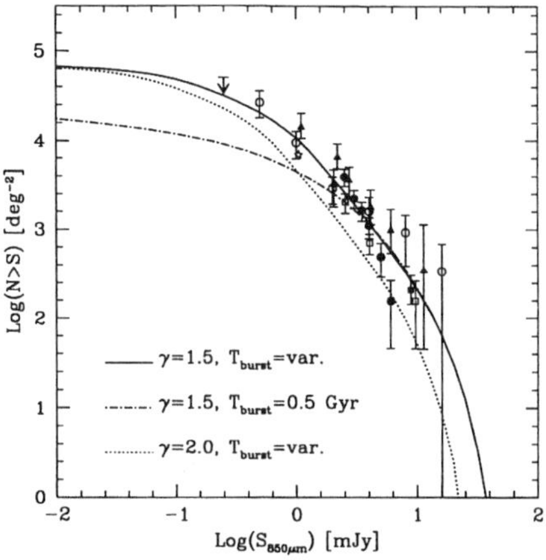

Figure 3. The formation rate for spheroids derived from the QSO luminosity function evolution, coupled with plausible assumption for their SED evolution can easily reproduce the 850μm SCUBA counts. Adapted from Granato et al. (2001).

3. Prospects for the future

Although mm observations have had a strong impact on models for the formation and evolution of galaxies in the past few years, and are leading to a major revision of them, a systematic study of the high-z universe is almost impossible with present instruments, which are limited by a small accessible area and poor resolution. As a consequence, up to now relatively few objects have been detected, and their optical identification, which is at present a necessary step to obtain a spectroscopic redshift, is difficult. Also, the confusion limits at faint fluxes are severe: the confusion noise may be important at $\lesssim 2$ mJy for SCUBA (Blain 2000).

In summary, we have now only $\sim 10^2$ sub-mm selected sources mostly without a key information: a reliable redshift. For most of the sources we have only "photometric" redshifts (or lower limits), based on ratios of fluxes in two different sub-mm bands or on sub-mm to radio ratios.

The detection rates for surveys to be performed with future IR and submillimetric facilities (e.g. SIRTF, LMT, ALMA, Herschel-FIRST) are expected to increase by orders of magnitude with respect to present ones (e.g. figure 4 in Blain 1999). A fundamental point is that these large samples could be of little use to further constrain the history of star formation in the universe, unless good estimates of the redshifts will be available. On one hand, it will be possible to combine information coming from surveys at different wavelengths, if properly planned, to estimate photometric redshifts (see also Silva et al. contribution to these proceedings). On the other hand the development of the "redshift machine" at LMT and ALMA, based on an order of magnitude increase of the bandwidth of sub-mm detectors, will allow to determine the spectroscopic redshift of the sources directly from CO transitions, without the need of the optical identification (see Carrasco contribution).

Acknowledgments

We would like to thank the organizers of this meeting for their hospitality and for financial support. GLG has been in part supported also by ASI contract 1/R/27/00 and by SAGG, LS by Cofin99. We also acknowledge the contribution of our GALFORM collaborators (Cedric Lacey, Carlton Baugh, Shaun Cole and Carlos Frenk) to the work presented here. Special thanks are due to our closest collaborators Alessandro Bressan, Gigi Danese, Gianfranco De Zotti and Pasquale Panuzzo.

References

Blain, A. W., 1999, in ASP Conf Series 191, Photometric Redshifts and High Redshift Galaxies, eds. R. Weymann, L. Storrie-Lombardi, M. Sawicki, & R. Brunner (San Francisco: ASP), p. 255.

Blain, A. W., 2000, astro-ph/0011479.

Cole, S., Lacey, C. G., Baugh, C. M. & Frenk, C. S. 2000, MNRAS, 319, 168.

Eggen, O. J., Lynden-Bell, D., & Sandage, A.R. 1962, ApJ, 136, 748.

Fixsen, D. J., Dwek, E., Mather, J. C., Bennett, C. L., & Shafer, R.A. 1998, ApJ, 508, 123.

Genzel, R., & Cesarsky, C. J. 2000, ARA&A, 38, 761.

Granato, G. L., Danese, L., & Franceschini, A. 1997, ApJ, 486, 147.

Granato, G. L. Lacey, C. G., Silva, L., Bressan, A., Baugh, C. M., Cole, S., & Frenk, C. 2000, ApJ, 542, 710.

Granato, G. L., Silva, L., Monaco, P., Panuzzo, P., De Zotti, G., & Danese, L. 2001, MNRAS, in press.

Hauser, M. G., et al. 1998, ApJ, 508, 25.

Hughes, D. H., et al. 1998, Nature, 394, 241.

Kawara, K., et al. 1998, A&A, 336, L9.

Larson, R. B. 1975, MNRAS, 173, 671.

Meurer, G. R., Heckman, T. M., & Calzetti, D. 1999, ApJ, 521, 64.

Meurer, G. R. Heckman, T. M., Seibert, M., Goldader, J., Calzetti, D., Sanders, D., & Steidel, C. C. 2000, in Cold Gas and Dust at High Redshift, ed. D. J. Wilner, in press (see also astro-ph/0011201).

Oliver, S. J., et al. 1997, MNRAS, 289, 471.

Puget, J.-L., Abergel, A., Bernard, J.-P., Boulanger, F., Burton, W. B., Desert, F.-X., & Hartmann, D. 1996, A&A, 308, L5.

Puget, J.-L., et al. 1999, A&A, 345, 29.

Rodighiero, G., Granato, G.L., Franceschini, A., Fasano, G., & Silva, L. 2000, A&A, 364, 517.

Sanders, D. B., & Mirabel I. F. 1996, ARA&A, 34, 749.

Silva, L., 1999, SISSA Ph.D. Thesis, (see the Web addresses *http://grana.pd.astro.it*, or *http://www.sissa.it/ap/PhD_Astro.html*).

Silva, L., Granato, G. L., Bressan, A., & Danese, L. 1998, ApJ, 509, 103.

Smail, I., Ivison, R.J., & Blain, A.W. 1997, ApJ, 490, L5.

Thomas, D., Greggio, L., & Bender R. 1999, MNRAS, 302, 537.

ON THE CALIBRATION OF STAR FORMATION RATES

Daniel Rosa-González, Elena Terlevich
Instituto Nacional de Astrofísica, Optica y Electrónica, Luis Enrique Erro 1, 72840 Tonantzintla, Puebla, México

and

Roberto Terlevich
Institute of Astronomy, Madingley Road, CB3 OHA Cambridge, UK, and Instituto Nacional de Astrofísica, Optica y Electrónica, Puebla, México

Abstract

A central problem in the "Madau–Lilly" diagram, i.e. the evolution of the Star Formation Rates (SFR) with redshift, is that the SFR is estimated using several different estimators at different redshift ranges that potentially can have systematic differences among them. We will address the question: How much of the observed change of the SFR with redshift may be due to systematic differences in the SFR calibrators. We have used a well calibrated sample of star forming galaxies in the local Universe to establish a self–consistent system of SFR estimators. The application of the new system to a high redshift sample shows an increase of the SFR by a factor of 14 from the local Universe to redshift one and a slow decrease from redshift one to five with a median value of approximately 0.3 solar masses per year per cubic megaparsec (for $h = 0.5$, Ω_M=1. and Λ=0).

Introduction

The global history of the star formation of the Universe seems to have a noticeable increase from $z \sim 0$ to $z \sim 2$ (Lilly et al. 1995). At $z > 3$ the behavior of the global SFR is unclear (e.g. Madau et al. 1996; Steidel et al. 1999). The evolution of the SFR is estimated using different methods for different redshift ranges. The Hα luminosity is used for galaxies with redshift up to 0.4, the [OII]λ3727 luminosity for $0.4 > z > 1.0$ and the UV continuum luminosity for galaxies with $z > 2$. The transformation between the observed luminosities to the current star formation is done using expressions derived from spectra constructed using population synthesis methods (e.g. Leitherer &

M. Chávez et al. (eds.),
New Quests in Stellar Astrophysics: The Link Between Stars and Cosmology, 265–268.

Heckman 1995). In addition the resulting SFR estimates are dependent on the reddening/extinction corrections.
In order to understand and quantify the possible systematic differences for the different SFR estimators, we have compiled a sample of nearby well studied galaxies for which good data is available in Hα, [OII]λ3727, UV continuum and FIR luminosities.

1. The Reference Galaxy Sample

To build the reference galaxy sample we selected from the literature all star-forming galaxies with well determined fluxes in Hα, Hβ, [OII]λ3727, FIR and the ultraviolet continuum. The resulting sample consist in galaxies classified either as HII galaxies, Starburst Galaxies (SB) or Blue Compact galaxies (BCG). The galaxies are located at large galactic latitudes ($|b| > 25$). A more complete information of the individual galaxies are published in Rosa–González et al. (2001).

2. Corrections to the Observed Fluxes

We investigated the effect on the observed fluxes due to dust extinction and the presence of underlying absorption.

2.1 Dust extinction Corrections to the Observed Fluxes

Calzetti and collaborators developed an empirical method to estimate the UV extinction (Calzetti et al. 1994). They proposed an alternative empirical law where the extinction correction is obtained by the comparison of different IUE spectra from a sample of nearby starburst galaxies. The obtained extinction curve could be considered such as an average of the different dust distributions described by Calzetti et al. (1994). In order to correct the ultraviolet flux using this procedure it is necessary to estimate the Av from the observed Hα/ Hβ ratios and then apply the extinction law to the observed ultraviolet fluxes.

2.2 Dust extinction corrections to the Emission Lines Fluxes

Extinction affects the emission lines in different degrees depending on wavelength. Correction is usually obtained from the observed ratio of Balmer lines, the intrinsic ratio, and an interstellar extinction curve. The effect due to the dust can be expressed by a vector in the log(F(Hα)/F(Hβ)) vs. log(F(Hγ)/F(Hβ)) plane, where F(Hα), F(Hβ) and F(Hγ) are the fluxes at the different wavelengths.

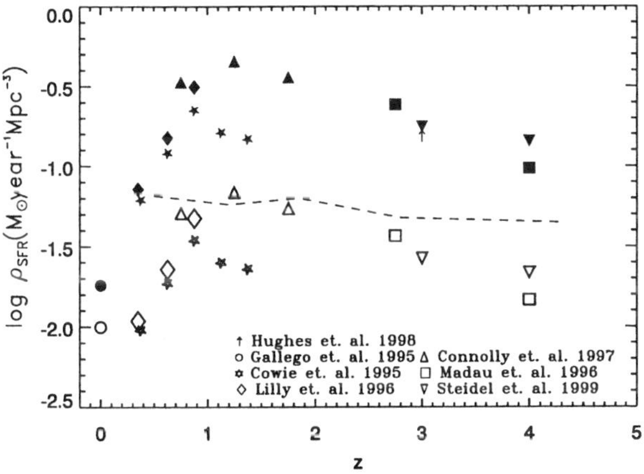

Figure 1. The SFR density as a function of the redshift. The open values are the published SFR given by the different authors rescaled to the SFR relations given in this work. The open circle is the SFR given by the H α emission line corrected by extinction from Gallego et al. (1995) sample. The open stars are the SFR given by the [OII]λ3727 from the sample of Cowie et al. (1995) without any correction. The SFR given by the UV continuum without any correction are from the samples of Connolly et al. (1997) (triangles) Madau et al. (1996) (squares) and Steidel et al. (1999) (upside down triangles). Solid symbols represent the statistical corrections to the data as of our analysis. The upward arrow is the lower limit given by Hughes et al. (1998) based on submm observations of the Hubble Deep Field. Dashed line is the lower limit by Chapman et al. (2001).

2.3 Underlying absorption corrections to the Emission Lines Fluxes

The presence in most galaxies of the sample of a population young and intermediate age stars is evidenced by the presence of absorption Balmer absorption lines.

The effect of the absorption lines in the observed emission lines can be represented also by a vector in the log(F(Hα)/F(Hβ)) vs. log(F(Hα)/F(Hβ)) plane. This vector can be combined with the extinction vector due to dust to estimate the value of the visual extinction.

3. Star Formation Rates

For each galaxy, we calculate the SFR by using the calibrations given by Kennicutt (1998) where a Salpeter IMF [N(m)$\propto m^{-2.35}$] with masses varying from 0.1 to 100 M$_\odot$, solar metallicity and a continuous star formation mode, i.e. constant SFR are assumed. For the case of the SFR(FIR) the scape of UV photons can be quantified by the parameter ϵ which can be estimated through the observed 60μm and ultraviolet luminosities. The SFR given by the FIR can be considered as an upper limit (see Fig. 1).

4. Results and Conclusions

After correct simultaneously by the presence of dust and the underlying population we find that the SFR given by the Hα is 40% lower than the SFR given by the FIR. The SFR given by the [OII]λ3727 forbidden line is only 20% lower than the SFR given by the FIR. If the MW extinction curve is adopted then the SFR given by the UV continuum is in average 2 times higher than the SFR given by the FIR. It is necessary to combine the visual extinction obtained by the Balmer decrement with Calzetti's law to obtain similar results with the UV continuum and the FIR.

Applying the statistical corrections found in this work to galaxies at high redshift we agree with the existence of a peak in the SFR density about $z = 1$ with an steep increase of the SFR from $z = 0$ to $z = 1$ (a factor of about 14) and a slow decay for $z > 1$.

References

Calzetti, D., Kinney, A. L., & Storchi-Bergmann, T. 1994, ApJ, 429, 582.

Chapman, S. C., Richards, E., Lewis, G., Wilson G., & Barger A. 2001, ApJ, 548, 147.

Connolly, A. J., Szalay, A. S., Dickinson, M., Subbarao, M. U., & Brunner, R. J. 1997, ApJ, 486, L11.

Cowie, L. L., Hu, E. M. & Songaila, A. 1995, Nature, 377, 603.

Gallego, J., Zamorano, J., Aragon-Salamanca, A., & Rego, M. 1995, ApJL, 455, L1.

Guzmán, R., Gallego, J., Koo, D. C., Phillips, A. C., Lowenthal, J. D., Faber, S. M., Illingworth, G. D., & Vogt, N. P. 1997, ApJ, 489, 559.

Howarth, I. D. 1983, MNRAS, 203, 301.

Hughes, D. H. et al. 1998, Nature, 394, 241.

Kennicutt, R. C., J. 1998, ARA&A, 36, 189.

Leitherer, C., & Heckman, T. M. 1995, ApJS, 96, 9.

Lilly, S. J., Tresse, L., Hammer, F., Crampton, D., & Le Fevre, O. 1995, ApJ, 455, 108.

Madau, P. , Ferguson, H. C. , Dickinson, M. E. , Giavalisco, M. , Steidel, C. C., & Fruchter, A. 1996, MNRAS, 283, 1388.

Rosa–González, D., Terlevich, E., & Terlevich, R. 2001, submitted to MNRAS.

Seaton, M. J. 1979, MNRAS, 187, 73.

Steidel, C.C., Adelberger, K.L., Giavalisco, M., Dickinson, M., & Pettini, M. 1999, ApJ, 519, 1.

EXCURSIONS INTO THE EVOLUTION OF EARLY-TYPE GALAXIES

Omar López-Cruz*
Instituto Nacional de Astrofísica, Optica y Electrónica (INAOE) & DAUG , México

David Schade
CADC, Herzberg Institute of Astrophysics (HIA), Canada

L. Felipe Barrientos
Departamento de Astron. & Astrofís. P. Universidad Católica de Chile (PUC), Chile

Michael D. Gladders[†] & H. K. C. Yee
Department of Astronomy and Astrophysics, University of Toronto (U of T), Canada

and

Tadayuki Kodama
Department of Astronomy, University of Tokyo, Japan

Abstract Recent observations have revealed that early-type galaxies (ETG) in clusters comprise an old galaxy population that is evolving passively. We review some recent observations from the ground and from the HST that show that ETG have undergone a significant amount of luminosity evolution. This evolution is traced by two projections of the fundamental plane (FP): the size-magnitude relation (SMR) and the color-magnitude relation (CMR). We will briefly discuss the relevance of all these results in the context of the universality of the IMF.

*Fulbright Visiting Scholar, Carnegie Observatories (OCIW), USA
[†]Carnegie Fellow, Carnegie Observatories (OCIW), USA

M. Chávez et al. (eds.),
New Quests in Stellar Astrophysics: The Link Between Stars and Cosmology, 269–276.

1. Introduction

The opening line of Walter Baade's lectures on the evolution of stars and galaxies (Baade, 1963) reads: "The study of the evolution of stars and galaxies is in a stage of rapid development, and we are still at the beginning". Almost half a century later the former statement is still timely, as advances in technology have opened new windows for the exploration of galaxy evolution. The first clue that galaxies change through cosmic time is that galaxies at high redshift (z) appear systematically brighter than those at lower z. We may identify two reasons for this brightening: it could be that either the stellar content was intrinsically brighter at an earlier epoch or that galaxies appear brighter due to the properties of the space-time. We are fortunate that those issues happened to be the main running themes of this rather interesting conference. Clearly, if we want to understand the brightening of high-z galaxies, we need to understand the role of stellar evolution in galaxies as well as the propagation of light in different cosmological models. With these elements in mind let us explore the evolution of galaxy cluster ETG and its possible connections with the evolution of field ETG. We note that ETG were once considered the simplest stellar systems. Nevertheless, ETG, despite their apparently simple appearance, pose some difficult problems to models of galaxy formation and evolution. Perhaps the most remarkable is the inexistence of a comprehensive theory of stellar formation in dynamically hot systems (Silk, 1999).

Clusters of galaxies provide us with probes to study the effects of the environment on the evolution of galaxies (Oemler, 1974; Dressler 1984). Indeed, it has been found that clusters provide a hostile environment for some galaxies (e.g., dwarf galaxies: López-Cruz et al., 1997; Hilker, 1998; and low surface brightness galaxies: Gregg & West, 1998). In contrast, ETG flourish in such rich environments: this is just a paraphrase of the density-morphology relation (Dressler, 1980). Interestingly, the physical parameters of these galaxies, such as colors, size (e.g., the effective radius R), luminosity (L), velocity dispersion (σ), surface brightness (I), etc., have the same pairwise correlations originally found for field ETG (e.g., $L \propto \sigma^4$). The correlations among these parameters are synthesized by the fundamental plane ($R \propto \sigma^{1.4\pm0.15} I^{-0.9\pm0.1}$; FP; Djorgovski & Davis, 1987). Nevertheless, the effects of the environment may be indicated by an augmented tightness in the correlations for cluster galaxies (e.g., Schade et al., 1999; Tragger, Faber, & Dressler, 2001).

In this review we present the results of some apparently disconnected evolutionary studies. We have explored two projections of the fundamental plane. These studies have led us to conclude that ETG in the central regions (the inner $\approx 1\,h_{50}^{-1}$ Mpc)[1] of rich clusters conform to an old population that has been

[1] A Hubble constant of $H_o = 50\,h_{50}$ Mpc km^{-1} s^{-1} is used throughout.

evolving passively since $z \sim 3$. We do not intend to present a comprehensive review (see elsewhere, e.g., Kormendy & Djorgovski 1989; Buzzoni, Renzini, & Serrano, 1995; Arnaboldi, Da Costa, & Saha, 1996; Cepa & Carral, 1998; and Ellis 2001); however, we do intend to arrive at a unifying view on the evolution of ETG.

2. The Size-Magnitude Relation

The work of Barrientos, Schade, & López-Cruz (1996) was the first study from a series that compared ETG in an intermediate-z cluster (CL 0939 + 4713 at $z = 0.41$, observed with the HST) with those of a nearby rich cluster (the Coma cluster at z=0.023, observed with the KPNO 0.9m) in a bidimensional space that is defined by axes: size versus integrated magnitude. The size is defined as the core radius in a bidimensional de Vaucouleurs fit to the surface brightness distribution, and the integrated magnitude was computed by the integration of the best fit. The resulting size-magnitude relation (SMR) is expected as it conforms one of the projections of the FP (Djorgovski & Davis 1987). When the k-corrected SMR of Coma and CL 0939 + 4713 are compared, an offset of $\Delta M_B = 0.72 \pm 0.24 + 5 \log h_{50}$ is detected. This offset indicates that the galaxies in CL 0939 + 4713 are brighter than the ones in Coma. We can explain this result either by assuming that the size remained fixed and the change was due to the evolution of the underlying stellar population (passive evolution) or by assuming that both the size and the magnitude were changing due to mergers. Although mergers could have been important at the time of cluster formation (Schweizer, 2001); the merger hypothesis is not favored because in the central regions of rich clusters the large cluster velocity dispersions ($\sigma_{cluster} \geq 750\ \mathrm{km\,s^{-1}}$) would make merging inefficient. Moreover, the general *merger rate*[2] is very low at $z < 1$. Hence, we can conclude that the observed changes are due to passive evolution. Similar amount of luminosity evolution was reported by previous observations in clusters around quasars (Yee & Green, 1987).

Changes in luminosity can easily be explained with the aid of the single burst model (c.f. Tinsley, 1980; Buzzoni, 1995; van Dokkum et al., 1998) where a burst of star formation occurs at an epoch $t = t_\circ$. It can be shown that the expected luminosity evolution is described by a power law $L(t) \propto \left[\frac{1}{t - t_\circ}\right]^\kappa$, with $\kappa = [1.3 - 0.27(s-1)] > 0$ where $x = (s-1)$ is the slope of the IMF; it has been found that κ depends on the metallicity and the passband[3]. Hence,

[2] For example *merger rate* $\propto (1+z)^{2.3\pm0.7}$ (Patton et al., 2002)

[3] A useful approximation valid at low z is provided by an empty universe model ($\Omega_\circ = 0$). A simple relation results for the luminosity evolution as a function of z: $L(z) \propto \left[\frac{(1+z)(1+z_\circ)H_\circ}{(z_\circ - z)}\right]^\kappa$, where $z_0[\approx 3] > z$ is the galaxy formation redshift.

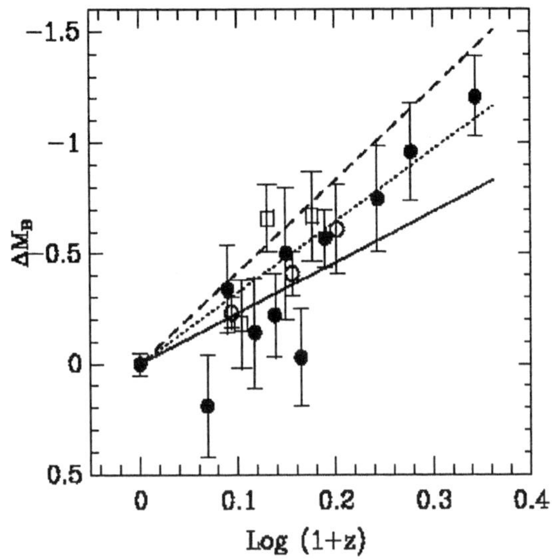

Figure 1. The luminosity shift ΔM_B including revised results from Schade et al. (1996a) and from Barrientos, Schade, & López-Cruz (1996). Filled symbols are for clusters in the present study, open circles are cluster points from Schade et al. (1996b) (including 2 clusters in common that have been offset in redshift slightly for clarity) and open squares are field galaxies from Schade et al. (1996a). The lines are from the models of Buzzoni (1995) for IMF power-law indices of $s = 3.35$ (solid line), $s = 2.35$ (short dashed), and $s = 1.35$ (long dashed). The models assume a present-day age of 15 Gyr and we show the theoretical $\Delta M(V)$ whereas the data is in M_B where the evolution would be ~ 0.15 mag larger by $z = 1$.

as t becomes larger than $t_\circ$ the luminosity decreases, in agreement with the expected dimming in luminosity as stars evolve off the giant branch. Further studies (Schade et al., 1996b; Schade, Barrientos, & López-Cruz, 1997) have included more clusters and a larger number of galaxies and have confirmed and extended the early result of Barrientos, Schade, & López-Cruz. Figure 1 taken from Schade, Barrientos, & López-Cruz (1997) summarizes the results from our previous studies. It is found that $\Delta M \propto z$, in addition the direct comparison with the models of Buzzoni (1995) has given us a constraint on s. The four most deviant points in this figure are produced by poor sampling. Therefore, we can safely conclude that our observations are in agreement with models with $s = 2.35$, i.e., $x = 1.35$ the Salpeter IMF (Salpeter, 1955).

3. The Color-Magnitude Relation

The first pairwise relationship discovered among the physical properties of ETG was the color-magnitude relation (CMR). It was originally discovered by Baum (1959) as a trend in which the brightest galaxies are also the reddest. In the same paper Baum, with the aid of a simple population synthesis model, demonstrated that ETG were dominated by old *Population I* stars, contrary to the then popular belief that globular clusters and ETG had the same stellar make up. Baum's observations and their interpretation signified a turning point in our understanding of ETG. The work of Sandage (1972), Faber (1973), and Visvanathan & Sandage (1977) showed that the CMR can be parameterized by a straight line within a luminosity interval of about 8 magnitudes. This remarkable property suggests that ETG from giants to dwarves, with the exception of some brightest cluster galaxies (BCG)[4], have shared a common history of star formation, even when dwarfs and giants are structurally and dynamically different: a dwarf elliptical galaxy is not a scaled-down giant ETG (Ferguson & Binggeli, 1994). This difference in dynamical behavior is further suggested by the break away of dwarves from the FP and the SMR. The other important property is the universality of the CMR at least for clusters at low redshift (e.g., Sandage & Visvanathan, 1978; Bower, Lucey, & Ellis, 1992; López-Cruz & Yee, 2001). The CMR was first explained by Arimoto & Yoshii (1987) in terms of the combined effects of dissipative galaxy formation and metallicity.

Three constraints on the last epoch of strong galaxy formation in ETG can be derived from the study of the CMR. Bower, Lucey, & Ellis (1992) showed that the dispersion of the CMR is a good indicator of the age spread in the stellar population, since newer episodes would increase the dispersion about the CMR. We have found that the dispersion is smaller than the limit of our observations for galaxies in the central regions of clusters at lower redshift (the LOCOS sample, López-Cruz, 1997; López-Cruz 2001). The second constraint is provided by the absolute color evolution of ETG, this can be measured by changes of the zero point of the CMR (see Stanford, Eisenhardt, & Dickinson, 1998). The third constraint is provided by the slope of the CMR: in the classical scenarios (e.g., Kodama & Arimoto, 1997) star formation is regulated by supernova driven winds. In these models the strongest changes in the stellar population properties occur during the first 5 Gyrs after formation. Hence if we are able to identify evolution in the slope of the CMR, we could in principle constrain the epoch of the last major starburst. We have attempted such a test using archive HST observations and LOCOS data (Gladders et al., 1998). The merit of this test is that the slope, being a ratio of colors, is independent of the calibrations and therefore we are provided by an unbiased estimator that can be readily compared with the models.

The comparison of our observed CMR slopes to a suite of four models with different formation epochs and cosmologies is presented in Figure 2. The overall slope of the red sequence seems to be consistent with passive evolution for all clusters, implying that the populations we have sampled have similar evolution-

[4]Sometimes BCG in cooling flow clusters show very blue cores (e.g., McNamara & O'Connell, 1992, López-Cruz, 1997) hence, some BCG deviate from the CMR

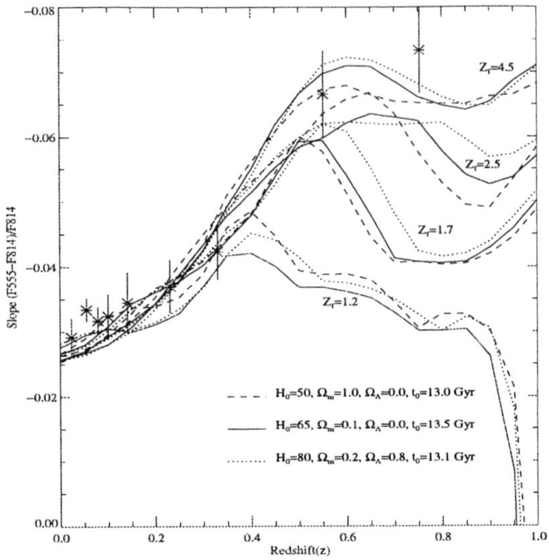

Figure 2. This figure has been taken from Gladders et al. (1998). The predicted slopes for formation models (Kodama 1997, Kodama & Arimoto 1997) using three different cosmologies to map age onto redshift. Note that while the details of the turn-over in the slope for a model at a given formation redshift changes slightly in different cosmologies, the change is not significant enough to affect the conclusion that at least some clusters form at $z > 2$.

ary histories and time-scales. The observed slopes of the higher redshift clusters are perfectly consistent with a high formation redshift, and a conservative limit of $z \geq 2$ can be set. Note that up to redshifts of z≈0.4, observations of the slope of the red sequence have little power to distinguish between formation epochs with z≥1.0. The change in slope at z≤0.4 is exclusively due to blue-shifting of the rest-frame band-passes. However, beyond z≈0.5, the expected slopes for the lowest-formation-epoch models rapidly diverge from the observed slopes. This turnover is a result of the differing metalicities in the ETG population as a function of mass; at ages younger than $\sim$ 5 Gyr, the color evolution of the stellar population in a metal-rich ETG is significantly different from that of a metal-poorer elliptical (Kodama, 1997; Kodama & Arimoto 1997). From Figure 2 it is evident that that the slope of the CMR provides a good marker on the last episode of star formation in cluster ETG that is almost independent of the cosmology.

4. Discussion & Conclusions

The two studies that have been presented here strongly suggest that ETG in clusters are evolving passively. Also Pahre, Djorgovski, & de Carvalho (1996) and Lubin & Sandage (2001) using the Tolman test have shown indirect evidence that also suggest that ETG are evolving passively.

We have used the poor-man's approach to the FP in order to address the evolution of ETG. Can we expect to detect the same kind of evolution using the FP itself? The answer is yes: there are many FP studies for cluster galaxies at low and intermediate z that have agreed that ETG are evolving passively (e.g., Jørgensen, Franx, & Kjærgard, 1996; van Dokkum et al., 1998).

Figure 1 also reveals that the best agreement between the models and the observations is reached with a Salpeter IMF. This is remarkable since Salpeter (1955) defined the IMF for the solar neighborhood. Field ETG at higher redshifts also reveal a Salpeter IMF (Schade et al., 1999). Star formation being such a complex physical process has all the characteristic of self-organized criticality (see Melnick & Selman, 2000). Our results along with those of Melnick & Selman further supports the universality of the IMF.

Our general conclusion, which also agrees with many other studies (see Peebles, 2002), is that cluster ETG coevally formed at $z > 3$ and have been evolving passively since then. Moreover, a single burst with a Salpeter IMF suffices to explain the observations presented in this review.

Acknowledgments

OLC research is supported in part by CONACyT-México Young Researcher Grant No. J32098-E.

References

Arimoto, N. & Yoshii, Y. 1987, A&A, 173, 23.

Arnaboldi, M., Da Costa, G.S., & Saha (eds.) 1996, ASP Conf. Series 116, The Nature of Elliptical Galaxies (San Francisco: ASP).

Baade, W, 1963, in The Evolution of Stars and Galaxies, ed. C. Payne-Gaposchkin (Cambridge: Harvard University Press).

Baum, W. A. 1959, PASP, 71, 106.

Barrientos, L. F., Schade, D. & López-Cruz, O. 1996, ApJ, 460, L89.

Bower, R. G., Lucey, J. R., & Ellis, R. S. 1992, MNRAS, 254, 601.

Buzzoni, A. 1995, ApJS, 98, 69.

Buzzoni, A., Renzini, A., & Serrano, A. (eds.) 1995, ASP Conf. Series 86, Fresh Views of Elliptical Galaxies (San Francisco: ASP).

Cepa, J. & Carral, P. (eds.) 1998, ASP Conf. Series 163, Star Formation in Early-Type Galaxies (San Francisco: ASP).

Djorgovski, S. & Davis, M. 1987, ApJ, 313, 59.

Dressler, A. 1980, ApJ, 236, 351.

Dressler, A. 1984 ARAA, 22, 185.

Ellis, R.S. 2001, astro–ph/0102056.

Faber. S, 1973, ApJ, 179, 731.

Ferguson, H. C. & Binggeli, B. 1994, A&A Rev., 6, 67.

Gladders, M. D., López-Cruz. O, Yee, H. K. C. & Kodama, T. 1998, ApJ, 501 , 571.

Gregg, M. D. & West, M. J. 1998, Nature, 396, 549.
Hilker, M. 1998, Ph.D. Thesis, Sternw. Bonn.
Jørgensen, I., Franx, M. & Kjærgard, P. 1996, MNRAS, 280 , 167
Kodama, T. 1997, Ph.D., University of Tokyo.
Kodama, T. & Arimoto, N. 1997, A&A, 320, 41.
Kormendy, J. & Djorgovski, S. 1989, ARAA, 27, 235.
López-Cruz, O. 1997, Ph.D. Thesis, University of Toronto.
López-Cruz, O., Yee, H.K.C, Brown, J.P., et al. 1997, ApJ, 475, L97.
López-Cruz, O. 2001, Rev. Mex. A&A Conf. Ser., 11, 183.
López-Cruz. O. & Yee, H.K.C 2002, in preparation
Lubin, L. M. & Sandage, A. 2001, AJ, 122, 1084.
McNamara, B. R. & O'Connell, R. W. 1992, ApJ,393, 579.
Melnick, J. & Selman, F. J. 2000, in ASP Conf. Series 215, Cosmic Evolution and Galaxy Formation, eds. J. Franco et al. (San Francisco: ASP), p. 159.
Oemler, A. J. 1974, ApJ, 194, 1.
Pahre, M. A., Djorgovski, S. G., & de Carvalho, R. R. 1996, ApJ, 456, L79.
Patton, D. R., Pritchet, C. J., Carlberg, R. G., et al. 2002, ApJ, 565, 208.
Peebles, J.P.E 2002, astro–ph/0201015.
Salpeter, E. E. 1955, ApJ, 121, 161.
Sandage, A. 1972, ApJ, 176, 21.
Sandage, A. & Visvanathan, N. 1978, ApJ, 223, 707.
Schade, D., Lilly, S. J., Le Fevre, et al. 1996a, ApJ, 464, 79.
Schade, D., Carlberg, R. G., Yee, H. K. C., et al. 1996b, ApJ, 464, L63.
Schade, D., Barrientos, L. F., & López-Cruz, O. 1997, ApJ, 477, L17.
Schade, D., Lilly, S. J., Crampton, D., et al. 1999, ApJ, 525, 31.
Schweizer, F. 2001, Phil. Trans. R. Soc. Lond., 358, 2063.
Silk, J. 1999, in Formation of Structure in the Universe, eds. A. Dekel & J.P. Ostriker (Cambridge: University Press), p. 98.
Stanford, S. A., Eisenhardt, P. R. & Dickinson, M. 1998, ApJ, 492, 461.
Tinsley, B. M. 1980, Fund. Cosmic Phys., 5, 287.
Tragger, S.C., Faber, S.M., & Dressler, A. 2001, astro–ph/0104404.
van Dokkum, P. G., Franx, M., Kelson, D. D. et al. 1998, ApJ, 500, 714.
Visvanathan, N. & Sandage, A. 1977, ApJ, 216, 214.
Yee, H. K. C. & Green, R. F. 1987, ApJ, 319, 28.

OPTICAL IDENTIFICATION OF X-RAY SOURCES IN A HIGH X-RAY FLUX SENSITIVITY AREA FROM THE RASS

R. Müjica
Instituto Nacional de Astrofísica, Optica y Electrónica, Puebla, Mexico

F. J. Zickgraf
Hamburger Sternawarte, Germany

V. Chavushyan, Y. Juárez, A. Serrano
Instituto Nacional de Astrofísica, Optica y Electrónica, Puebla, Mexico

I. Appenzeller, and J. Krautter
LSW-Heidelberg, Germany

Abstract

We have optically identified a complete sub-sample of ROSAT All-Sky Survey (RASS) X-ray sources contained in 6 study areas. Of the original 12° × 12° area near the NEP only one quarter was observed until now in order to keep the number of sources on a manageable size. This area is of particular interest because the RASS integration time is about a factor of ten longer than in the other areas and therefore a factor of 3-4 deeper in flux. We have started to observe the RASS sources in the remaining 3/4 of this area. First results are presented.

Introduction

ROSAT detected about 60,000 sources during the all-sky survey, i.e. about 1.5 sources per square degree. Complete, spectroscopically identified subsamples of the RASS are extremely important because, first, it is impossible to do spectroscopy on all sources, and second it is the only way to achieve an unbiased sample of optically identified X-ray sources, in order to carry out statistical studies and to obtain identification criteria to extrapole towards all X-ray sources. We have optically identified a complete sub-sample of RASS X-ray sources (Appenzeller et al. 1998, Paper II) contained in 6 study areas. Originally each area measured 12° × 12° in size. Due to the large number of X-ray sources in

M. Chávez et al. (eds.),
New Quests in Stellar Astrophysics: The Link Between Stars and Cosmology, 277–278.

two of the six areas, one close to the North Galactic Pole (NGP, area IV) and one near the North Ecliptic Pole (NEP area V), the original sample was reduced in size by observing only part of the latter two areas (Zickgraf et al. 1997, Paper I) in order to get a manageable number of sources. The statistical analysis was done by Krautter et al. (1999, Paper III).

1. The Area

The sample was redefined in terms of the minimum fluxes or countrates: 0.01cts/s in area V and 0.03 cts/sec in the other areas. The samples in the NEP and NGP were also reduced in size to smaller sub-areas. The resulting areas were called Va (western quarter of field V) and IVac (western half of area IV). Area Va was later extended to the east to obtain a field of $6.2^\circ \times 6^\circ$. Area V contained 671 sources, 183 were included in area Va. The goal of the present project is to identify the x-ray sources in the remaining 3/4 of the original area, that contains $\sim$ 300 sources with countrate $\geq$ 0.01cts/sec, following the same ID criteria as in the first part of the project (see Paper I).

2. Observations

The observations were carried out at the 2.1m telescope of the Guillermo Haro Observatory, operated by INAOE. The LFOSC focal reducer was used, it allows direct CCD imaging, filter photometry, and multi-object spectroscopy. A setup covering the spectral range 4200-9000 Åwith a dispersion of 8.2 Å/pix was adopted. Usually, all possible counterparts within a circle of about 50" to 60" radius around the X-ray position were observed.

3. Preliminary Results

We started to observe all the sources inside the error circle of the RASS positions in the remaining 3/4 of area V with the same limiting count-rate (0.01 cts/sec). Since May 1999 we have observed more than 100 RASS positions and we obtained in this way more than 300 spectra of candidate optical counterparts. We are following the method for optical identification as explained in detail in Paper I. The summary of new identifications is as follows: 22 Stars, 13 AGN, 1 Cluster and 2 Multiple. We have other 50 positions already observed, but not analyzed or requiring complementary observations.

References

Appenzeller et al. 1998, ApJS, 117, 319, Paper II.
Krautter et al. 1999, A&A, 350, 743, Paper III.
Zickgraf et al. 1997, A&AS, 123, 103, Paper I.

INTERMEDIATE RESOLUTION Hβ SPECTROSCOPY AND PHOTOMETRIC MONITORING OF 3C 390.3

J. R. Valdés, L. Carrasco, V. H. Chavushyan, F. Legrand
Instituto Nacional de Astrofísica, Optica y Electrónica, Apartado Postal 51, C.P. 72000, Puebla, México

A. I. Shapovalova, A. N. Burenkov, V. V. Vlasuyk, V. P. Mikhailov, O. I. Spiridonova
Special Astrophysical Observatory of the Russian AS, Nizhnij Arkhyz, Karachaevo-Cherkesia, 369167, Russia

A. M. Dumont, S. Collin
DAEC, Observatorie de Paris, section de Meudon, Place Jansen, F-92195 Meudon, France

V. T. Doroshenko, V. M. Lyuty, N. G. Bochkarev
Sternberg Astronomical Institute, University of Moscow, Universitetskij Prospect 13, Moscow 119899, Russia

O. Kurtanidze, and M. G. Nikolashvili
Abastumani Astrophysical Observatory, Georgian AS, Mt. Kanobili, 383762, Abastumani, Georgia

1. Observations

The BVR photometry of 3C 390.3 was carried out at three different observatories i.e. SAO RAS in North Caucasus (Russia), the Crimean Laboratory of the Sternberg Astronomical Institute (CL SAI, Russia) and the Abastumani Astrophysical Observatory (AAO) in Georgia. The observations at the SAO RAS were obtained with the 1 m and 60 cm Zeiss telescopes equipped with an offset guided automatic photometer. The VRI photometry at the Crimean Laboratory was obtained with the 60 cm telescope equipped with a CCD camera. BVRI observations at the AAO were obtained with the 70 cm menisk telescope equipped with a CCD camera.
Spectra of 3C 390.3 were obtained with the 6 m and 1 m telescopes of the SAO RAS and at INAOE's 2.1 m telescope of the "Guillermo Haro Observatory"

M. Chávez et al. (eds.),
New Quests in Stellar Astrophysics: The Link Between Stars and Cosmology, 279–280.

(GHO). The spectra were obtained with long slit spectrographs equipped with CCD detector arrays. The typical wavelength interval covered is from 4000 Å to 7500 Å. The spectral resolution varied between 4 and 15 Å.

2. Main results

Some of our main conclusions on the 4-year (1995–1999) spectroscopic and broad-band BVRI photometric monitoring of 3C 390.3 are:
1. The historical light curve in B shows a large increase in brightness during 1970–1971 followed by a gradual decrease until 1982.
2. The broad Hβ and continuum fluxes vary by a factor of about three during the time interval 1995–1999.
3. From a cross-correlation analysis, a delay of $\approx$ 100 days in the response of the Hβ emission with respect to the optical continuum changes is obtained from the observations at any given time. Yet, for observations obtained between the June 1997 and 1999, a period in which the sampling was better, a shorter lag of 33 to 36 days appears jointly with the one of $\approx$ 100 days.
4. The observed flux for the blue and red wings of the profile varied quasi-simultaneously. A similar result was obtained by Dietrich et al., (1998) from monitoring data for 3C 390.3 during 1994–1995. These results exclude the presence of considerable radial motions in the BLR.
5.- The changes of the radial velocity of the blue bump relative to the Hβ narrow emission in 1995 until Oct. 1999 follow the same trend found by Eracleous et al., (1997). Our results provide further support for the dismissal of the binary BH hypothesis for 3C 390.3, on the basis of the masses required.
6. The difference profiles of the Hβ broad emission revealed that the radial velocities of the blue and red bumps in the profiles varied, showing an increase in absolute value by about 2000 km s^{-1} during the observed period. The peak velocity shifts are also expected to result from the variable accretion rate near the black hole.
7. We have fitted the broad Hβ line profiles in the framework of an accretion disk model. We obtained a satisfactory agreement for a disk with an inclination angle of about 25° and with the region of maximum emission located at about 200 R_g. Our results do not support either the models of outflowing biconical gas streams or those of supermassive binary BHs. Instead, we conclude that our results favor the accretion disk model.

References

Dietrich, M., et al., 1998, ApJS, 115, 185.

Eracleous, M., Halpern, J. P., Gilbert, A. M., Newman, J. A., & Filippenko, A. V. 1997, ApJ, 490, 216.

VIII

NEW INSIGHTS FROM RADIO AND MILLIMETER SPECTRAL WINDOWS

FROM PROTOSTARS TO GALAXIES WITH GTM/LMT

Luis Carrasco, and Alberto Carramiñana
on behalf of the LMTO science team

Instituto Nacional de Astrofísica, Optica y Electrónica,
Luis Enrique Erro 1, Tonantzintla, Puebla 72840, México

Abstract

The general plan and status of the Gran Telescopio Milimétrico – Large Millimeter Telescope jointly with some of the potential scientific projects to be carried out with this instrument are discussed in this paper.

Introduction

Millimeter wave astronomy is presently undergoing a rapid expansion in terms of ground-based and space-borne facilities. Historically, much theoretical effort has been devoted to the understanding of star formation. The telescopes and satellites presently under development will have the sensitivity and resolution to test the understanding of star formation in our galaxy, in nearby galaxies and in the early universe. Observations of molecules in the Milky Way reveal chemical processes related to the formation history of our galaxy. The CO molecule can also be traced in nearby galaxies to prove both the dynamics and chemistry of cold gas. And deep observations of cold gas and dust at high redshift will tell us how the first stars and the first galaxies formed.

The new generation of far IR and millimeter telescopes will ensure wide wavelength coverage, sensitivity and spatial resolution. Space facilities, like *FIRST* or *STIRF*, are needed to observe in the medium and far IR bands unreachable from the ground. And while the anticipatedly acclaimed Atacama Large Millimeter Array (ALMA) project advances, large single dishes will provide essential clues to the understanding of the formation of stars and galaxies. In that respect, the Gran Telescopio Milimétrico / Large Millimeter Telescope (GTM/LMT) will build on the findings of present millimeter telescopes, like IRAM, JCMT and SEST.

M. Chávez et al. (eds.),
New Quests in Stellar Astrophysics: The Link Between Stars and Cosmology, 283–294.

1. El Gran Telescopio Milimétrico

The Gran Telescopio Milimétrico / Large Millimeter Telescope (GTM/LMT) is a bi-national project between the INAOE and the University of Massachusetts. The main goal is the construction of a 50 meter aperture antenna, designed to have a total surface error $\epsilon \approx 70\mu m$. Hence this instrument will have a good performance at frequencies up to 350 GHz. The construction site is located in the volcán Sierra Negra[1], eight kilometers SW from Citlaltepetl, the highest mountain in México. The high altitude of the site, about 4600 m above sea-level, guarantees a low amount of water vapor in the atmosphere, while its latitude (18° 59" North) allows excellent access to the Galactic Center. Although conceived primarily as a large aperture single-dish instrument, its unique geographical location intermediate between US-Canada and South American radio telescopes might eventually prove to be strategic for VLBI observations. Project news and developments can be found at our web site: *www.lmt.org*.

1.1 Engineering

The design of the GTM/LMT consists of an open air antenna with a segmented active primary surface, composed of aluminum covered carbon fiber panels of about 5m each (see Fig. 1). Each panel will be mounted on computer controlled actuators to ensure an optimal figure, canceling deformations due to gravity, thermal gradients and wind. The beamsize of the GTM/LMT will be about 5" at 300 GHz, which requires a pointing accuracy of about 1". The GTM/LMT will be able to perform under wind with speeds up to 10 m/s, a value larger than the higher quartile of the wind speed statistical distribution for the site.

1.2 Instrumentation

The initial instrumentation plan for the GTM/LMT contemplates a set of instruments adequate to use the antenna to its full capacities right from the commissioning phase. The instruments envisaged are:

- **SEQUOIA**, a focal plane array for observations in the 3mm band, which will allow GTM/LMT to perform fast imaging with high sensitivity over large scales with full resolution of the telescope. SEQUIOA is currently operating in the 14m millimeter telescope of FCRAO, with a 4×4 element configuration. The instrument will be upgraded to a 32 pixel array which will cover a 96"$\times$96" field with 12" spatial resolution on the GTM/LMT. The receiver has low noise temperatures ($\sim$ 60K), a dewar orientable along any coordinate system and angle. Its input has 15 GHz bandwidth,

[1] a correction from the former nomenclature "Cerro La Negra".

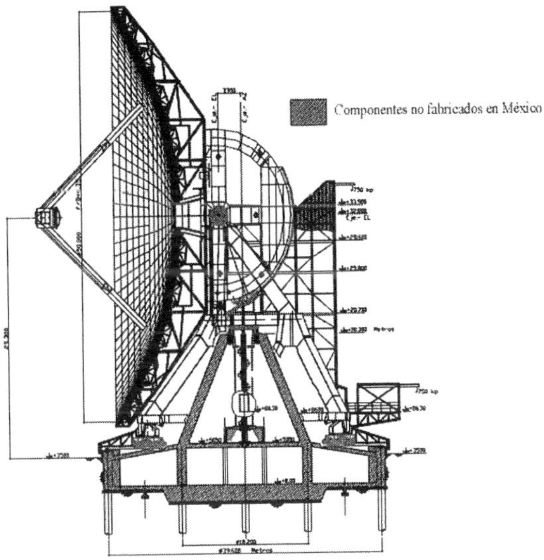

Figure 1. Design of the GTM/LMT antenna. Note the segmented structure of the primary surface. The concrete piles and foundation cone have now been poured. The alidade has been built and trial assembled. Transportation to the site will start within the next few months.

allowing multi-line observations in the 85-100 GHz or 100-115 GHz windows (i.e., molecules like N_2H^+, CS, HCN, HCO^+, ^{13}CO, $C^{18}O$).

- **BOLOCAM** is a bolometer based focal plane array for continuum observations currently under construction. It is designed around a monolithic wafer containing 151 silicon nitride micromesh ("spider-web") bolometers arranged in a hexagonal pattern with a center-to-center spacing of 5 mm. These detectors are developed from the same technology used in the ground-based SuZIE, AKBAR and BOLOMAT receivers, in the MAXIMA, ARCHEOPS and BOOMERANG balloon-borne experiments and planned for use in the *PLANCK* and *FIRST* satellites. They provide photon noise limited performance at $\lambda = 1$ mm from a base temperature of 300 mK. Science with BOLOCAM includes studies of the dust in the Galaxy, extragalactic surveys for galaxy counts and distributions, and the Sunyaev-Zel'dovich effect.

- The **"Redshift-Machine"**: current surveys are unable to identify the submm sources with their optical/IR/radio counterparts without ambiguity. Thus, the redshift distribution, luminosities and star-formation history of submm selected galaxies at $z > 1$ are unclear and, despite the improved resolution of future surveys conducted with the LMT/BOLOCAM, this situation will remain true. The GTM/LMT will have a wide-band spectrograph, currently denoted as the "redshift machine", designed to

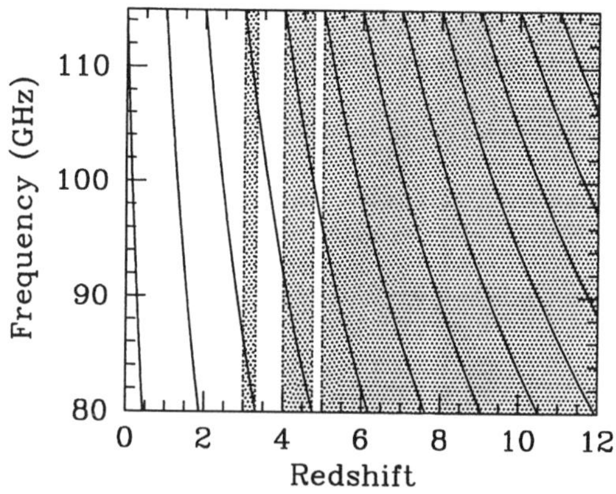

Figure 2. Frequency of different CO transition lines at high redshift. The curved lines represent the frequency *versus* redshift position of each CO line from $J = 1 \rightarrow 0$ to $J = 12 \rightarrow 11$. The shaded region indicates the redshift intervals where at least two lines fall inside the spectral window of the redshift machine.

search for lines in the redshifted CO ladder. At high-z the frequency separation of adjacent J-transitions of the CO molecule is $\delta\nu \simeq 115\,\mathrm{GHz}/(1+z)$, i.e. at $z > 2$ two adjacent CO spectral lines will be separated by $\delta\nu < 35\,\mathrm{GHz}$. Figure 2 shows the expected frequencies for the CO line ladder at various redshifts. We see that for $\delta\nu < 35$ GHz, we will be able to detect one or more lines simultaneously depending on z. Unambiguous redshift measurements are possible whenever at least two spectral lines fall inside the spectral window, which corresponds to $3.0 \leq z \leq 3.31$, $4.0 \leq z \leq 4.75$ and $z \geq 5.0$. Outside these intervals, the detection of a single line gives at most five possible options for the redshift.

- **Commissioning receivers:** for the characterization of GTM/LMT performance and initial science both 3mm and 1.3mm receivers are considered. Highly sensitive receivers with excellent science capabilities, including a high frequency receiver with full spectroscopic capability, will ensure that initial scientific projects with GTM/LMT will be carried on as soon as possible.

- **Support instrumentation** includes a wave-front sensor to measure the tilt of the incoming wavefront so that atmosphere-induced pointing errors may be removed and a holography receiver to measure the antenna surface at initial setting.

The LMT Observatory (LMTO) instrumentation team is presently developing a second generation instrumentation plan which will incorporate highly sensitive monolithic array cameras.

Table 1. Atmospheric Opacity at 215 GHz. La Negra Site period: 1997-2001

Month	1st Q	Median	3rd Q
Jan	0.053	0.073	0.108
Feb	0.062	0.092	0.141
Mar	0.085	0.133	0.198
Apr	0.091	0.140	0.224
May	0.124	0.211	0.327
Jun	0.250	0.347	0.535
Jul	0.273	0.397	0.634
Ago	0.315	0.476	0.697
Sep	0.197	0.286	0.424
Oct	0.083	0.213	0.431
Nov	0.083	0.135	0.204
Dec	0.073	0.985	0.151

1.3 The Site

The site for the GTM/LMT was selected amongst a large number of high altitude peaks in México based mainly on the water vapor content of the atmosphere. Radiometers measuring the opacity at 215 GHz were placed on about half a dozen preselected potential sites, their data shows that altitude is the predominant factor for high-transparency, even more than the geographical position. Sierra Negra, the highest of the sites surveyed, often shows an opacity value at 215 GHz below 0.10, specially during the winter months, indicative of submillimeter conditions. Table 1 lists the median and quartiles of the statistical distribution function for the atmospheric opacity at the site.
The site has the advantage of a relatively easy access (two hours from the city of Puebla). Wind statistics, critical for the performance of GTM/LMT at high frequencies, indicate a median of 6 m/s and a third quartile of 9 m/s. The weather at the site has both seasonal and diurnal modulations, with winter conditions better than summer conditions and nights better than days.
More recently, the site is being monitored for optical astronomy. Seeing measurements carried out with a Differential Image Motion Monitor (DIMM) indicate a median value of 0.6", with nightly medians rarely reaching 0.9". A report on preliminary optical seeing measurements at La Negra can be found in Carrasco, Carramiñana & Avilés (2002).

1.4 Construction Status

Site development tasks are being carried out since late 1997 and the telescope is under construction. Figures 3 and 4 show the present status of the civil works in the mountain and the plant trial assembly of the alidade in Tlalnepantla, México.

Figure 3. GTM/LMT foundation status, July 2001

Figure 4. GTM/LMT Alidade trial assembly, July 2001

2. Science with the Large Millimeter Telescope

The LMTO science group has contemplated a wide range of research topics for the LMT, ranging from solar system studies to high-redshift surveys. A relevant and interesting example of combined optical and millimeter observations of an object with consequences for the understanding of the formation of the solar system can be found in Jewitt, Aussel & Evans (2001).

2.1 Protostars

A broad and important topic of research is the study of the velocity fields in protostellar regions. In general it is not possible to separate the different types of motions, namely infall, rotation, outflows and turbulence. A specific goal for SEQUOIA observations on the GTM/LMT is the definitive identification of a *collapsing* protostar signature, which requires the direct observation of collapsing or infalling molecular gas, with the "classic," asymmetric, blue line profiles as observational signatures. Absorption line infall signatures will also be searched for.

As a specific example, GTM/LMT can study molecular outflows in young stellar objects. Although a lot is known after two decades of study of molecular outflows, many questions still remain unanswered: how do outflows evolve with time? What is the time variability of the outflow phenomenon? What is the linear extent of outflows? Can outflows trigger the birth of successive generations of stars? etc... The LMT will be able to study shocks driven by outflows. Protostellar shocks can be used as a clock to study the the time-dependent chemistry of outflows. Bally and Devine (1997) find that HH flows are very extended (pc scales). Large scale maps can be made with the LMT to determine the linear extent of the corresponding molecular outflows. This study would have considerable implications for the impact of outflows on the ISM, triggered star formation and the generation of interstellar turbulence.

2.2 Galactic ISM studies

Several topics of ISM studies with LMT have been identified. Here we list a few examples:

- Astrochemistry studies of the ISM have traditionally been carried out at a single position in the sky, but comparison of maps of different species shows interesting variations within molecular clouds. Focal plane array observations permit chemical variations within clouds to be studied and work at FCRAO has suggested that a detailed relation between cloud evolution and chemical abundance gradients exists. Another particular interest is the role of interface regions in the chemical balance of the clouds.

- The structure and equilibrium of molecular clouds are clearly related to turbulent motions at various spatial scales (Heyer, 1999). The resolution of the LMT will enable the study of turbulence with maps with a large spatial dynamic range, to address open questions regarding interstellar turbulence: what is the energy spectrum $E(k)$ for turbulent motions in the ISM? What is the decay rate of kinetic energy? Where and how is kinetic energy injected and dissipated? How is kinetic energy redistributed between scales? Interface regions apparently play an important role. LMT's resolution will enable studies of star formation sites in the galactic molecular ring and will enable comparison with photometric surveys.

- Galactic Center studies focus frequently on the circumnuclear disk (SgrA). Within the Galactic center region one can also target objects with extreme physical conditions. Broad bandwidths are required ($\sim 300\,\mathrm{km/s}$) to study the kinematics of the region.

- Magnetic fields are believed to play an important role in the evolution of the ISM. Polarization measurements of background starlight are used to map the magnetic field geometry. Near-infrared polarization studies of background stars present some problems: relative polarization does not increase with increased extinction; only a small subset of grains is responsible for producing the polarization of background starlight, and these are underrepresented in the dense interiors of clouds; polarization measurements of background stars do not trace the magnetic field in the interior of the cloud. BOLOCAM at GTM/LMT has the sensitivity to measure the emission from magnetically aligned dust grains, arising from the high column density interiors of clouds, and to map out the magnetic field geometry in these regions through millimeter and submillimeter polarization measurements.

2.3 Molecular gas in galaxies

Three types of studies of nearby galaxies can be identified: (a) large scale mapping of the spatial distribution of molecular gas; (b) patterns and kinematics of star formation regions in spiral arms; (c) line ratio determination of excitation conditions. Single dish observations will continue to be essential for these studies, since the filled aperture provides the best possible surface brightness sensitivity.

As an example, the passage of density waves across spiral arms predicts different spatial distributions of CO (molecular gas), HI (photodissociated gas) and Hα (stars) (Fig. 5), according to the classical sequence of star formation (CO $\rightarrow$ HI $\rightarrow$ IR $\rightarrow$ UV). However, there are open questions requiring more sensitive observations, like in the case of M83, where the CO distribution does not seem to match the dust distribution. Additional physical processes, like UV or cosmic-ray heating, have been studied. GTM/LMT single dish observations -matching the OVRO observations in spatial resolution, but with much higher sensitivity-will result in more detailed comparison of the distribution of the different gas components. An important clue is given by their motion. Measurements in M51 and M83 confirm the expected streaming directions of motion, but with velocities twice as large as expected. SEQUOIA on LMT will map one footprint with $\Delta v = 1\,\mathrm{km/s}$ in 20 hours.

On the largest scales, LMT will have the resolution and sensitivity for mapping cold dust in nearby galaxies, like M51 and NGC 4565. Dust, associated with atomic hydrogen, provides an independent estimate of ISM mass and serves

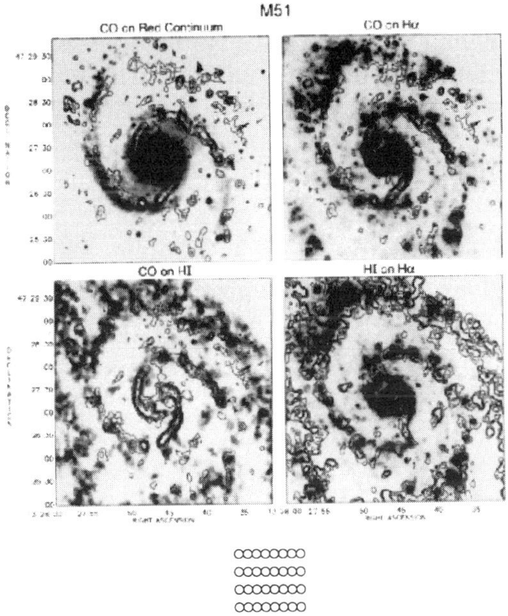

Figure 5. Comparative CO-red continuum, CO-Hα, CO-HI and HI-Hα maps of M51, illustrating the different positions of molecular gas, neutral gas and stars (from Rand, Kulkarni & Rice 1992). The SEQUOIA array is schematically shown at the same scale, on the right, for comparison.

to calibrate CO as mass tracer. Inside the Galaxy, BOLOCAM at GTM/LMT will be able to map giant molecular clouds with beam sizes $\leq$ 14" at flux densities of a few mJy in a 1 sec integration. This enables identification and study of density structures, with $\gtrsim$ 10 magnitudes of visual extinction, within molecular clouds at high resolution such as filaments $< 0.14\,\mathrm{pc} \times 7\,\mathrm{pc}$ in the Orion molecular cloud, and cold dense dust protostellar condensations in NGC 2024. The identification of several of these condensations will provide information about the IMF.

2.4 Spectroscopy of high-z galaxies

One of the major goals of observational cosmology is to acquire empirical data with sufficient diagnostic power to constrain and develop the theoretical modeling of the high-redshift universe, ultimately leading to an understanding of the processes by which galaxies form and subsequently evolve. New bolometer arrays on the largest submm and mm-telescopes now offer a unique view of the high-redshift universe and the key challenge in this field is to quantify the star formation history of galaxies and clusters, particularly those at redshifts > 1. The SCUBA JCMT observations of the HDF indicate the presence of a population of millimeter galaxies, probably precursors of present day dusty starbursts

(Hughes et al. 1998). LMT can map large regions of the sky with both depth and coverage providing statistically valuable samples. Large area deep millimeter surveys are prone to be a key project for the LMT. A 200 hour survey with BOLOCAM could cover a 10 deg^2 region at 1 mJy, providing several 10^4 sources. The immediate follow-up study would then be CO redshift measurements with the "redshift machine".

2.5 The S-Z effect

The Sunyaev-Zel'dovich effect (S-Z; Sunyaev-Zel'dovich 1972) is a distortion in the spectrum of the Cosmic Microwave Background (CMB) due to its interaction with the hot intracluster plasma, gravitationally bound in rich clusters of galaxies. The S-Z effect has two components, a thermal component due to the random motion of electrons in the intracluster plasma and a kinetic component due to the bulk motion of the cluster with respect to the CMB co-moving frame. Measurements of the thermal S-Z effect, combined with measurements of the surface brightness and spectrum of the X-ray emission from the intracluster plasma can be used to determine the Hubble constant and baryon mass fraction of the X-ray emitting gas. Cluster number counts in blank sky surveys can provide a determination of Ω_{matter} and probe cluster formation and evolution. Measurements of the kinetic S-Z effect give estimates of the radial component of individual cluster peculiar velocities with an accuracy of a few hundred km/s. A large sample of these measurements can be used to map the velocity field and mass distribution in the universe on the largest scales.

The brightness of the S-Z thermal effect changes dramatically across the millimeter-wave bands. The non-relativistic spectrum has a decrement in intensity with a peak at $\nu = 2.255kT/h = 128$ GHz, an increment in intensity with a peak at $\nu = 6.508kT/h = 369$ GHz, and a null where the intensity of the CMB through the cluster is equal to the unperturbed intensity at $\nu = 3.83kT/h = 217$ GHz. The distinct shape of this spectrum allows S-Z signals to be distinguished from local sources of emission such as the atmosphere with measurements at millimeter wavelengths.

The S-Z kinetic effect has the same spectral signature as primordial CMB fluctuations. Theoretically, measurements in any two bands where the S-Z effect is detected allow the separation of the S-Z kinetic and thermal effects. However, at frequencies < 100 GHz or > 280 GHz, the kinetic spectrum is similar to the thermal spectrum. Also, for expected cluster velocities of ~ 500 km/s, the thermal effect is almost an order of magnitude brighter than the kinetic effect at most frequencies and estimates of the kinetic effect from differences between detections are limited by uncertainty in calibration in addition to statistical noise. Fortunately, the peak of the S-Z kinetic effect is close to the null

frequency of the thermal effect ($\sim$ 220 GHz). Observations at this frequency offer the best possibility of measuring the effect.
The BOLOCAM array is specifically designed for mapping low surface brightness sources like the S-Z effect from galaxy clusters and will be able to recover all of the flux from on-the-fly maps covering more than $10' \times 10'$ on the sky. A survey of known clusters with BOLOCAM will provide a data base of S-Z measurements with comparable sensitivity and resolution to existing X-ray images and complement lower frequency measurements from ground-based interferometers. Mounted on the LMT, BOLOCAM will have a FOV of 2' and a resolution of $\sim 10''$ at a frequency of 221 GHz and $\sim 7''$ resolution at 271 GHz. The projected sensitivity of BOLOCAM will allow a $> 10\sigma$ detection of the S-Z thermal effect in a 1 hour integration on suitable clusters. Measurements with BOLOCAM on the LMT at 221 GHz will reach a sensitivity to cluster peculiar velocity of < 300 km/s in a ten minute integration. Over the next few years, the X-ray satellites *CHANDRA*, *XMM-Newton* and *ASTRO-E* will provide more maps of X-ray emission from galaxy clusters with improved sensitivity, angular resolution ($0.5''$), and spectral resolution (0.012 keV). Observations of clusters and coordination with S-Z measurements is a high priority for these satellites.

Acknowledgments

We thank CONACYT, México for financial support of the GTM/LMT project and the LMTO science team for their intellectual contributions. GTM/LMT is a joint project of the Instituto Nacional de Astrofísica, Óptica y Electrónica and the University of Massachusetts.

References

Bally, J., & Devine, D., 1997, in IAU Symposium No. 182, Herbig-Haro Flows and the Birth of Stars, eds. B. Reipurth & C. Bertout (Dordrecht: Kluwer), p. 29.

Carrasco, B. E., Carramiñana, A., & Avilés, J. L. 2002, in ASP Conf. Ser., Astronomical Site Evaluation in the Visible and Radio Range, (San Francisco: ASP), in press.

Heyer, M. 1999, in Interstellar Turbulence, eds. J. Franco & A. Carramiñana, (Cambridge: Cambridge Univ. Press), p. 198.

Hughes, D. H. et al. 1998, Nature, 394, 241.

Jewitt, D., Aussel, H., & Evans, A. 2001, Nature, 411, 446.

Rand, R. J., Kulkarni, S., & Rice, W. 1992, ApJ 390, 66.

Sunyaev, R., & Zel'dovich, Ya. B. 1972, Comments on Astrophysics and Space Physics 4, 173.

SEARCHING FOR AN EARLY EPOCH OF STAR FORMATION WITH MM/SUB-MM BLANK FIELD SURVEYS

E. Chapin, D. H. Hughes, E. Gaztañaga, and I. Aretxaga
Instituto Nacional de Astrofísica, Optica y Electrónica, Puebla, Mexico

Abstract

If there exists an epoch of star-formation at $z > 3$ then we can claim that at least some of the modern day giant elliptical galaxies may have formed in a violent, short-lived (≤ 1 Gyr) starburst rather than evolving gradually through the merger of intermediate mass disks. A coordinated campaign of multi-wavelength blank field FIR–mm surveys using a combination of future telescopes will allow us to search for dust-enshrouded star-formation at these high redshifts. In addition we will be able to improve estimates of the parameters that describe the evolutionary history of the starburst galaxy population, and their luminosity function.

1. Evolution of the high-z sub-mm galaxy population

The mm to sub-mm band is ideally suited to searching for star formation because most of the radiation from these regions is thought to be absorbed by dust in the optical and re-emitted thermally in the sub-mm where the dust is optically thin. This belief is supported by the lack of optical counterparts to the sources detected in SCUBA (Hughes et al. 1998) and MAMBO (Bertoldi et al. 2000) surveys. Therefore if there *is* an epoch of star-formation at $z > 3$ during which the modern-day giant ellipticals were formed, then mm and sub-mm surveys provide the best way of conducting a relatively unbiased search.

The form of evolution in the starburst galaxy population may be determined from the number of sources as a function of limiting flux in any survey. A prescription for modeling source counts as a function of characteristic source SED, cosmology, luminosity evolution and clustering is described in Hughes & Gaztañaga (2000). A comparison of existing data with these models demonstrates that we lack the dynamic range in flux densities with modern instruments to successfully constrain the evolutionary parameters. At 850 μm with SCUBA for example it is not possible to scan large areas ($> 0.2\,\mathrm{deg}^2$) and detect significant numbers of high-z sources > 10 mJy. This bright flux density limit corresponds to star formation rates (SFR) $\geq 10^3\ \mathrm{M}_\odot/\mathrm{yr}$ which are required to

M. Chávez et al. (eds.),
New Quests in Stellar Astrophysics: The Link Between Stars and Cosmology, 295–296.

form massive ellipticals at $z > 3$. At faint flux densities the beam size yields a 3-σ confusion limit due to extragalactic sources of about 2 mJy (Hughes et al. 1998).

2. Redshifts and SFR

In order to determine the star-formation history of the universe, we need to perform multi-wavelength surveys to obtain photometric estimates of the redshifts and hence luminosities and SFRs. While the idea of using sub-mm colors is not new (Hughes et al. 1998; Carilli & Yun 1999) we have developed a Monte-Carlo simulation to evaluate confidence intervals for inferred redshifts using different combinations of colors. In this way we will be able to select statistically significant high-z candidate sources for future follow-up with a broad-band CO line spectrometer (Hughes 2000). A full description of the Monte-Carlo simulations is in preparation (Hughes et al. 2001).

3. Simulating Future Instruments

Over the next 5-10 years a variety of new telescopes such as BLAST (Devlin et al. 2001), the Large Millimeter Telescope, SIRTF and Herschel will be available to address the above problems of sensitivity and wavelength coverage. We have created a telescope mapping simulator to assist with the development and observing strategies for this new generation of instruments. The telescope simulator can model arbitrary focal plane arrays at different wavelengths, with variable beam sizes and detector characteristics. The arrays are scanned across an "input sky" created from a simulated catalogue of extragalactic sources (using a realistic model for the evolution of the luminosity function), and generate signal time-streams for each bolometer. These data may be analyzed using the reduction pipeline for the real instrument. Further information concerning the simulator may be found at: *http://www.inaoep.mx/~echapin/scansim.html*
This combination of simulations (source catalogues, sky maps, detector arrays and telescopes) provides a powerful tool to influence design and engineering decisions, examine scheduling issues and test the overall feasibility of the science case for future cosmological experiments.

References

Bertoldi, F., Menten, K. M., Kreysa, E., Carilli, C. L., & Owen, F. 2000, in Cold Gas and Dust at High Redshift, ed. D. J. Wilner (San Francisco: PASP).

Carilli, C.L., & Yun, M. 1999, ApJ, 513, L13.

Devlin, M.J., et al. (the BLAST project) 2001, astro-ph/0012327.

Hughes, D.H. 2000, in Clustering at High Redshift, eds. A. Mazure, O. Le Fèvre, & V. Le Brun (San Francisco: ASP), p. 81.

Hughes, D.H., & Gaztañaga, 2000, astro-ph/0004002.

Hughes, D.H. et al. 1998, Nature, 394, 241.

THE CREATION AND EVOLUTION OF GALAXIES IN THE UNIVERSE: MILLIMETER COSMOLOGY @ INAOE

E. Gaztañaga, D. H. Hughes, E. Chapin, and I. Aretxaga
INAOE, Astrofísica, Tonantzintla, Apdo. Postal 216 y 51, Puebla 7200, México

Abstract

Ongoing efforts at INAOE are focusing to develop a new group of research center around millimeter cosmology. To illustrate the scientific case for such a group, we present a sequence of real and simulated images of galaxy evolution, starting from CMB structures through a stage of dust obscured star formation to the more recent epoch where galaxies are seen at optical wavelengths.

Introduction

The primary optical and UV radiation emitted by the first generations of young, massive stars in galaxies is absorbed by dust and re-radiated at FIR wavelengths. In distant galaxies this rest-frame FIR emission is redshifted to longer wavelengths due to the expansion of the Universe. The formation of galaxies is therefore expected to be a spectacular luminous event at submm and mm wavelengths (eg see astro-ph/0004002 and references therein). The Large Millimeter Telescope (LMT, see *http://www.lmtgtm.org*) INAOE/Umass, will look back 14,000 million years to epochs when galaxies and stars were forming in the early Universe.

The Cosmic Background Radiation (CMB) spectrum also peaks at millimeter wavelengths. Observations of angular CMB temperature fluctuations (measured from satellites, such as COBE, or balloons , such as QMAP, MAXIMA or Boomerang) correspond to the earliest structures in the Universe. The LMT will be able map these background fluctuations on scales and with amplitudes that have since grown to become the galaxies and clusters that we observe in the present day Universe.

M. Chávez et al. (eds.),
New Quests in Stellar Astrophysics: The Link Between Stars and Cosmology, 297–298.

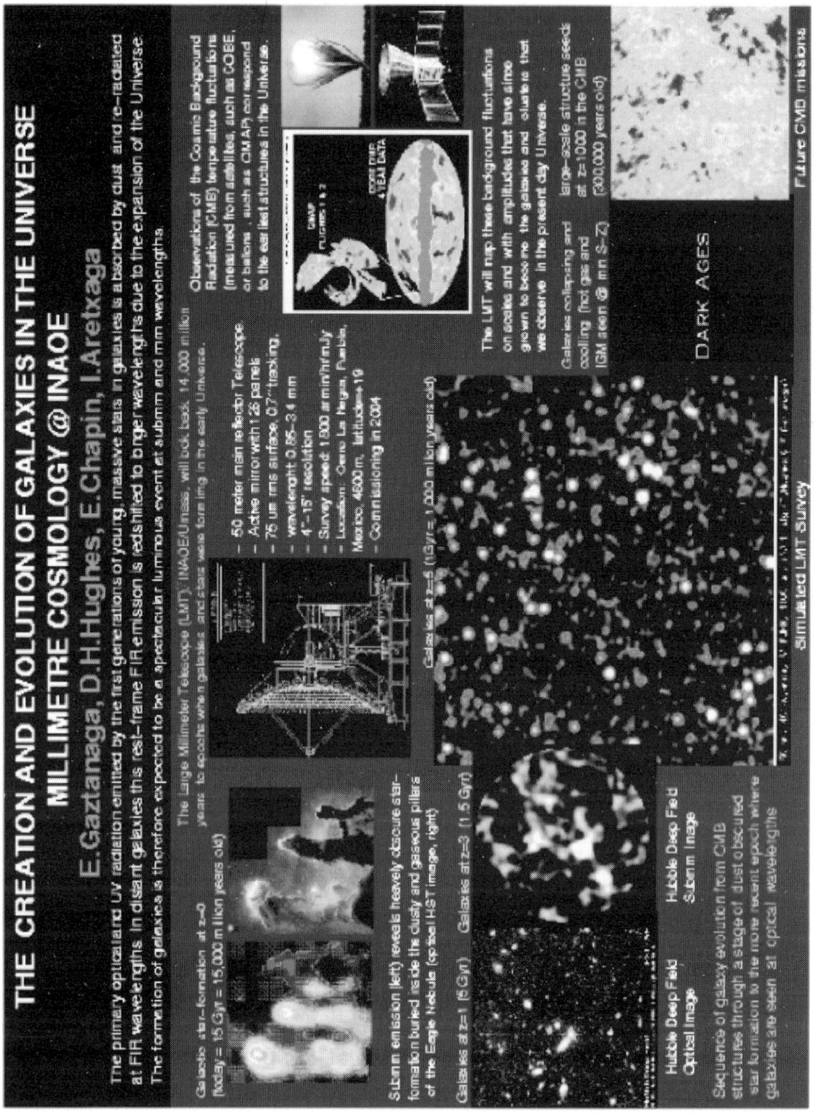

Figure 1. Poster presented at the conference. For the full color version see: http://www.inaoep.mx/∼gazta/mmposter.gif

21-CM EMISSION FROM THE DAMPED LYMAN-α ABSORBER SBS 1543+593

D. V. Bowen

Princeton Observatory, USA

E. Brinks

Depto. de Astronomía, Guanajuato, Mexico

W. Huchtmeier

Max-Plank-Institute für Radioastronomie, Germany

T. M. Tripp, and E. B. Jenkins

Princeton Observatory, USA

Abstract

We detect 21-cm emission from the Low Surface Brightness (LSB) galaxy SBS 1543+593, which gives rise to a Damped-Lyα (DLyα) absorption line in the spectrum of the background QSO HS 1543+5921.

We recently discovered (Bowen et al. 2001a) that the nearby LSB galaxy SBS 1543+593 (see Fig. 1 and *http://www.astro.princeton.edu/~dvb/lsb.htm* for further details) gives rise to a DLyα system in the spectrum of the background QSO HS 1543+5921 ($z_e = 0.807$). The line of sight to the QSO passes within 2.4″ ($\equiv 0.3h_{75}^{-1}$ kpc) from the center of the galaxy[1], and an H I column density of N(H I)= 20.35 dex is measured along the sightline to the quasar. The galaxy has a central surface brightness of $\mu_B(0) = 23.2$ mag arcsec^{-2} and an absolute magnitude of $M_R = -16.5$.

The 21-cm line emission observations were made 22nd January 2001, using the 100 m radio telescope at Effelsberg. The 21-cm emission detected from SBS 1543+539 is shown as a function of heliocentric velocity in Fig. 2 (Bowen

[1] $h_{75} = H_0/75$ km s^{-1} Mpc^{-1}, where H_0 is the Hubble constant, and $q_0 = 0.0$

M. Chávez et al. (eds.),
New Quests in Stellar Astrophysics: The Link Between Stars and Cosmology, 299–302.

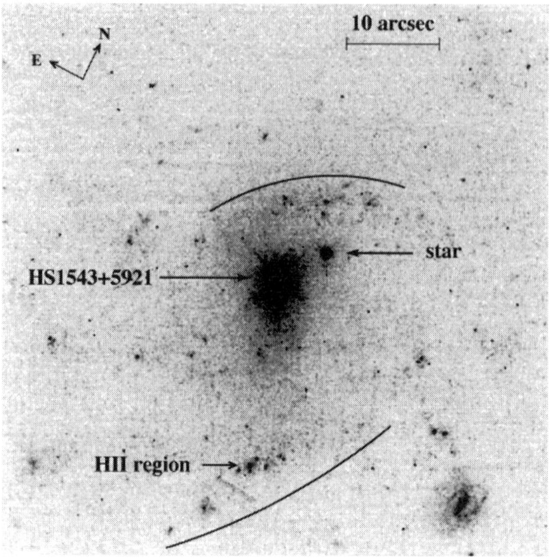

Figure 1. Reproduction of a 804 sec coadded HST STIS image of SBS 1543+593 taken with no filter. Pixels were originally 0.05 arcsec square but the image has been smoothed by a 3 × 3 pixel boxcar. The H II region whose redshift was measured by Reimers & Hagen (1998) is marked.

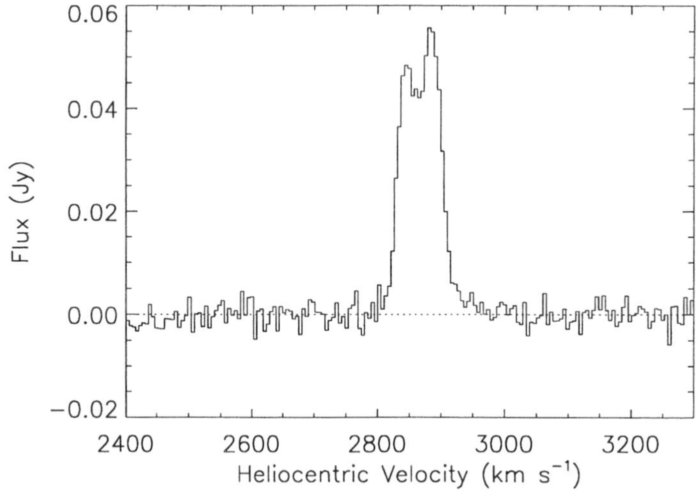

Figure 2. 21-cm emission from SBS 1543+593. The bulk of the emission arises at 2868 km s^{-1}, with two peaks arising from the rotating, inclined disk.

et al. 2001b). The center of the emission measured from the FWHM of the profile is 2868 ± 2 km s^{-1} ($cz = 0.0096$) and shows the classic double-horn peaks, with maxima at 2843.5 and 2884.0 km s^{-1}. Such profile shapes are well understood to arise from the disks of spiral galaxies with flat rotation curves and exponential gas distributions. From these data, we derive an H I mass of $M_{\rm HI} = 1.3 \times 10^9\ M_\odot$, and calculate a lower limit to the dynamical mass of $M_{\rm dyn} \geq 2.8 \times 10^9\ h_{75}^{-1}\ M_\odot$. The H I mass derived for SBS 1543+593 is about 1/5 that of the Milky Way, and the H I mass-to-light ratio is $M_{\rm HI}/L \sim 4$, consistent with other LSB galaxies of similar luminosity.

Comparisons of $M_{\rm HI}$ with limits obtained towards two other $z \sim 0.1$ DLyα systems show that SBS 1543+593 would not have been detected in those observations. Hence LSB galaxies similar to SBS 1543+593 can be responsible for DLyα systems at even modest redshifts without being detectable from their 21-cm emission.

References

Bowen, D. V., Tripp, T. M., & Jenkins, E. B. 2001, AJ, 121, 1456.

Bowen, D. V., Huchtmeier, W., Brinks, E., Tripp, T. M., & Jenkins, E. B. 2001, A&A, 372, 820.

Reimers, D., & Hagen, H. 1998, A&A, 329, L25.

COSMOLOGICAL PARAMETERS AND THE BARYON DENSITY FROM CMB AND GALAXY FLUCTUATIONS

E. Gaztañaga
INAOE, Astrofísica, Tonantzintla, Apdo. Postal 216 y 51, Puebla 7200, Mexico

J.Barriga
IEEC-ICE/CSIC, Edf. Nexus-201 - c/ Gran Capità 2-4, 08034 Barcelona, Spain

M. G. Santos, and S. Sarkar
Theoretical Physics, University of Oxford, 1 Keble Road, OX1 3NP, UK

Abstract

Recent observations of the CMB fluctuations (eg Boomerang and Maxima) show an anomalously low second acoustic peak, which has been interpreted as an indication of a high primordial Big-Bang nucleosynthesis (BBN) baryon density. We reexamine the prediction by Adams et al (1997) that this would be the consequence of a 'step' in the primordial spectrum induced by a spontaneous symmetry breaking phase transition during primordial inflation. We demonstrate that a deviation from scale-invariance around $k \sim 0.1h$ Mpc^{-1} can simultaneously explain both the feature identified earlier in the APM galaxy power spectrum as well the recent CMB anisotropy data, with a baryon density consistent with the BBN value.

Introduction

It is commonly assumed by astronomers that inflation predicts a scale-invariant 'Harrison-Zeldovich' (H-Z) spectrum of scalar density perturbations, $P^0(k) \propto k^n$ with $n = 1$. This is an essential ingredient when fitting cosmological parameters, such as the baryon density Ω_B, to observations of large scale structure or CMB fluctuations. Relaxing this assumption, produces quite different constraints. Adams et al. (1997) realized that the spectrum may not be *scale-free* because the rapid cooling of the universe during primordial inflation can result in spontaneous symmetry breaking phase transitions which may interrupt

M. Chávez et al. (eds.),
New Quests in Stellar Astrophysics: The Link Between Stars and Cosmology, 303–308.

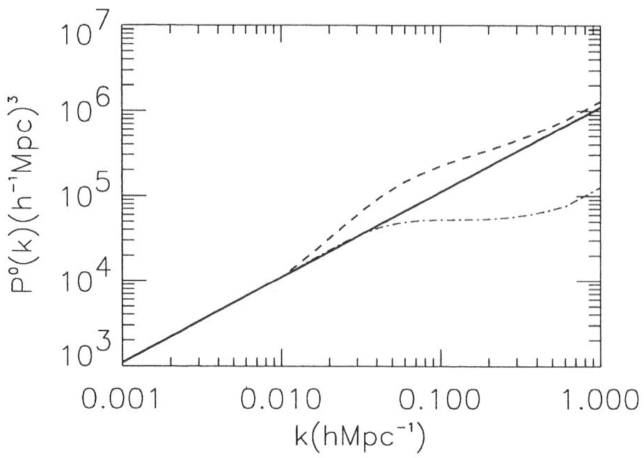

Figure 1. Reconstruction of the primordial density power spectrum from the APM data, adopting a CDM shape parameter Γ of 0.5 (dot-dashed line) and 0.2 (dashed line). A Harrison-Zeldovich spectrum (full line) is also shown.

the slow roll of the inflation field for brief periods. In order for such spectral features to be observable in the LSS or CMB, it is of course necessary that they occur within the last $\sim$ 50 e-folds of inflation, corresponding to spatial scales going up to the present Hubble radius $H_0^{-1} \sim 3000h^{-1}$Mpc. Since the density perturbation can be observed on scales from the Hubble radius down to $\sim 1h^{-1}$Mpc, corresponding to about 8 e-folds of expansion, it would be not unreasonable to expect at least one such spectral break to be seen today.

1. Reconstructing the primordial spectrum

On the linear scales $0.01h^{-1}\text{Mpc} \lesssim k \lesssim 0.6h\text{Mpc}^{-1}$, the APM galaxy power spectrum is an *unbiased* (or moderately linearly biased) tracer of the mass. The linear power spectrum recovered under this assumption from $P_{\text{APM}}(k)$ is well fitted in this range by $P_{\text{lin}}(k) \simeq \frac{7\times10^5 k\,(h^{-1}\text{Mpc})^3}{[1+(k/k_c)^2]^{1.6}}$ (Baugh & Gaztañaga 1996) where $k_c = 150(H_0/c) \simeq 0.05\,h\text{Mpc}^{-1}$. If we assume a CDM transfer function $T(k)$ with a 'shape' given by our cosmological parameters: eg $\Gamma = \Omega_m h e^{-[\Omega_B(1+\sqrt{2h}/\Omega_m)-0.06]}$, the primordial spectrum of matter fluctuations, can then be reconstructed as: $P^0(k) = P_{\text{lin}}(k)/T^2(k)$. Figure 1 shows the recovered primordial spectrum for two choices of Γ corresponding to the sCDM model ($\Gamma = 0.5$) and a low density variant $\Gamma = 0.2$. Outside the APM measurements in k we have assume a H-Z spectrum. Note how the $\Gamma = 0.5$ reconstruction has significantly less power than a scale-free H-Z spectrum on scales $k \gtrsim 0.1h\text{Mpc}^{-1}$, while the $\Gamma = 0.2$ reconstruction is closer to a H-Z spectrum but has relatively more power.

Ω_Λ	h	k_{start}	A/B	σ_8	χ^2
0.0	0.50	0.07	4.3-6.5	0.65-0.73	11.9
0.2	0.55	0.06	3.6-5.6	0.69-0.78	10.0
0.3	0.60	0.06	3.4-5.0	0.76-0.85	10.3
0.4	0.65	0.05	3.1-4.6	0.77-0.87	13.1
0.5	0.70	0.05	2.6-4.1	0.81-0.93	18.8
0.6	0.75	0.05	2.0-2.8	0.90-0.99	29.7

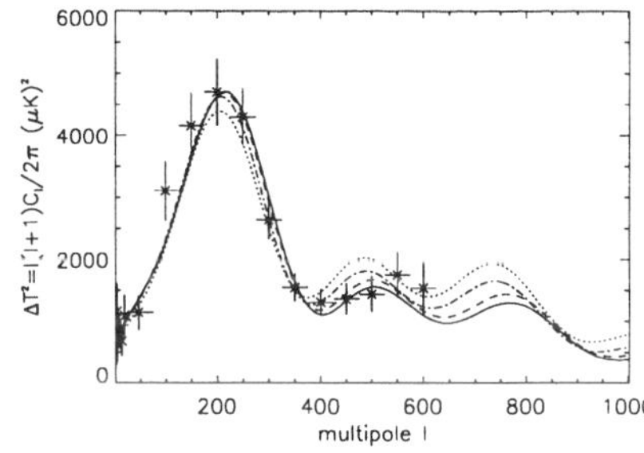

Figure 2. Left: Parameters for best fits to COBE+Boomerang+APM data with the BBN constraint on the baryon density. Right: CMB angular power spectra for the best-fit models with $\Omega_\Lambda = 0$, $\Omega_\Lambda = 0.2$, $\Omega_\Lambda = 0.4$ and $\Omega_\Lambda = 0.6$ (continuous, dashed, dot-dashed and dotted line). All models have the BBN baryon density $\Omega_B = 0.019/h^2$.

The above results suggest to parameterize the "step" in the primordial power spectrum as:

$$P^0(k) = \begin{cases} Ak, & k \leq k_{\text{start}} \\ Ck^\alpha, & k_{\text{end}} \leq k \leq k_{\text{start}} \\ Bk, & k \geq k_{\text{end}} \end{cases}$$

where k_{start} and k_{end} mark the beginning and end of the break from H-Z spectra with amplitudes A and B. (The values of C and α are specified by the other parameters.) In the multiple inflation model (Adams et al. 1997), the actual form of the spectrum during the phase transition is difficult to calculate since the usual 'slow-roll' conditions are violated. However a robust expectation is that $\ln(k_{\text{end}}/k_{\text{start}}) \sim 1$ because the field undergoing the symmetry-breaking phase transition evolves exponentially fast to its minimum. The ratio of the amplitudes A/B is determined by the (unknown) super-potential couplings of the field undergoing the phase transition but is expected to exceed unity (i.e. there is a decrease in the power). Using the above parameterization for $P^0(k)$, Barriga et al. (2001) have fit recent CMB data of Boomerang (de Bernardis et al. 2000) and Maxima-1 (Hanany et al. 2000) as well as to the measured APM spectrum and COBE. The left panel of Fig. 2 shows the best fit parameters after imposing the BBN constraint and also requiring that $0.5 \leq \ln(k_{\text{end}}/k_{\text{start}}) \leq 2$. We see that the data now prefer lower values of Ω_Λ (and h). In particular the value $\Omega_\Lambda \sim 0.7$ favored by the SN Ia data (Perlmutter et al. 1999; Riess et al. 1998) is *not* permitted. The right panel of Fig. 2 shows the fit to the CMB data for these models.

2. Discussion

Allowing the primordial spectrum not to be scale-invariant has interesting consequences. In our analysis we find that high values of Ω_m in a flat universe (and therefore low values of Ω_Λ) are compatible with CMB and LSS data. The standard interpretation of LSS data (eg APM $P(k)$ and cluster abundances) favors of a low $\Gamma \sim \Omega_m h \sim 0.2$ for a H-Z spectrum. But these observations can also be explained with larger values of Ω_m if we allow for a break in the primordial spectrum. Our best fit values are shown in the left panel of Fig.2. The required spectral break decreases with increasing Ω_Λ but one cannot do without such a break. The value $\Omega_\Lambda \sim 0.7$ favored by the SN Ia data (Perlmutter et al. 1999; Riess et al. 1998) is *not* permitted, while it is possible to have a universe with no cosmological constant and $\Omega_m \simeq 1$. We note that the small angular-scale CMB anisotropy data (de Bernadis et al. 2000; Hanany et al. 2000) favor a deviation from scale-invariance in the range $k \sim (0.06 - 0.6)h\text{Mpc}^{-1}$ *independently* of the similar previous indication from studies of LSS, in particular the APM survey.

The hypothesis of a primordial density perturbation with broken scale-invariance is eminently falsifiable. The ongoing 2DF and SDSS redshift surveys can confirm or rule out such a feature in the power spectrum of galaxy clustering, while the forthcoming MAP mission will determine whether *all* the secondary acoustic peaks in the CMB angular spectrum are indeed suppressed as expected. By contrast the alternative hypothesis of a baryon density $\sim 70\%$ higher than the BBN value predicts a *boosted* third peak Tegmark & Zaldarriaga 2000), so this would be a definitive test distinguishing between the two possibilities. [1]

Broken scale-invariance has a natural explanation in a phase transition occurring during inflation as expected in super-symmetric theories. If established this would provide the first direct connection between astronomical data and physics at very high energies. [The reader is referred to Barriga et al. (2001) for further details, references and acknowledgments].

References

Barriga J., Gaztañaga E., Santos M., & Sarkar S. 2001, MNRAS, 324, 977.

de Bernadis P. et al. (the Boomerang project), 2000, Nature, 404, 955.

Adams J.A., Ross, G.G., & Sarkar, S. 1997, Nucl. Phys., B503, 405.

Baugh, C.M., & Gaztañaga, E. 1996, MNRAS, 280, L37.

Hanany, S. et al. (the MAXIMA project), 2000, ApJ, 545, L5.

Perlmutter S. et al. 1999, ApJ, 517, 565.

Riess A.G. et al., 1998, AJ, 116, 1009.

Tegmark, M., & Zaldarriaga, M., 2000, ApJ, 544, 30.

[1] After this work was completed, the Boomerang (astro-ph/0104460) and MAXIMA (astro-ph/0104459) collaborations have presented new data analysis. Their new fits, assuming a H-Z spectrum, now allow a baryon density consistent with the BBN value. This does not contradict our predictions. In fact, note how the new Boomerang data show a low third acoustic peak, just as predicted by Fig. 2.

List of Participants

Name	Pages	Email
Abia, Carlos	59	cabia@ugr.es
Albarran, Marna		marna@inaoep.mx
Aragón Calvo, Miguel Angel		maac@inaoep.mx
Avila-Rees, Vladimir	83	avila@astroscu.unam.mx
Bertone, Emanuele	27, 39	bertone@merate.mi.astro.it
Blom, Harry		harry.blom@wkap.nl
Boesgaard, Ann Merchant		boes@ifa.hawaii.edu
Bonifacio, Piercarlo	77	bonifaci@ts.astro.it
Bressan, Alessandro	159, 211, 167, 171, 175	bressan@pd.astro.it
Brinks, Elias	299	ebrinks@astro.ugto.mx
Buzzoni, Alberto	189, 27, 39, 103,153	buzzoni@tng.iac.es
Cardona, Octavio	29, 41	ocardona@inaoep.mx
Carigi, Leticia	91	carigi@astroscu.unam.mx
Carrasco, Luis	283, 279	carrasco@inaoep.mx
Cardwell, Andrew	147	cardwell@ll.iac.es
Castillo-Domínguez, Edgar		cade@inaoep.mx
Cellone, Sergio	153	scellone@fcaglp.unlp.edu.ar
Chapin, Edward	295, 297	echapin@inaoep.mx
Chávez, Arturo		achavez@cucei.udg.mx
Chávez, Miguel	41, 39, 103, 27	mchavez@inaoep.mx
Chieffi, Alessandro	47, 199	achieffi@ias.rm.cnr.it
Croft, Rupert	249	rcroft@cfa.harvard.edu
de Alba-Martínez, Durruti J.		dalba@udgserv.cencar.udg.mx
del Río, M. Soledad		sole@inaoep.mx
Demarque, Pierre	111	demarque@astro.yale.edu
di Benedetto, Paolo G.		pdibene@ifctr.mi.cnr.it
Ellison, Sara	207	sellison@eso.org
Ferguson, Annette	131	ferguson@astro.rug.nl
Franco Hernández, Ramiro		quantumrem@yahoo.com
Gallardo del Angel, Agustín		delfin-x@inaoep.mx
García-Malacon Yamil A.		yamil@inaoep.mx
Gaztañaga, Enrique	303, 297, 215, 295	gazta@inaoep.mx
Gnedin, Nick		gnedin@casa.colorado.edu
Gómez-Carranza, Juan Carlos		jcgc@inaoep.mx
Granato, Gian Luigi	257, 175	granato@pd.astro.it
Grimaldi-Urbiola, Cesar A.		grimaldi@inaoep.mx
Guichard, José		jguich@inaoep.mx
Gutierrez-Reséndiz, Fidel		fidel@astro.ugto.mx
Guzmán Jiménez, Violeta		violetaguzmanj@hotmail.co
Hauschildt, Peter	15	yeti@hal.physast.uga.edu
Hernández-Doring, Xavier	69,83,127	xavier@astroscu.unam.mx
Hernández-Utrera, Oscar		dhuos@inaoep.mx
Hubrig, Swetlana	33	shubrig@eso.org
Juárez, Juan Antonio		jantonio@inaoep.mx

Juszkiewicz, Roman		roman@camk.edu.pl
Kanbur, Shashi	143	shashi@fcrao1.astro.umass.edu
Koenigsberger, Gloria	135	gloria@astroscu.unam.mx
Kopp, Angela		angela.j.kopp@lawrence.edu
Kurucz, Robert	3	rkurucz@cfa.harvard.edu
Limongi, Marco	199, 47, 59	marco@nemo.mporzio.astro.it
Lipovka, Anton	73	aal@cajeme.cifus.uson.mx
López-Cruz, Omar	269	omar@inaoep.mx
Malagnini, Maria Lucia	103	malagnini@ts.astro.it
Marleau, Francine		marleau@maia.stmarys.ca
Mayya, Divakara	167, 155	ydm@inaoep.mx
Mercado, Abelardo		abelmh@inaoep.mx
Moitinho, Andre	119	andrem@astrosen.unam.mx
Molaro, Paolo	241	molaro@ts.astro.it
Montaña, Alfredo		montana@inaoep.mx
Mújica, Raúl	277	rmujica@inaoep.mx
Nakamura, Fumitaka	63	nakamura@astro.berkeley.edu
Nepote, Juan Miguel		lafisicarifa@yahoo.com
Ontiveros-González, Guillermo		memomemo36@hotmail.com
Orozco-Aguilera, Teresa		toa@inaoep.mx
Ortiz-Palacios, Martha		ahtramop@inaoep.mx
Ortuño-Encinas, Andrea		aortuno@fismat1.fcfm.buap.mx
Panuzzo, Pasquale	171	panuzzo@sissa.it
Peimbert, Manuel		peimbert@astroscu.unam.mx
Pettini, Max	229	pettini@ast.cam.ac.uk
Porras, Alicia	121	aporras@inaoep.mx
Porras, Rosario		rosariop@inaoep.mx
Przybilla, Norbert	139	nob@usm.uni-muenchen.de
Puerari, Ivanio	253,151	puerari@inaoep.mx
Ramírez, Luis		luisrg@inaoep.mx
Ramírez-Ruiz, Enrico	223	enrico@ast.cam.ac.uk
Rodríguez, Ary		ary@inaoep.mx
Rodríguez-Merino, Lino H.	39	lino@inaoep.mx
Romano, Roberto	155	rromano@inaoep.mx
Rosa-González, Daniel	265	danrosa@inaoep.mx
Rosales-Ortega, Fernando F.		fer_fab@hotmail.com
Rozas, Maite	147	maite@astrosen.unam.mx
Sandóval-González, Leonardo		leonardo@inaoep.mx
Saucedo, Julio	73	jsaucedo@cosmos.cifus.uson.mx
Schaerer, Daniel	185	schaerer@ast.obs-mip.fr
Schiavon, Ricardo	181	ripisc@ucolick.org
Schlattl, Helmut	55	hs@astro.livjm.ac.uk
Silva, Laura	175, 171, 257	silva@ts.astro.it
Szeifert, Thomas		tszeifer@eso.org
Taylor, Benjamin	95	taylorb@physc3.byu.edu
Tinoco-Villa Monica Emilia		tinocofisica@hotmail.com

Torres-Papaqui, Juan Pablo	37	papaqui@inaoep.mx
Torres-Tonche, Pablo		ptorres@ccip.udg.mx
Urrutia, Eugenio		eurrutia@upaep.mx
Valdés, Jose Ramón	279, 167	jvaldes@inaoep.mx
Valdez-Gutierrez, Margarita	151	mago@inaoep.mx
Valls-Gabaud, David	127	dvg@ast.obs-mip.fr
Villalobos, Salvador		chasobel@yahoo.com
Zoccali, Manuela	107	mzoccali@eso.org

Previously published in Astrophysics and Space Science Library book series:

- **Volume 273**: **Lunar Gravimetry**
  Author: Rune Floberghagen
  Hardbound, ISBN 1-4020-0544-X, April 2002
- **Volume 271**: **Astronomy-inspired Atomic and Molecular Physics**
  Author: A.R.P. Rau
  Hardbound, ISBN 1-4020-0467-2, March 2002
- **Volume 269**:**Mechanics of Turbulence of Multicomponent Gases**
  Authors: Mikhail Ya. Marov, Aleksander V. Kolesnichenko
  Hardbound, ISBN 1-4020-0103-7, December 2001
- **Volume 268**:**Multielement System Design in Astronomy and Radio Science**
  Authors: Lazarus E. Kopilovich, Leonid G. Sodin
  Hardbound, ISBN 1-4020-0069-3, November 2001
- **Volume 267: The Nature of Unidentified Galactic High-Energy Gamma-Ray Sources**
  Editors: Alberto Carramiñana, Olaf Reimer, David J. Thompson
  Hardbound, ISBN 1-4020-0010-3, October 2001
- **Volume 266: Organizations and Strategies in Astronomy II**
  Editor: André Heck
  Hardbound, ISBN 0-7923-7172-0, October 2001
- **Volume 265: Post-AGB Objects as a Phase of Stellar Evolution**
  Editors: R. Szczerba, S.K. Górny
  Hardbound, ISBN 0-7923-7145-3, July 2001
- **Volume 264: The Influence of Binaries on Stellar Population Studies**
  Editor: Dany Vanbeveren
  Hardbound, ISBN 0-7923-7104-6, July 2001
- **Volume 262: Whistler Phenomena**
  Short Impulse Propagation
  Authors: Csaba Ferencz, Orsolya E. Ferencz, Dániel Hamar, János Lichtenberger
  Hardbound, ISBN 0-7923-6995-5, June 2001
- **Volume 261: Collisional Processes in the Solar System**
  Editors: Mikhail Ya. Marov, Hans Rickman
  Hardbound, ISBN 0-7923-6946-7, May 2001
- **Volume 260: Solar Cosmic Rays**
  Author: Leonty I. Miroshnichenko
  Hardbound, ISBN 0-7923-6928-9, May 2001
- **Volume 259: The Dynamic Sun**
  Editors: Arnold Hanslmeier, Mauro Messerotti, Astrid Veronig
  Hardbound, ISBN 0-7923-6915-7, May 2001
- **Volume 258: Electrohydrodynamics in Dusty and Dirty Plasmas**
  Gravito-Electrodynamics and EHD
  Author: Hiroshi Kikuchi
  Hardbound, ISBN 0-7923-6822-3, June 2001
- **Volume 257: Stellar Pulsation - Nonlinear Studies**
  Editors: Mine Takeuti, Dimitar D. Sasselov
  Hardbound, ISBN 0-7923-6818-5, March 2001

- **Volume 256: Organizations and Strategies in Astronomy**
  Editor: André Heck
  Hardbound, ISBN 0-7923-6671-9, November 2000
- **Volume 255: The Evolution of the Milky Way**
  Stars versus Clusters
  Editors: Francesca Matteucci, Franco Giovannelli
  Hardbound, ISBN 0-7923-6679-4, January 2001
- **Volume 254: Stellar Astrophysics**
  Editors: K.S. Cheng, Hoi Fung Chau, Kwing Lam Chan, Kam Ching Leung
  Hardbound, ISBN 0-7923-6659-X, November 2000
- **Volume 253: The Chemical Evolution of the Galaxy**
  Author: Francesca Matteucci
  Hardbound, ISBN 0-7923-6552-6, May 2001
- **Volume 252: Optical Detectors for Astronomy II**
  State-of-the-art at the Turn of the Millennium
  Editors: Paola Amico, James W. Beletic
  Hardbound, ISBN 0-7923-6536-4, December 2000
- **Volume 251: Cosmic Plasma Physics**
  Author: Boris V. Somov
  Hardbound, ISBN 0-7923-6512-7, September 2000
- **Volume 250: Information Handling in Astronomy**
  Editor: André Heck
  Hardbound, ISBN 0-7923-6494-5, October 2000
- **Volume 249: The Neutral Upper Atmosphere**
  Author: S.N. Ghosh
  Hardbound, ISBN 0-7923-6434-1, (in production)
- **Volume 247: Large Scale Structure Formation**
  Editors: Reza Mansouri, Robert Brandenberger
  Hardbound, ISBN 0-7923-6411-2, August 2000
- **Volume 246: The Legacy of J.C. Kapteyn**
  Studies on Kapteyn and the Development of Modern Astronomy
  Editors: Piet C. van der Kruit, Klaas van Berkel
  Hardbound, ISBN 0-7923-6393-0, August 2000
- **Volume 245: Waves in Dusty Space Plasmas**
  Author: Frank Verheest
  Hardbound, ISBN 0-7923-6232-2, April 2000
- **Volume 244: The Universe**
  Visions and Perspectives
  Editors: Naresh Dadhich, Ajit Kembhavi
  Hardbound, ISBN 0-7923-6210-1, August 2000
- **Volume 243: Solar Polarization**
  Editors: K.N. Nagendra, Jan Olof Stenflo
  Hardbound, ISBN 0-7923-5814-7, July 1999

Missing volume numbers have not yet been published.
For further information about this book series we refer you to the following web site: http://www.wkap.nl/prod/s/ASSL

To contact the Publishing Editor for new book proposals:
Dr. Harry (J.J.) Blom: harry.blom@wkap.nl